FORTRAN WITH ENGINEERING APPLICATIONS

D. D. McCracken

FORTRAN with Engineering Applications
A Guide to FORTRAN IV Programming
A Guide to COBOL Programming
A Guide to ALGOL Programming
A Guide to IBM 1401 Programming
A Guide to FORTRAN Programming
Digital Computer Programming

D. D. McCracken and W. S. Dorn

Numerical Methods and FORTRAN Programming

F. J. Gruenberger and D. D. McCracken

Introduction to Electronic Computers:
Problem Solving with the IBM 1620

D. D. McCracken, H. Weiss, and T. H. Lee

Programming Business Computers

FORTRAN

WITH ENGINEERING APPLICATIONS

Daniel D. McCracken

McCRACKEN ASSOCIATES, INC.

JOHN WILEY & SONS, INC., NEW YORK · LONDON · SYDNEY

Library of Congress Catalog Card Number: 67–17343

Printed in the United States of America

ISBN 0 471 58236 0

To

Charles,
Judith,
Cynthia,
Virginia,
Rachel,
Aliza,
and
Thomas

PREFACE

This book is written for the person who wants to get a rapid grasp of the use of a computer in the solution of problems in engineering and related disciplines. The essentials of FORTRAN are presented in a form that can be mastered in a few hours of careful reading and practice. The reader who wishes to go beyond the fundamentals will also find his needs met; among the 29 case studies are several that apply computer techniques of an intermediate level of sophistication to a variety of realistic and interesting engineering problems.

It is anticipated that the text will be helpful in a number of different situations.

1. The book is suited to individual study in school or industry.

2. It can be used as the text for a course in the engineering, science, or mathematics area. One semester hour would provide enough time to cover a good sampling of the case studies and allow students time to run several problems on a computer.

3. It can be used as a supplementary text in some other course, one not directly connected with computing. Covering the fundamentals of FORTRAN will take only a few hours and this "lost" time will, in many cases, be more than recovered by improved learning in the primary subject matter. Students will then be able to use the knowledge gained in all subsequent courses.

4. One attractive possibility combines elements of (2) and (3). A short course early in the student's career would cover the essentials of programming at a level in keeping with his background and interests at the time. In subsequent courses instructors could use that knowledge and build on it by referring to specific case studies for additional programming techniques or for illustrations of approaches to the solution of engineering problems.

5. Students in an industrial course in FORTRAN programming will find the book useful as an elaboration of the necessarily rather terse descriptions found in many FORTRAN manuals and as a guide to the ways in which FORTRAN and computers can be applied to the solution of realistic problems.

In all cases the value of the book is enhanced by the inclusion of a large number of graded exercises, with answers to a selected set. These exercises induce thorough mastery of the fundamentals and provide an insight into methods and applications that often go considerably beyond those presented in the text.

Of particular importance is the case study approach. Although the examples make only a minimum demand on the reader's background, they nevertheless place each new programming idea in the context of a meaningful application. They demonstrate how to go about setting up a program; they show how the program actually works by reproducing real output from test runs for every case study. One case study is the medium for a discussion of program checkout and another surveys the critical subject of error analysis.

The introduction on page 1 outlines the case studies and suggests selections of them to meet special needs.

It is a pleasure to acknowledge the many and varied contributions of T. R. Bashkow of Columbia University, W. S. Dorn and R. F. Lever of the International Business Machines Corporation, C. H. Davidson and D. L. Dietmeyer of the University of Wisconsin, F. F. Kuo of the University of Hawaii, O. O. Pardee of Syracuse University, and H. H. Skilling of Stanford University.

I must single out the contributions of Fred J. Gruenberger of Informatics, Inc., who has been a faithful reviewer of all my books. He has consistently encouraged my best efforts and firmly blue-penciled my carelessness and in this capacity has been a major formative influence on my work.

Finally, I wish to express appreciation to my editors at John Wiley and Sons, Charles B. Stoll and Walker G. Stone, for their support in ways too numerous to mention. In more than 10 years of close association they have given nothing but good advice and warm encouragement. It has been a pleasure to work with them.

DANIEL D. McCRACKEN

Ossining, New York
March 1967

CONTENTS

FORTRAN WITH ENGINEERING APPLICATIONS

ATTENTION, READER:
HOW TO USE THIS BOOK EFFECTIVELY

To obtain optimum value from a study of this book the reader should be aware of a few matters concerning its content and organization.

1. There is not just one version of FORTRAN; there are dozens. The FORTRAN language described here is intended to be in compliance with the language adopted as a standard by the United States of America Standards Institute (formerly the American Standards Association), although for clarity no attempt has been made to discuss every feature of the standard. Furthermore, no attempt has been made to point out all the variations between the standard and the other FORTRAN languages; to do so would hopelessly clutter the exposition. Most readers will find that the differences are not great and that with the aid of a manual on their own FORTRAN the presentation here will be readily adapted to the particular situation.

The word "FORTRAN" is used here as in the United States of America Standards Institute document; it refers to the full language, sometimes also called FORTRAN IV. The USASI uses "Basic FORTRAN" to designate a more limited language sometimes called FORTRAN II or simply FORTRAN. Elimination of this deplorable confusion of terminology is best served, it is felt, by adhering henceforth to the decision of the United States of America Standards Institute.

2. The speed and depth of learning will be enhanced by actual practice in preparing and running programs—early and often. The material in the case studies has been developed in the hope that the reader will have the advantage of machine practice, beginning as early as Case Study 1 (the exercises on page 6 are suitable for a first practice run). But the presentation is not restricted to this usage; the reader without access to a computer will find the book suited to his needs.

3. The first dozen case studies develop a nucleus of knowledge about FORTRAN and computing that will allow the beginner to tackle realistic applications. Some readers might choose to omit Case Study 11 on complex numbers and 13 on logical operations as not relevant to their needs. Case Study 10 on program checkout and 12 on error analysis are much broader than their titles might first seem to indicate; they should definitely be included in any study plan.

The reader going beyond a bare minimum should study Case Studies 14 to 16, which develop the central combination of the DO statement and subscripted variables. Case Studies 17 to 23 introduce ideas which are an essential part of the "equipment" of the student but can be postponed, if necessary, until the need for them arises.

Case Studies 24 to 29 integrate the material of the earlier examples in applications that involve a variety of programming techniques in a framework of problems which are interesting in their own right. They are, in general, longer than the earlier ones. It is possible that not all readers will choose to cover all of the last six case studies in full detail, but every reader should study one or two of them, at least, to get an idea of the full scope of the task of going from problem specification to final solution.

1

CASE STUDY 1 RESISTANCE OF A WIRE: PROGRAM BASICS

Let us begin our study of programming by considering an example of a program. Suppose that we wish to compute the resistance of a piece of copper wire, given its length and cross-sectional area. The required formula is

$$R = 0.0000081456 \frac{L}{A}$$

where R = resistance, ohms
L = length, feet
A = area, square inches

The constant in the formula, 0.0000081456, is the resistivity of standard annealed copper at 20°C, for length in feet and area in square inches.

The program that we wish to display is required to read a data card that contains numerical values for the length and area; it is to compute the resistance; it should then print the value of length, area, and resistance.

Figure 1.1 shows a FORTRAN program to do this job, written on FORTRAN coding paper. Let us examine this program in some detail, to begin to see what the elements of the FORTRAN language are and how a program is constructed. With this much background the student will be able to write meaningful practice programs of his own.

The first three lines are called *comment lines* because the letter C appears in position 1. They are for the information of the human reader; they have no effect on the computer. Free use of comments is strongly en-

couraged, for identification and to make clear the purpose of the various parts of a complicated program. The third comment line simply provides spacing to make it easier to read the program.

Observe here that we are approaching two problems at once: the problem of making a computer program easy for a *person* to read, and the problem of telling the computer exactly what we want it to do. We approach human understandability by means of frequent comments, by choice of meaningful variable names, and by certain matters of spacing in the writing of the program. We approach the problem of making certain that the computer is precisely instructed in its task by following various rules for writing *statements*.

The line beginning REAL is the first statement of our program. We see that a statement begins in position 7. (The 1 and 2 on two of the statements are *statement numbers*.) The function of the REAL statement in this program is to specify to the computer that the three variables named L, A, and R are of the real type. This means that they are able to take on any positive or negative value in the approximate range of 10^{-40} to 10^{+40}, or be zero. It may be noted that the variable names used in the program are the same as the names appearing in the formula; this will not always be possible, as we shall see later. We shall also learn later that FORTRAN provides four types of variables in addition to REAL: INTEGER, DOUBLE PRECISION, COMPLEX, and LOGICAL.

3

```
C CASE STUDY 1
C A PROGRAM TO COMPUTE RESISTANCE OF A WIRE FROM LENGTH AND AREA
C
      REAL L, A, R
      READ (5, 1) L, A
  1   FORMAT (8F10.0)
      R = 0.00000081456 * L / A
      WRITE (6, 2) L, A, R
  2   FORMAT (1H , 1P10E13.5)
      STOP
      END
```

Figure 1.1. An example of a FORTRAN program. A data card containing the length and area of a piece of copper wire is read by the program, which then computes and prints the resistance of the wire.

The next statement, READ, calls for a data card to be read by the computer. The action of the READ statement is to read values for the variables named in the statement, L and A, from a punched card. The 5 in parentheses identifies an input unit that holds the card, and the 1 is the statement number of the FORMAT statement that comes next.

The FORMAT statement describes the format, or arrangement, of the data values on the card, as follows. The first data value is punched anywhere in columns 1–10; a decimal point must be punched; a minus sign is the first punch in columns 1–10 if the number is negative. The second number on the card is punched in the same format, but in columns 11–20, etc., up to a maximum of eight values. The value punched in columns 1–10 will be assigned to the first variable named in the READ statement, the value in columns 11–20 will be assigned to the second variable named in the READ statement, and so on.

We shall see later that great flexibility in the arrangement of data values (and printed results) is provided by the various features of the FORMAT statement that we shall study at the appropriate time. We accordingly defer until a little later a consideration of the exact meaning of the symbols in parentheses in the FORMAT statement.

Figure 1.2 shows a data card that a student might prepare for reading by this program. The values used, as may be seen in the printing at the top of the card, are 1000.0 and 0.0005047. The piece of wire is thus described as being 1000 feet in length and as having a cross-sectional area of 0.0005047 square inches. (This is the area of B. & S. Gauge 22 wire.) The printing at the top of the card was done by the *card punch* on which the card was prepared; the lines separating the *fields* (groups of columns) were drawn by the author. The computer, of course, senses only the punched holes.

Now we come to the statement that calls for the actual computation of the resistance of the wire described by the data card. An *arithmetic assignment statement* calls for the computation of the value of the expression on the right and the assignment of that value to the variable named on the left. Let us examine this important statement in some detail.

The left side of an assignment statement must always consist of just one variable name; this variable is assigned a value as a result of executing the statement. The equal sign identifies the statement to the computer as an assignment statement. Now we see an *expression* corresponding quite closely to the formula. Multiplication is signified in FORTRAN by an asterisk and division by a slash. In later examples we shall see that addition and subtraction are designated by the usual plus and minus signs and that exponentiation is represented by two asterisks.

Let us emphasize once again what the arithmetic assignment statement calls for: the expression on the right is evaluated by use of whatever values have been assigned to the variables appearing in

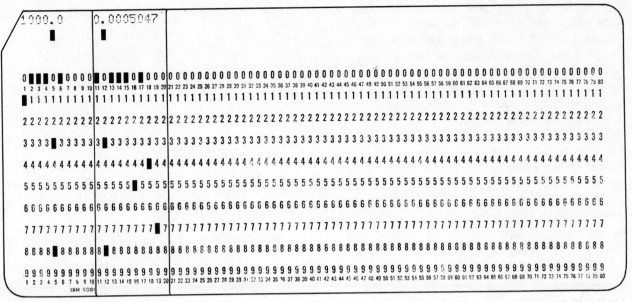

Figure 1.2. An example of a data card for reading by the program in Figure 1.1. The wire is described as 1000 feet long and as having an area of 0.0005047 square inches. (B. & S. gauge 22.)

the expression; the newly computed expression value is assigned to the variable named on the left.

The WRITE statement calls for the data and the answer to be printed. The 6 identifies an output unit and the 2 is the statement number of the FORMAT statement that follows, much as in READ. The three variables for which we want values printed are listed in the order in which we want the values to appear on the page.

The FORMAT statement is a little different this time. It specifies single spacing of the paper in the printer and permits a maximum of 10 numbers on one line, each of which will have five digits after the decimal point and be printed in "exponent form," that is, as a number times a power of 10.

The STOP statement is required to inform the computer that processing should cease at this point.

The END statement indicates that no more statements follow, that is, that this is the end of the program. It may appear in this example that STOP and END duplicate each other, but in fact they do not. END must *always* be the final statement in the program, whereas STOP may frequently appear elsewhere than just before the END statement. Sometimes, a STOP is not required. We shall see examples of these variations in later case studies.

When this program was submitted for computer execution with the data card of Figure 1.2, it produced the line of output shown in Figure 1.3. We see that the numbers have been printed in a somewhat different form from the way they were entered

on the data card. This form is called exponent form, as noted; we are to read a number such as 1.00000E 03, for instance, as meaning $1.00000 \cdot 10^3$, or 1000. The second number is $5.04700 \cdot 10^{-4}$, or 0.0005047.

This form of printing numbers permits us to write programs without knowing in advance how large or how small the results may be. Later we shall learn how to present results in the form in which they were punched on the data card, if we so choose.

While we are exploring the basic ideas of programming the pattern of this program will be repeated several times. We may therefore profitably emphasize the framework and the function of each part.

1. Comments. Use where and as needed to identify and explain the program.

2. REAL. Name all variables used in the program.

3. READ. Using the sample form READ (5,1) list the variables for which values are to be read, by writing the variable names in the same order as the values are punched on the card.

4. FORMAT. Including the statement number 1 in front, write this line exactly as in the sample program.

1.00000E 03   5.04700E-04   1.61395E 01

Figure 1.3. The line of output produced when the program in Figure 1.1 was executed with the data card in Figure 1.2. Note the exponent form of the values: 1.61395E 01, for instance, means $1.61395 \cdot 10^1 = 16.1395$.

5. Arithmetic assignment statement(s). One or more statements to compute the results, or answers, desired. Each arithmetic assignment statement consists of a single variable name on the left, an equal sign, and an expression on the right. All variables named in the expression on the right must already have been given values by reading a card or by executing some other assignment statement.

6. WRITE. Using the sample form WRITE (6,2) list the variables for which values are to be printed in the order in which they are to appear across the page.

7. FORMAT. Write the second FORMAT statement exactly as it appears in the sample, including the statement number.

8. STOP.

9. END.

EXERCISES

In each of the following exercises you are to prepare a complete program, according to the outline just given, and show how your data card might appear. The programs you write will be complete with the exception of a few "control cards" that your instructor will supply if you are to run the programs on the computer. For the sake of easier checking of the programs in a class situation, specific data values are suggested, although the reader will realize that the programs should operate correctly with any data values.

Answers to starred exercises are given at the back of the book.

*1. A card contains values of the base B and the altitude H of a triangle. Compute the area A of the triangle from the formula

$$A = BH/2$$

Print the values of B, H, and A, in that order. Remember to write the constant 2 with a decimal point in the program. On your data card let B = 10.0 and H = 16.5.

2. Write a program to read from a data card a temperature in degrees centigrade, compute the equivalent Farenheit temperature, and print both, using the formula

$$F = 1.8C + 32$$

Let C = 22.5.

*3. A card contains values of a resistance and a voltage, in that order. The program is to compute current from Ohm's law,

$$i = \frac{v}{R}$$

where i = current, amperes
  v = voltage, volts
  R = resistance, ohms

Print the values of voltage, resistance, and current, in that order. On your data card let $R = 4700$, $v = 8.37$.

4. A card contains values of voltage, resistance, and current, as measured in an experiment. A program is to read the card, compute the theoretical current based on an application of Ohm's law, then print the three data items and the computed result, in the order: voltage, resistance, experimental current, theoretical current. Let $v = 2.24$, $R = 6420$, $i_{exp} = 0.00033$.

*5. The capacitance of a parallel plate capacitor with air dielectric is given by

$$C = 0.000008855 \frac{A}{s}$$

where C = capacitance microfarads
  A = area, square meters
  s = separation, meters

Let $A = 0.43$ and $s = 0.0005$.

6. The capacitance of a parallel plate capacitor is given by

$$C = \frac{0.2244KA}{s}$$

where C = capacitance, picofarads
  K = dielectric constant, dimensionless
  A = area, square inches
  s = separation, inches

A program is to read area, separation, and dielectric constant, in that order; compute the capacitance; print the three data items in the same order as they appeared on the card, together with the computed capacitance. Let $K = 3.5$, $A = 2.5$, and $s = 0.01$.

CASE STUDY 2 RESISTANCE OF A CIRCULAR WIRE: VALUES OF VARIABLES; INTEGER AND REAL CONSTANTS

Our second example of a program is a slight extension of the task of the first. We can accordingly consolidate our knowledge of program structure, while amplifying one or two points.

We start again with a piece of standard annealed copper wire at 20°C, but this time we know that the wire is of circular cross section, and instead of area we are given the diameter. The program is required to produce not only the resistance but also the area. The mathematical statement of the task can thus be presented as two formulas:

$$A = \frac{\pi}{4} d^2$$

$$R = 8.1456 \cdot 10^{-6} \frac{L}{A}$$

where A = area, square inches
d = diameter, inches
L = length, feet
R = resistance, ohms

The program is shown in Figure 2.1.

We note the similarities with the preceding program. The REAL statement lists the additional variable. Incidentally, no significance is attached to the order in which variables are named in the REAL statement. The READ and FORMAT statements are familiar.

The arithmetic assignment statement for computing the cross-sectional area demonstrates two new items. First there is the exponentiation operator—the double asterisk.

The command to the computer is to raise the value of d to the second power. Second, we have our first example of an *integer constant*—one that is permitted to take on only integer (whole number) values and is written without a decimal point.

Statements in a FORTRAN program are executed in the sequence in which they are written unless we direct otherwise with the GO TO and IF statements that we shall encounter later. Therefore once a value has been assigned to a variable by an assignment statement we may immediately use that variable in a subsequent statement, knowing that the variable has the value just assigned to it. This we do in the statement for computing the resistance: the A in this statement will have the value just given to it by the preceding statement.

We also see here that real constants in a program may be written in exponent form in a manner just like we have seen on output. Thus, when we write 8.1456E-6, we intend $8.1456 \cdot 10^{-6}$. If the exponent is positive instead of negative, we omit the plus sign. Examples of other real constants written in exponent form and an equivalent way of writing the same number without an exponent are

12.08E5	1208000.
1.E7	10000000.0
1.E-6	0.000001
.75311E-4	0.000075311

```
C CASE STUDY 2
C A PROGRAM TO COMPUTE RESISTANCE OF A CIRCULAR WIRE,
C GIVEN LENGTH AND DIAMETER
C
      REAL L, A, D, R
      READ (5, 1) L, D
    1 FORMAT (8F10.0)
      A = 0.78539816 * D ** 2
      R = 8.1456E-6 * L / A
      WRITE (6, 2) L, D, A, R
    2 FORMAT (1H , 1P10E13.5)
      STOP
      END
```

Figure 2.1. A FORTRAN program to compute the resistance of a circular wire, given the length and diameter.

Let us now summarize the characteristics of real and integer constants. An integer constant must be a whole number: zero or any positive or negative number of 10 or fewer decimal digits.* An integer constant must always be written *without* a decimal point. A real constant, on the other hand, may be a whole number or have a fractional part. It may be zero or any positive or negative number in the approximate range of 10^{-40} to 10^{+40}. A real constant must always be written *with* a decimal point. For very large or very small numbers we may use the exponent form, but still with a decimal point in the part that precedes the E. The part after the E, which is called the exponent, must not have a decimal point.

The rest of the program contains no new ideas. A possible data card is shown in Figure 2.2; the

*The exact limits of sizes for both integer and real constants vary from one computing system to the next.

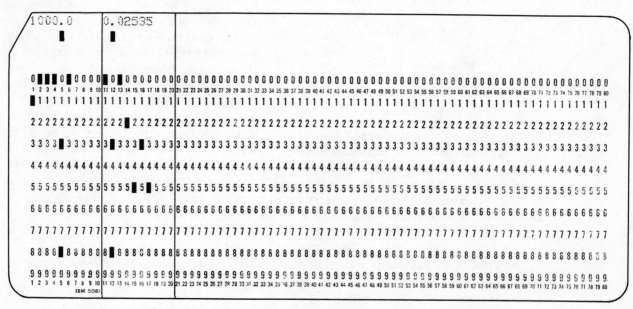

Figure 2.2. A data card that could be used with the program in Figure 2.1.

1.00000E 03   2.53500E-02   5.04715E-04   1.61390E 01

Figure 2.3. The line of output produced when the program in Figure 2.1 was executed with the data card in Figure 2.2.

value for the diameter is again that of B. & S. Gauge 22. The line of output produced when the program was run is shown in Figure 2.3.

EXERCISES

*1. Write the following numbers as FORTRAN real constants:

256   2.56 $-43,000$ 10^{12} 0.000000492 -10 -10^{-16}

2. Write the following as FORTRAN real constants:

16   4.59016 $-10,000$ 10^{17} 0.000006 -1 -10^{-10}

*3. All of the following are unacceptable as real constants. Why?

87,654.3 $+987$ $9.2E + 87$ $7E - 9$

4. All of the following are unacceptable as real constants. Why?

$-100,000$ $1E - 55$ $2.379427 - E12$ $2E + 5.1$

*5. Do the following pairs of real constants represent the same number in each case?

a. 16.9 $+16.9$
b. 23000.   2.3E4
c. 0.000007 $.7E - 5$
d. 1.0   1.
e. .906E5 $+906.0E + 2$

*6. Some of the following are unacceptable as integer constants. Identify the errors.

$+234$ $-234.$ 23,400   1E12 $+1000000000000$ $+10000$

7. Some of the following are unacceptable as integer constants. Identify the errors.

-16.5 16000   16,000.   2.E12.5   0.01

*8. A data card contains values of force, mass, and time, in that order. Write a program to compute acceleration and position; then print time, force, mass, acceleration, and position, in that order. The required formulas, assuming that the mass was at rest at time zero, are

$$a = \frac{F}{m}$$

$$x = \frac{1}{2}at^2$$

where F = force, pounds
m = mass, slugs
a = acceleration, feet/second^2
t = time, seconds
x = distance moved since time $t = 0$, feet

9. A data card contains two values: the radius of a sphere and the density of the material of which it is made. Compute the volume of the sphere; then compute its weight, which is the product of volume and density. Print, in order, radius, density, weight, and volume.

*10. A card contains values for frequency in cycles per second and inductance. The program is to compute angular frequency and reactance according to

$$\omega = 2\pi f$$

$$X = 10^{-6}\omega L$$

where ω = angular frequency, radians per second
f = frequency, cycles per second
X = inductive reactance, ohms
L = inductance, microhenrys

11. A card contains a value for the length of the side of a cube of copper. The program is to print the length, the weight of the cube, the number of gram-molecules (moles) of copper in the cube, and the number of atoms of copper in the cube, using the formulas

$$W = 8.89l^3$$

$$N = \frac{W}{63.57}$$

$$A = 6.023 \cdot 10^{23}N$$

where
l = side of cube, centimeters
W = weight of cube, grams
8.89 = density of copper
N = number of moles
63.57 = atomic weight of copper
A = number of copper atoms in cube
$6.023 \cdot 10^{23}$ = Avogadro's number

CASE STUDY 3 TENSION IN A ROPE ON A PULLEY; PARENTHESES; HIERARCHY OF ARITHMETIC OPERATORS

Figure 3.1 describes two unequal masses connected by a weightless rope passing over a frictionless and massless pulley. The formula for the tension in the rope is

$$T = \frac{2m_1 m_2}{m_1 + m_2} g$$

where m_1, m_2 = mass, slugs

g = gravitational constant, 32.2 ft/sec^2

T = tension, pounds

The data card this time contains just the values of the two masses; we take the value of g to be a constant and write it in the program. We are to compute the tension and print it along with the values of the two masses.

The program shown in Figure 3.2 is based on familiar ideas, with two exceptions that we wish to study: we see that a variable name need not be just a single letter and we see parentheses in the arithmetic assignment statement.

The new twist in naming is made necessary by the fact that there are two masses, which must have separate names. We have made an obvious choice and used M1 and M2. Fortran permits a variable name to have one to six letters or digits, the first of which must be a letter. We can thus devise names such as X45, VECTOR, Z, FOURTH, ROOT3, KLMNOP, A12345, ALPHA, ZCUBED, TORQUE, and FARAD. On the other hand, here are some combinations that are illegal as FORTRAN variable names: ZSQUARED (too many letters), 12AT7 (does not start with a letter), A + B (contains something other than a letter or a digit), X3.4 (contains something other than a letter or a digit).

We are completely free to invent variable names within the limits of the simple formation rules (i.e., one to six letters or digits, the first being a letter). Most programmers try to choose names that remind them of the meaning of the variable, such as ZCUBED or FARAD, but this is a matter of individual discretion. In the program at hand the choice of M1 and M2 was only one of many possibilities. We might, for instance, have written MONE and MTWO, or FIRSTM and SECNDM.

In this book we usually follow the practice of naming every variable in a *type statement*, that is, REAL, INTEGER, COMPLEX, LOGICAL, or DOUBLE PRECISION, in which case no meaning is attached to the initial letter of a variable name. We should note, however, that this practice is not actually mandatory. If a variable is not mentioned in any type statement, it is automatically taken to be either real or integer, according to the following rule: if the first letter is I, J, K, L, M, or N, it is integer and otherwise real. We prefer not to begin by becoming dependent on this convention because of the problems that arise if a variable that was intended to be real is absentmindedly assigned the name of an integer variable. This mistake is remarkably easy to make, considering that certain choices of symbol, such as I for current and M for mass, are so familiar.

The parentheses are used in arithmetic expressions to convey the intentions of the programmer, much as in ordinary algebraic notation. Without the parentheses, the product would be divided by M1 and M2 would be added to the quotient because division is a "stronger" operation than addition. FORTRAN and the familiar mathematical notation agree in this regard.

In other words, we use parentheses in an arithmetic expression to dictate that we wish a grouping other than that contained in the *hierarchy of operators* applied in the absence of parentheses, which is that all exponentiations are performed first, then all multiplications and divisions, then all additions and subtractions.

For example, $(A+B) * C$ means to add first and then multiply; without the parentheses, $A+B * C$ means to multiply first and then add. $A/(B+C)$ forces the addition to be done first; without parentheses we get A divided by B first and the quotient added to C.

Or consider the expression $(A+B)/(C-D)$ and note what is meant by the four possible variations, with and without parentheses:

$$(A + B)/(C - D) \qquad \frac{A + B}{C - D}$$

$$A + B/(C - D) \qquad A + \frac{B}{(C - D)}$$

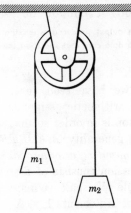

Figure 3.1. The weight and pulley arrangement used as a basis for Case Study 3.

$$(A + B)/C - D \qquad \frac{A + B}{C} - D$$

$$A + B/C - D \qquad A + \frac{B}{C} - D$$

Certain expressions, usually considered ambiguous in mathematical notation, are permitted in FORTRAN, but should be used with care. For a prime example, does $A/B * C$ mean $\frac{A}{BC}$ or $\frac{A}{B} C$? The second expression is the answer because in the absence of parentheses it is evaluated from left to right. We shall not dwell on this point, however, because a completely general description of what can happen is altogether too complicated to be worth our trouble, particularly when one can hardly ever get into trouble by observing the dictum, *when in doubt, parenthesize.*

```
C CASE STUDY 3
C A PROGRAM TO COMPUTE THE TENSION IN PULLEY ROPE SUPPORTING TWO MASSES
C
      REAL M1, M2, T
      READ (5, 1) M1, M2
    1 FORMAT (8F10.0)
      T = 64.4 * M1 * M2 / (M1 + M2)
      WRITE (6, 2) M1, M2, T
    2 FORMAT (1H , 1P10E13.5)
      STOP
      END
```

Figure 3.2. A program to compute the tension in a pulley rope.

2.00000E 00   1.00000E 00   4.29333E 01

Figure 3.3. The output produced when the program in Figure 3.2 was run with a data card that specified the masses as 2 and 1 slugs.

Now that we have seen several examples of expressions in arithmetic assignment statements, a formal definition is in order so that we may use the concept in full generality. *A FORTRAN expression is a rule for computing a numerical value.* In many cases an expression consists of a single constant or a single variable. We shall see many examples later, for instance, of statements like $X = 12.0$ in which "the expression on the right" is the single constant 12.0. All of the examples so far, however, have involved combining two or more constants and/or variables with the arithmetic operators $+$, $-$, $*$, $/$, and $**$.

FORTRAN requires that real and integer quantities never be "mixed" (combined) in an expression, with one exception: a real quantity can be raised to an integer power. Thus an expression like $X**2$ is permitted, but $X + 2$ is not. The reason for this rule is that real and integer quantities are stored and operated on within the computer in quite different ways in most machines.*

Table 3.1 lists some examples of correct and incorrect expressions to illustrate the rules we have discussed.

The WRITE statement again contains nothing new. A line of output is exhibited in Figure 3.3.

* In a few FORTRAN systems "mixed mode" expressions *are* permitted.

EXERCISES

1. Which of the following are acceptable variable names? In the absence of a type statement (REAL, etc.) which would be the names of real variables and which the names of integer variables? G, GAMMA, GAMMA-423, I, IJK, IJK, J79-14, LARGE, R(2)16, BT07TH, 2N173, 6CA7, CDC160, DELTA, KAPPA, EPSILON, A1.4, A1P4, FORTRAN, ALGOL.

2. Same as Exercise 1. K, I12G, CAT, X+2, XPLUS2, NEXT, 42G, LAST, MU, A*B, X1.4, (X61), GAMMA81, AI, IA, X12, 1X2, XTHIRD, XFOURTH, IBM7094, IBM360, MODEL40, COBOL, PL/I.

3. Following are a number of mathematical expressions and corresponding FORTRAN expressions, each of which contains at least one error. Point out the errors and write the correct expressions.

a. $(x + y)^4$ — X + Y**4

*b. $\dfrac{x + 2}{y + 4}$ — X + 2.0/Y + 4.0

c. $\dfrac{a \cdot b}{c + 2}$ — AB/(C + 2.)

d. $-\dfrac{(-x + y - 16)}{y^3}$ — $-(-$X + Y $- 16)/$ Y**3

*e. $\left(\dfrac{x + a + \pi}{2z}\right)^2$ — (X + A + 3.1415927)/ (2.0*Z)**2

f. $\left(\dfrac{x}{y}\right)^{n-1}$ — (X/Y)**N $- 1$

*g. $\left(\dfrac{x}{y}\right)^{r-1}$ — (X/Y)**(R $- 1$)

h. $\dfrac{a}{b} + \dfrac{c \cdot d}{f \cdot g \cdot h}$ — A/B + CD/FGH

i. $(a + b)(c + d)$ — A + B*C + D

TABLE 3.1

Mathematical Notation	Correct Expression	Incorrect Expression
$a \cdot b$	A*B	AB (no operation)
$a \cdot (-b)$	A*($-$B) or $-$A*B	A* $-$ B (two operations side by side)
$a + 2$	A + 2.0	A + 2 (mixed integer and real)
$-(a + b)$	$-$(A + B) or $-$A $-$ B	$-$A + B or $-+$A + B
a^{i+2}	A**(I + 2)	A**I + 2 ($= a^i + 2$, and is mixed integer and real)
$a^{b+2} \cdot c$	A**(B + 2.0)*C	A**B + 2.0*C ($= a^b + 2 \cdot c$)
$\dfrac{a \cdot b}{c \cdot d}$	A*B/(C*D) or A/C*B/D	A*B/C*D $\left(= \dfrac{a \cdot b}{c} \cdot d\right)$
$\left(\dfrac{a + b}{c}\right)^{2.5}$	((A + B)/C)**2.5	(A + B)/C**2.5 $\left(= \dfrac{a + b}{c^{2.5}}\right)$
$a[x + b(x + c)]$	A*(X + B*(X + C))	A(X + B(X + C)) (missing operators)
$\dfrac{a}{1 + \dfrac{b}{(2.7 + c)}}$	A/(1.0 + B/(2.7 + C))	A/(1.0 + B/2.7 + C)

j. $a + bx + cx^2 + dx^3$ A + X(B + X
                                    (C + DX)

which can be rewritten

$a + x[b + x(c + dx)]$

k. $\dfrac{1{,}600{,}042x + 10^5}{4{,}309{,}992x + 10^5}$ (1,600,042X + 1E5)/
                                        (4,309,992X + 1E5)

l. $\dfrac{1}{a^2}\left(\dfrac{r}{10}\right)^a$ 1/A**2*(R/10)**A

4. Write FORTRAN expressions corresponding to each of the following mathematical expressions.

*a. $x + y^3$

b. $(x + y)^3$

c. x^4

*d. $a + \dfrac{b}{c}$

e. $\dfrac{a + b}{c}$

*f. $a + \dfrac{b}{c + d}$

g. $\dfrac{a + b}{c + d}$

*h. $\left(\dfrac{a + b}{c + d}\right)^2 + x^2$

i. $\dfrac{a + b}{c + \dfrac{d}{e + f}}$

*j. $1 + x + \dfrac{x^2}{2!} + \dfrac{x^3}{3!}$

*k. $\left(\dfrac{x}{y}\right)^{g-1}$

l. $\dfrac{\dfrac{a}{b} - 1}{g\left(\dfrac{g}{d} - 1\right)}$

*5. The combined resistance of two resistances R1 and R2 in parallel is given by

$$R = \frac{1}{1/R_1 + 1/R_2}$$

Write a complete program to read values of R_1 and R_2 from a card, compute R, and print all three resistance values. For numeric values, try $R_1 = 30$, $R_2 = 60$, for which the result is 20.

6. A certain series-parallel combination has a combined resistance given by

$$R = R_1 + \frac{R_2 R_3}{R_2 + R_3}$$

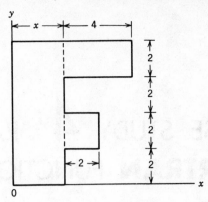

Figure 3.4. The figure for which a center of mass is to be found in Exercise 8.

Write a complete program to read values of R_1, R_2, and R_3, compute R, and print all four values.

7. Three masses, m_1, m_2, and m_3, are separated by distances r_{12}, r_{13}, and r_{23}. If G is the universal gravitational constant, the binding energy holding the mass particles together is given by

$$G\left(\frac{m_1 m_2}{r_{12}} + \frac{m_1 m_3}{r_{13}} + \frac{m_2 m_3}{r_{23}}\right)$$

Write a program to read values of m_1, m_2, m_3, r_{12}, r_{13}, and r_{23}; then compute and print the binding energy along with the data values. For mass in kilograms and distance in meters, $G = 6.67 \cdot 10^{-11}$ newton-meter^2/kg^2.

*8. Figure 3.4 displays the letter F which we suppose has been cut from sheet metal; the dimension marked X is the only variable. The x-coordinate of the center of mass of this figure is given by

$$\bar{x} = \frac{\sum\limits_{i=1}^{3} m_i x_i}{\sum\limits_{i=1}^{3} m_i}$$

where the m's are the masses of the three rectangles into which the figure can be decomposed, as sketched, and the x's are the distances of the centers of mass of these rectangles from the y-axis. The mass units divide out; so we can replace mass with area. For instance, the vertical bar has a mass (area) of $8x$ and its center of mass is at a distance of $x/2$ from the y-axis. Write the formula for the x-coordinate, simplifying as much as possible; then prepare a program that reads X and prints X and XBAR.

9. The y-coordinate of the center of mass of the letter F in Figure 3.4 is given by

$$\bar{y} = \frac{\sum\limits_{i=1}^{3} m_i y_i}{\sum\limits_{i=1}^{3} m_i}$$

with notation as suggested in Exercise 8. Write a program that reads X and prints X and YBAR.

CASE STUDY 4 RADIOACTIVE DECAY; FORTRAN FUNCTIONS

The rate of emission of particles by a substance undergoing a single type of radioactive disintegration is given by

$$R = R_0 e^{-\lambda t}$$

in which

$$\lambda = 0.693/T$$

where R = rate emission at time t, any convenient units

R_0 = rate of emission at time zero, same units as R

λ = disintegration constant, seconds^{-1}

t = time, any convenient units

T = half-life, same units as t

We wish to write a program that will read values of R_0, t, and T, compute λ and R, and print all five values. To do so we must be able to compute an exponential, a function that is readily provided by FORTRAN's mathematical functions. Let us look briefly at this matter before writing the program.

FORTRAN includes a number of common mathematical functions, such as square root, absolute value, sine, cosine, arctangent, logarithm, and the exponential we need in this case study. The exact list of functions available depends to some extent on the version of FORTRAN being used and on the particular computer installation. All FORTRAN systems, however, have the functions named.

Every function has a preassigned name. The names of those we shall use at this time are given in Table 4.1.

TABLE 4.1

Mathematical Function	FORTRAN NAME
Exponential	EXP
Natural logarithm	ALOG
Common logarithm	ALOG10
Sine of an angle in radians	SIN
Cosine of an angle in radians	COS
Hyperbolic tangent	TANH
Square root	SQRT
Arctangent; angle computed in radians	ATAN
Absolute value	ABS

Using a FORTRAN function is a simple matter of writing its name and following it with parentheses enclosing an expression. This directs FORTRAN to evaluate the expression and to evaluate the function at that value. In our case, for example, all that is necessary to get the desired exponential is to write

EXP (—LAMBDA*T)

in the appropriate place in an arithmetic assignment statement. LAMBDA and T must already have been given values by preceding statements—but that is true of any variable appearing in any expression.

The names of functions may not be used as variable names. The reader can perhaps imagine the difficulty of distinguishing between, say, SQRT used as a variable name and SQRT used as a function name in an illegal statement such as

RESULT = SQRT + SQRT(X)

It is worth emphasizing that the *argument* of a function (the expression in parentheses) may be *any* expression. Naturally, if the function is designed to accept a real argument, as in all the examples so far, the expression must have a real value, which is only reasonable. If the argument is a single variable, we may find something as simple as COS(X) in using a function. On the other hand, in computing one of the roots of a quadratic equation from the familiar formula, we might see a function call like this:

$$SQRT(B**2 - 4.0*A*C)$$

The value supplied by a function may then be used in an expression in the same way that a variable can be used. For example, we might write

$$RHO = COS(X)$$

Or if we are computing one of the roots of a quadratic we might have

$$ROOT1 =$$
$$(-B + SQRT(B**2 - 4.0*A*C))/(2.0*A)$$

The first set of parentheses defines the numerator in the formula

$$root_1 = \frac{-b + \sqrt{b^2 - 4ac}}{2a}$$

The second set encloses the argument of the square root function. The third set forces the entire product, $2a$, to be interpreted as being in the denominator.

The argument of a function may very well involve another function. Suppose, for instance, that we need to write a statement to compute V from

$$V = \frac{1}{\cos X} + \log \left| \tan \frac{X}{2} \right|$$

Because we have no function to compute the tangent, we must use a trigonometric identity:

$$\tan \theta = \frac{\sin \theta}{\cos \theta}$$

The statement to compute V could then be the following:

$$V = 1.0/COS(X)$$
$$+ ALOG(ABS(SIN(X/2.0)/COS(X/2.0)))$$

This one statement has the same effect as the next four statements, executed in the order written.

$$Y = X/2.0$$
$$TAN = SIN(Y)/COS(Y)$$
$$ABSVAL = ABS(TAN)$$
$$V = 1.0/COS(X) + ALOG(ABSVAL)$$

The latter method is perhaps a little easier to read, but otherwise the choice is largely arbitrary.

Quite a number of other functions are available in FORTRAN, including those that accept arguments of other types than real and those that supply function values of some other type such as complex, integer, or double precision. A list of the functions supplied with almost all FORTRANs is given in Appendix 2.

```
C CASE STUDY 4
C A PROGRAM TO FIND THE RATE OF EMISSION OF RADIOACTIVITY OF A SUBSTANCE
C
      REAL RO, TH, T, LAMBDA, R
      READ (5, 1) RO, TH, T
    1 FORMAT (8F10.0)
      LAMBDA = 0.693 / TH
      R = RO * EXP(-LAMBDA * T)
      WRITE (6, 2) RO, TH, T, LAMBDA, R
    2 FORMAT (1H , 1P10E13.5)
      STOP
      END
```

Figure 4.1. A program to find the rate of emission of radioactivity of a substance.

4.37000E 02   1.45000E 01   3.90000E 01   4.77931E−02   6.77620E 01

Figure 4.2. The output of the program in Figure 4.1.

The program shown in Figure 4.1 is actually quite simple, once the general idea of a function is clear. LAMBDA is used as an intermediate variable, computed before the statement for getting the radio-activity. If the problem specifications had not called for LAMBDA to be printed, there would not have been any particular need for computing it separately. We might have written

$$R = RO * EXP((-0.693/TH) * T)$$

The parentheses around −0.693/TH are not actually necessary but were inserted to make doubly sure that the *fraction* rather than just the denominator (TH), would be multiplied by what follows. As noted, this is what would have happened without the parentheses, but we are following the "when in doubt, parenthesize" advice. Figure 4.2 exhibits a line of output.

EXERCISES

1. Each of the following arithmetic assignment statements contains at least one error. Identify them.

    a. −V = A + B
    b. 4 = I
    c. V − 3.96 = X**1.67
    d. X = (A + 6)**2
    e. A*X**2 + B*X + C
    f. K6 = I**A
    g. Z2 = A* − B + C**4
    h. X = Y + 2.0 = Z + 9.0
    i. R = 16.9X + AB

2. Write arithmetic assignment statements to do the following:

*a. Add 2 to the current value of the variable named BETA; make the sum the new value of a variable named DELTA.

b. Subtract the value of a variable named B from the value of a variable named A, square the difference, and assign it as the new value of W.

*c. Square A, add to the square of B, and make the new value of C the square root of the sum.

*d. A variable named R is to have its present value replaced by the square root of 2.

e. Multiply THETA by π and store the cosine of the product as the new value of RHO.

f. Add the values of F and G, divide by the sum of the values of R and S, and square the quotient; assign this result to P.

*g. Multiply the cosine of two times X by the square root of one half of X; set Y equal to the result.

*h. Increase the present value of G by 2 and replace the present value of G with the sum.

i. Multiply the present value of A by −1.0 and replace the present value of A with the product.

j. Assign to OMEGA the value of 2π.

k. Assign to the variable named D a value 1.1 times as great as the present value of the variable named D.

3. Write arithmetic assignment statements to compute the values of the following formulas. Use the letters and names shown for variable names.

*a. $$AREA = 2 \cdot P \cdot R \cdot \sin\left(\frac{\pi}{P}\right)$$

b. $$CHORD = 2R \sin \frac{A}{2}$$

*c. $$ARC = 2\sqrt{Y^2 + \frac{4X^2}{3}}$$

d. $$s = \frac{-\cos^4 x}{x}$$

*e. $$s = \frac{-\cos^{p+1} x}{p + 1}$$

*f. $$g = \frac{1}{2} \log \frac{1 + \sin x}{1 - \sin x}$$

g. $$R = \frac{\sin^3 x \cos^2 x}{5} + \frac{2}{15} \sin^3 x$$

h. $$D = \log|\sec x + \tan x|$$

*i. $$e = x \arctan \frac{x}{a} - \frac{a}{2} \log (a^2 + x^2)$$

j. $$f = -\frac{\pi}{2} \log |x| + \frac{a}{x} - \frac{a^3}{9x^3}$$

k. $$Z = -\frac{1}{\sqrt{x^2 - a^2}} - \frac{2a^2}{3(\sqrt{x^2 - a^2})^3}$$

*l. $$Q = \left(\frac{2}{\pi x}\right)^{1/2} \sin x$$

m. $$B = \frac{e^{x/\sqrt{2}} \cos (\sqrt{x/2} + \pi/8)}{\sqrt{2\pi x}}$$

*n. $$Y = (2\pi)^{1/2} x^{x+1} e^{-x}$$

o. $$t = a \cdot e^{-\sqrt{w/2p \cdot x}}$$

*4. For the circuit shown in Figure 4.3, in which $R = 50{,}000$ ohms, and $C = 1.0$ mfd, it is given that $i = 1$ ma (constant) and that $v(0) = 10$ volts. The voltage as a function of time is given by

$$v = 50 - 40e^{-20t}$$

The voltage v can be computed as soon as a value of t is known; write a program to read a value of t, compute v, and print both. (This exercise is based on Problem 2–31b in H. H. Skilling, *Electrical Engineering Circuits*, 2nd ed., Wiley, 1965, p. 78.)

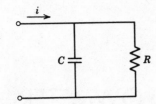

Figure 4.3.   The circuit in Exercise 4.

5. For the circuit in Exercise 4, given that $i = 10^{-3}$ cos $10^5 t$, the forced component v_f of v is given by

$$v_f = 0.01 \cos\left(10^5 t - \frac{\pi}{2}\right)$$

Write a program to read a value of t, compute v_f, and print both. (Based on Problem 2–31c in Skilling, *loc. cit.*)

6. Figure 4.4 describes a conical pendulum. It can be shown that the *period* τ of this pendulum is given by

$$\tau = 2\pi \sqrt{\frac{(L \cos\theta)}{g}}$$

where g is the gravitational constant, 32.2 ft/sec². Write a program that reads a card containing values of L and θ *in degrees* and prints the values of L, θ, and the period.

7. Figure 4.5 represents a projectile being fired into the air with an initial velocity v_0 at an angle of θ_0 with the horizontal. Neglecting air resistance, we describe the projectile at some time t later by

velocity component along the horizontal,

$$v_x = v_0 \cos\theta_0 \quad \text{(independent of time)}$$

velocity component along the vertical,

$$v_y = v_0 \sin\theta_0 - gt$$

* See, for instance, David Halliday and Robert Resnick, *Physics for Students of Science and Engineering*, Wiley, 1960, pp. 102–103.

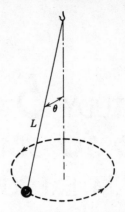

Figure 4.4.   The conical pendulum in Exercise 6.

magnitude of velocity along path of flight,

$$v = \sqrt{v_x{}^2 + v_y{}^2}$$

angle that the velocity vector makes with the horizontal,

$$\theta = \arctan\frac{v_y}{v_x}$$

Write a program that reads a card containing values of v_0, θ_0 (in degrees), and t, computes v_x, v_y, v, and θ (in degrees), and prints all seven values.

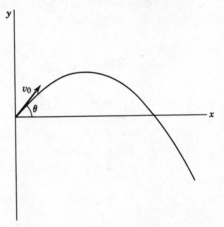

Figure 4.5.   The projectile trajectory in Exercise 7.

CASE STUDY 5 A VENTURI FLOWMETER; THE **GO TO** STATEMENT; PROGRAM EXECUTION

The Venturi meter is used to measure the flow of a liquid through a pipe, as sketched in Figure 5.1. A liquid of density ρ flows through a pipe of cross-sectional area A. At the throat the area is reduced to a and a manometer tube is attached. The manometer liquid, such as mercury, has a density ρ'. The rate of flow is given by

$$Q = aA \sqrt{\frac{2(\rho' - \rho)gh}{\rho(A^2 - a^2)}}$$

where
Q = flow rate, cubic feet per second
a, A = areas, square feet
ρ, ρ' = densities, any convenient consistent units
g = acceleration of gravity, feet per second^2
h = height difference, feet

In this case study we should like to develop a program that will prepare a table of flow rates as a function of height differences for fixed values of the areas and densities. The program will read one card that contains the areas and densities, print these values, then read many additional cards, each containing a height. For each height card the height and flow will be printed, which will permit us to make a plot or table for use with the flowmeter.

The program we need will involve two new features. It will have more than one READ statement, which is not actually very remarkable, and will have a provision for repeating a part of the program as many times as there are height cards. This facility is provided by the GO TO statement, as we can see in the program shown in Figure 5.2.

The first READ obtains values of the two areas and the two densities from one card. These four numbers will not change during the execution of the program, so we print them once at the top of the output page. Thus we find a WRITE statement in this program before any calculation is done; such an arrangement is not only permissible but usual. Now we find another READ statement—this one to get a value of height from another data card. The READ refers back to the same FORMAT statement used by the first READ, which is entirely legal: a FORMAT statement need not immediately follow the READ or WRITE that refers to it, and more than one READ or WRITE can reference the same FORMAT.

This second READ, it will be noted, has been given a statement number, 3, which identifies the statement when later we wish to tell the computer to go back and repeat part of the program.

Now we come to an assignment statement that will compute the flow rate, and then a WRITE to print the height and the flow. We do not print the values of the other four variables because they do not change. It would not be "wrong" to print them, too, but then every line would contain the same four values. The result would be a somewhat unattractive page of output.

Next we find the new statement: GO TO 3. The purpose is simple: return now to execute statement number 3. In other words, go back to read another height card, compute the flow, print height and flow, go back again, etc. etc. Thus we keep going back around the program for as long as there are height cards.

And what will happen when there are no more cards and the program tries to read another? At that point the computer system will detect the out-of-cards condition and act just as though a STOP had been encountered. Actually, as we shall see shortly, a special card is usually placed near the end of the deck to signal the "end of file," or no-more-data-cards. It is then, when the end-of-file card is read, that the computer stops the execution of the program.

It does not matter whether there is one height card or a hundred. However many there are, we will get that many lines of printing, plus the initial line with the fixed parameters. This kind of program structure is fairly typical.

Let us now follow through the process of getting

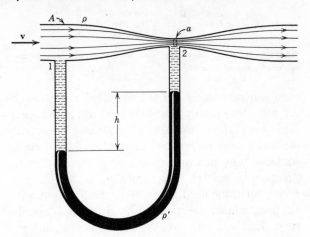

Figure 5.1. The Venturi flowmeter that is the basis in Case Study 5.

this program run on the computer. It is to be hoped that the reader will by now have already run several programs, but perhaps the operations carried out may not have been clear. In any case, now is a good time to describe the process.

The first action is to get the program punched

```
C CASE STUDY 5
C A PROGRAM TO COMPUTE THE FLOW RATE THROUGH A VENTURI FLOWMETER
      REAL A1, A2, RHO, RHOPR, H, Q
C THIS FIRST READ IS EXECUTED ONLY ONCE
      READ (5, 1) A1, A2, RHO, RHOPR
    1 FORMAT (8F10.0)
C WRITE A LINE CONTAINING THE UNCHANGING PARAMETERS
      WRITE (6, 2) A1, A2, RHO, RHOPR
    2 FORMAT (1H , 1P10E13.5)
C THIS SECOND READ IS EXECUTED ONCE FOR EACH HEIGHT CARD
    3 READ (5, 1) H
C COMPUTE THE FLOW FOR THIS HEIGHT
      Q = A1*A2*SQRT (64.4*(RHOPR-RHO)*H/(RHO*(A1**2-A2**2)))
C WRITE HEIGHT AND FLOW
      WRITE (6, 2) H, Q
C GO BACK TO READ ANOTHER HEIGHT CARD AND DO IT ALL OVER
      GO TO 3
      END
```

Figure 5.2. A program to compute the flow through the flowmeter in Figure 5.1.

Zero	0	The letter O	$\emptyset$
One	1	The letter I	I
Two	2	The letter Z	$\bar{Z}$

Figure 5.3. One acceptable way to distinguish certain easily confused pairs of characters.

onto cards by using a machine called a *card punch* or, more commonly, *keypunch*. Card punching is sometimes done by the programmer, especially student programmers, but more often it is done by someone else, in which case it is essential that the program be written very carefully, so that the key-punch operator need never be in doubt about what the programmer meant. Neat handwriting is mandatory. Further, some convention must be followed when writing easily confused pairs of characters, such as 1 and I or 0 and O. One acceptable method of making these distinctions is shown in Figure 5.3.

When the program has been punched and checked it is nearly ready for the computer. (We shall have a good deal more to say about checking in Case Study 10.) Before going to the computer, however, it is necessary in most computer installations to add a few *control cards* to the deck. The exact nature of the requisite control cards varies considerably from one computer installation to the next, so we shall

not give a detailed description of them. Their general purpose is to permit the programmer to describe the actions he wants performed on his deck, rather than to give written or verbal instructions to the computer operator. The control card method is much faster and less open to error.

This sketch of the use of control cards assumes that the computer we shall use is run under the control of an *operating system*, which is a master control program for running the computer between jobs, keeping track of the time used by each program, scheduling input and output, and many other functions. Not all FORTRANs are run under an operating system, but it is becoming rare to find a computer without an operating system that is capable of accepting the full FORTRAN language (also called FORTRAN IV). FORTRAN systems that are run without an operating system would tend to be Basic FORTRAN (also called FORTRAN II). We shall assume throughout that an operating system is employed, which is of some importance to the presentation. For instance, the practice of letting the FORTRAN program read the end-of-file card might be quite unacceptable without an operating system.

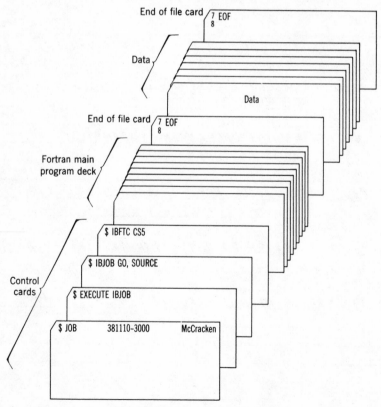

Figure 5.4. Schematic representation of the make-up of a deck containing control cards, a FORTRAN program, and program data.

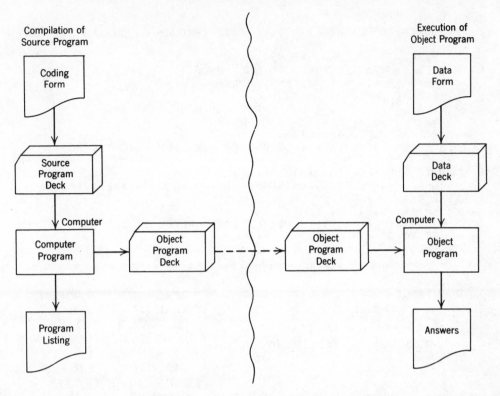

Figure 5.5. Schematic representation of the complete process of compiling and executing a FORTRAN program.

The reader who is in doubt about how his computer is used should ask, but the odds are much in favor of the existence of an operating system if full FORTRAN can be processed.

The deck we would submit to the computing center for the Venturi flowmeter case study would be as diagrammed in Figure 5.4. First there are a few control cards, then the deck as punched from the coding sheets, then an end-of-file card, then the data, then another end-of-file card.

When this deck is processed, two very distinct operations take place. First, the *source program* (what we have written) is *compiled* into an *object program*. The compilation is also called *translation*, which better indicates what happens: the FORTRAN language is "translated" into the very different language of the computer. After this is done, the object program is left in the computer, ready to be executed, as we shall ordinarily do things. The second action is to execute that object program, in the course of which the data cards, if any, will be read.

When a deck is submitted to be compiled and run all in one "pass" this way, the distinction between compilation and execution is of little importance to the programmer. It is also possible,

however, to ask (using appropriate control cards) for the object program to be punched out on another deck of cards, so that if the same program is to be run again it can be done without recompiling.

The situation is graphically described in Figure 5.5. The wavy line separates compilation of the source program from the execution of the object program. If we submit the deck with instructions to do both phases together, which is called "compile and go" in the trade, the wavy line in the diagram is merely an instant in time.

The diagram indicates that the compilation phase produces as output something called a program listing. This is simply a printed copy, produced by the computer's printer, of the source program as we wrote it. Because it is more compact and generally easier to read, the program listing becomes the primary program documentation, and the coding sheets can just as well be discarded. From now on we shall display all programs in the form of the program listings produced when the programs were tested. The listing for this case study appears as Figure 5.6.

Figure 5.7 shows the output produced when our program was executed. It may be noted that identification of the variables has been added "by

```
C CASE STUDY 5
C A PROGRAM TO COMPUTE THE FLOW RATE THROUGH A VENTURI FLOWMETER
C
      REAL A1, A2, RHO, RHOPR, H, Q
C THIS FIRST READ IS EXECUTED ONLY ONCE
      READ (5, 1) A1, A2, RHO, RHOPR
    1 FORMAT (8F10.0)
C WRITE A LINE CONTAINING THE UNCHANGING PARAMETERS
      WRITE (6, 2) A1, A2, RHO, RHOPR
    2 FORMAT (1H , 1P10E13.5)
C THIS SECOND READ IS EXECUTED ONCE FOR EACH HEIGHT CARD
    3 READ (5, 1) H
C COMPUTE THE FLOW FOR THIS HEIGHT
      Q = A1*A2*SQRT(64.4*(RHOPR-RHO)*H/(RHO*(A1**2-A2**2)))
C WRITE HEIGHT AND FLOW
      WRITE (6, 2) H, Q
C GO BACK TO READ ANOTHER HEIGHT CARD AND DO IT ALL OVER
      GO TO 3
      END
```

Figure 5.6. The program listing produced by the compiler as one of the outputs of running the program in Figure 5.2.

hand" to make it easier to study the results. We shall learn a little later (Case Study 9) how to include the production of headings in the program itself, so that the identifications can be made by the computer.

EXERCISES

1. Each of the following programs contains at least 10 errors. Identify them.

*a.   REAL, V, VM, W, T,
     READ (5, 1), VM, W, T,
 1 FORMAT (8F10.0)
     V = VM*COS WT
     WRITE (6, 2) VM, W, T, V,
     STOP

b.   REAL VM, R, W, T, PHI, IF
     READ (5, 1) VM, R, L, W, T
     PHI = ARCTAN(WL/R)
     IF = VM / SQRT(R**2 + W**2*L**2)
        COS(WT − P)
     WRITE (2, 6) VM, R, L, W, T, IF
 2 FORMAT (1H, 10F13.5)
     GO TO 3
     END

c.   REAL A, B, C, D
     READ A, B
     C = 1 + (A − 12B)/(A + 12./−B
     D = 1/(1 + (1 + (1 + (A + B − C)/6))**−3
     WRITE A, B, C, D
     STOP
     END

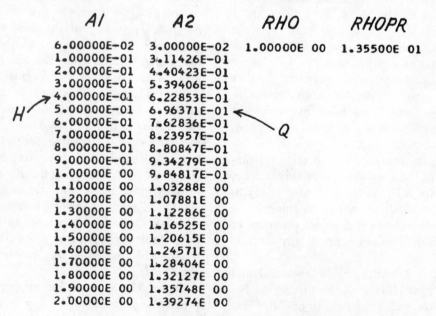

A1	A2	RHO	RHOPR
6.00000E-02	3.00000E-02	1.00000E 00	1.35500E 01
1.00000E-01	3.11426E-01		
2.00000E-01	4.40423E-01		
3.00000E-01	5.39406E-01		
4.00000E-01	6.22853E-01		
5.00000E-01	6.96371E-01		
6.00000E-01	7.62836E-01		
7.00000E-01	8.23957E-01		
8.00000E-01	8.80847E-01		
9.00000E-01	9.34279E-01		
1.00000E 00	9.84817E-01		
1.10000E 00	1.03288E 00		
1.20000E 00	1.07881E 00		
1.30000E 00	1.12286E 00		
1.40000E 00	1.16525E 00		
1.50000E 00	1.20615E 00		
1.60000E 00	1.24571E 00		
1.70000E 00	1.28404E 00		
1.80000E 00	1.32127E 00		
1.90000E 00	1.35748E 00		
2.00000E 00	1.39274E 00		

Figure 5.7. The output of the program in Figures 5.2 and 5.6.

*2. Write a program that reads a succession of data cards, each containing an angle x in radians. For each angle compute $(\sin x)/x$ and print this quantity with x. Use a sequence of values for x that approaches zero to demonstrate that as x approaches zero the quotient approaches 1.

3. Write a program that reads a succession of data cards, each containing an angle x in radians. For each angle compute $\sin x$ and $\sin^2 x$ and print these two quantities with x. Do not use SIN as the name of the value of the sine of x, for it is illegal to take the name of any FORTRAN function as a variable name. For data use a succession of x-values between zero and 2π. These functions come into play in computing the mean and root mean square of a sine function.

4. Modify the program of Exercise 3 so that it accepts an angle in degrees and converts to radians before computing the sine functions. This can be done either by a statement in the form of

$$XRAD = XDEG / 57.29578$$

between the READ and the assignment statement or by using an expression in the argument of the sine, leading to a statement like

$$S = SIN(X/57.29578)$$

Your data may now be a succession of angles between zero and 360.

5. Keeping in mind the hints in Exercise 4, modify the program in Figure 5.2 so that it will accept the height in inches and the two areas in square inches. The units on the densities are still immaterial, so long as both are expressed in the same units. The flow should still be computed in cubic feet per second. The numerical values printed on the output should be in the new units, just as they are read from the data card, but the computation must be done in consistent units. Either method of conversion suggested in Exercise 4 may be used.

*6. Modify the program you wrote for Exercise 4, Case Study 4, so that it reads a succession of cards with time values. If the times are chosen appropriately, you will be able to plot the current growth as a function of time.

7. Modify the program in Exercise 5, Case Study 4, to read a succession of time values and plot the results.

8. Modify the program in Exercise 7, Case Study 4, to read a succession of time values and plot the results.

*9. For the circuit shown in Figure 5.8, assuming zero stored energy at $t = 0$, the current is given by

$$i = K_1 e^{s_1 t} + K_2 e^{s_2 t}$$

where

$$-K_1 = K_2 = \frac{V}{2L\sqrt{\dfrac{R^2}{4L^2} - \dfrac{1}{LC}}}$$

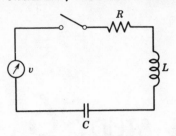

Figure 5.8. The circuit in Exercises 9 and 10.

$$S_1 = -\frac{R}{2L} - \sqrt{\frac{R^2}{4L^2} - \frac{1}{LC}}$$

$$S_2 = -\frac{R}{2L} + \sqrt{\frac{R^2}{4L^2} - \frac{1}{LC}}$$

(This formulation assumes that $R^2/4L^2 > 1/LC$. See Case Study 7 for a more general treatment.)

Write a program that reads a card containing values of V, R, L, and C, the fixed parameters, and prints them on a line at the start of a page. The program should then read a succession of cards, each of which contains a time value, compute the current for each, and for each print a line giving the time and current.

Try values of V = 2, R = 1000, L = 1, C = 10^{-5}, $t = 0.000, 0.002, 0.004, \ldots, 0.030$, and plot the results.

10. The circuit is the same as in Exercise 9, except that the applied voltage is $v = V_m \cos(\omega t)$. The total current (natural plus forced) is given by

$$i = V_m \cos(\omega t - \varphi) + Ae^{s_1 t} + Be^{s_2 t}$$

where

$$|Z| = \sqrt{R^2 + \left(\omega L - \frac{1}{\omega C}\right)^2}$$

$$\varphi = \tan^{-1} \frac{\omega L - \dfrac{1}{\omega C}}{R}$$

$$S_1 = -\frac{R}{2L} - \sqrt{\frac{R^2}{4L^2} - \frac{1}{LC}}$$

$$S_2 = -\frac{R}{2L} + \sqrt{\frac{R^2}{4L^2} - \frac{1}{LC}}$$

$$A = \frac{V_m}{S_1 - S_2}\left(\frac{1}{L} + \frac{S_2}{|Z|}\cos\varphi - \frac{\omega}{|Z|}\sin\varphi\right)$$

$$B = \frac{V_m}{S_1 - S_2}\left(\frac{1}{L} + \frac{S_1}{|Z|}\cos\varphi - \frac{\omega}{|Z|}\sin\varphi\right)$$

(This formulation assumes that $R^2/4L^2 > 1/LC$.) Write a program along the lines of Exercise 9 and use it to compute data for a plot of current versus time. Let $\omega = 400, V_m = 2$.

CASE STUDY 6 COLUMN DESIGN; THE LOGICAL **IF** STATEMENT

The safe loading of a certain type of load-bearing column is a function of the slimness ratio* of the column. A handbook gives two empirical formulas for the relation between the two in different ranges of the slimness ratio:

$$S = \begin{cases} 17{,}000 - 0.485R^2 & \text{for } R < 120 \\[2ex] \dfrac{18{,}000}{1 + \dfrac{R^2}{18{,}000}} & \text{for } R \geq 120 \end{cases}$$

where S = safe load, pounds per square inch
R = slimness ratio

We wish to write a program that will read a succession of slimness cards and compute and print the safe load for each. The problem, of course, is how to choose between the two formulas, which we shall see can be handled nicely by the *logical IF statement*.

The Logical IF is of the general form

$$\text{IF (e) S}$$

where *e* is a *logical expression* and S is any other statement except another logical IF or a DO (discussed in Case Study 15). The simplest form of logical expression is one that asks a question about two arithmetic expressions, as in our example: is RI less than or equal to 120? Or the question might be, is x greater than 12? Or is G**2 equal to H**3? We ask these questions on the basis of any of the following six *relational operators:*

*I am indebted to H. G. Heaslet for pointing out that "slenderness ratio" is the more common term.

Relational Operator	Meaning
.LT.	Less than
.LE.	Less than or equal to
.EQ.	Equal to
.NE.	Not equal to
.GT.	Greater than
.GE.	Greater than or equal to

The periods in these operators are required to distinguish them from possible variable names.

The action of the logical IF is as follows: if the logical expression is true, statement S is executed; if the logical expression is false, statement S is not executed. Either way, the next statement executed is the one following the IF, unless S was a GO TO and the expression was true.

We can see a simple example of the use of the logical IF statement in the column-design program in Figure 6.1. After reading a value of R we ask whether R is less than 120 and if so compute S from the appropriate formula for that case: S = 1.7E4 — 0.485 *R**2. If R is greater than or equal to 120, the assignment statement in the first IF will not be executed, but the one in the second IF will.

It may occur to some readers to wonder why a condition is needed in the second statement: if R is not less than 120, it must be greater than or equal to 120—there are no other possibilities. Why not write

IF (R .LT. 120.0) S = 1.7E4 — 0.485 * R**2
S = 1.8E4 / (1.0 + R*R/1.8E4)

```
C CASE STUDY 6
C COLUMN DESIGN
C
      REAL R, S
   23 READ (5, 1) R
    1 FORMAT (8F10.0)
      IF (R .LT. 120.0) S = 1.7E4 - 0.485 * R**2
      IF (R .GE. 120.0) S = 1.8E4 / (1.0 + R*R/1.8E4)
      WRITE (6, 2) R, S
    2 FORMAT (1H , 1P10E13.5)
      GO TO 23
      END
```

Figure 6.1. A program to compute the safe loading of a column.

The difficulty, of course, is that the second statement would *always* be executed, even if the answer on the first IF were "yes." Thus S would be computed from the first formula and recomputed from the second. The second result would destroy the first, and we would naturally not have the correct value for S.

The power of the logical IF is considerably increased by the combination of several relational expressions with the *logical operators* .AND., .OR., and .NOT.. Suppose, for instance, that we wanted to test R in the case study to establish that it is neither less than 20 nor greater than 225; we are to stop if the value is outside this range. We could write

IF (R .LT. 20.0 .OR. R .GT. 225.0) STOP

Or suppose the problem specifications were these: if the slimness ratio is less than 20.0 or greater than 225.0, set the safe load equal to zero to warn of the questionable data; otherwise, use the formulas given before:

IF (R .LT. 20.0 .OR. R .GT. 225.0) S = 0.0
IF (R .GE. 20.0 .AND. R .LT. 120.0)
    S = 1.7E4 − 0.485 * R**2
IF (R .GE. 120.0 .AND. R .LE. 225.0)
    S = 1.8E4 / (1.0 + R*R/1.8E4)

This combination does the required job. The .OR. in the first IF says to set S equal to zero if *either* of the relations is true. Actually, the .OR. is "satisfied" if either *or both* of the relations is true, but in this case it happens not to be possible for both to be true for any one value of R. In the second IF the complete logical expression is true if *both* relations are true: R must be greater than or equal to 20.0 *and* less than 120.0). This can be true only if R is within the range of 20 to 120, which is what we want. Without the .AND. to bring in the greater than or equal to 20, the formula would be evaluated even if R were less than 20. A similar comment applies to the last IF.

The .NOT. operator reverses the "truth value" of the expression to which it is applied. For instance, the logical IF statement

IF (.NOT. (X .LT. 12.)) R = X + 31.0

has the same effect as

IF (X .GE. 12.0) R = X + 31.1

because "not less than" means the same as "greater than or equal to."

In the expression .NOT. (X .LT. 12.0) the parentheses are required to inform the compiler that the .NOT. modifies the entire relational expression. Under certain circumstances the meaning would otherwise be ambiguous.

Logical expressions of considerable complexity can be built up by using the logical operators. Seldom, however, will it be necessary for us to delve into these complexities, and accordingly we now leave the subject. Case Study 13 takes up the matter again in connection with logical variables.

Figure 6.2 is the output of the program when it was run with a series of slimness ratios.

EXERCISES

1. In each of the following you are to write statements to carry out the actions described. You may regard these actions as a small part of a larger program; that is, you may assume that previous statements have given values to all variables and you need not write input or output statements.

*a. If a is greater than b, set x equal to 16.9, but if a is less than or equal to b, set x equal to 56.9.

b. If rho + theta $< 10^{-6}$, transfer to the statement numbered 156; otherwise do nothing.

*c. If rho + theta $< 10^{-6}$, transfer to statement 156; otherwise transfer to statement 762.

*d. Place whichever of the variables x and y is algebraically larger in BIG.

R	S
2.00000E 01	1.68060E 04
2.50000E 01	1.66969E 04
3.00000E 01	1.65635E 04
3.50000E 01	1.64059E 04
4.00000E 01	1.62240E 04
4.50000E 01	1.60179E 04
5.00000E 01	1.57875E 04
5.50000E 01	1.55329E 04
6.00000E 01	1.52540E 04
6.50000E 01	1.49509E 04
7.00000E 01	1.46235E 04
7.50000E 01	1.42719E 04
8.00000E 01	1.38960E 04
8.50000E 01	1.34959E 04
9.00000E 01	1.30715E 04
9.50000E 01	1.26229E 04
1.00000E 02	1.21500E 04
1.05000E 02	1.16529E 04
1.10000E 02	1.11315E 04
1.15000E 02	1.05859E 04
1.20000E 02	1.00000E 04
1.25000E 02	9.63569E 03
1.30000E 02	9.28367E 03
1.35000E 02	8.94410E 03
1.40000E 02	8.61702E 03
1.45000E 02	8.30237E 03
1.50000E 02	8.00000E 03
1.55000E 02	7.70970E 03
1.60000E 02	7.43119E 03
1.65000E 02	7.16418E 03
1.70000E 02	6.90832E 03
1.75000E 02	6.66324E 03
1.80000E 02	6.42857E 03
1.85000E 02	6.20393E 03
1.90000E 02	5.98891E 03
1.95000E 02	5.78313E 03
2.00000E 02	5.58621E 03

Figure 6.2. The output of the program in Figure 6.1.

e. Place whichever of the variables x, y, and z is algebraically largest in BIG3. (This can be done with only two IF statements. Establish which of x and y is larger, place it in a temporary location, and then compare this number with z to find the largest of the three.)

f. The variables named r and s may be positive or negative. Place the one that is larger *in absolute value* in BIGAB.

*g. An angle named THETA is known to be positive and less than 30 radians. Subtract 2π from THETA as many times as necessary to reduce it to an angle less than 2π; leave the reduced angle in THETA.

*h. If g and h are both negative, set SIGNS equal to -1; if both are positive, set SIGNS to $+1$; if they have different signs, set SIGNS to zero.

i. Y1, Y2, and Y3 are the ordinates of three points on a curve. If Y2 is a *local maximum*, that is Y2 > Y1 and Y2 > Y3, transfer to statement 456; otherwise transfer to statement 567.

*j. If $a < 0$ and $b > 0$, or if $c = 0$, set OMEGA equal to $\cos(x + 1.2)$; otherwise do nothing.

k. If $i = 1$ and $R < S$, transfer to statement 261;

if $i = 1$ and $R \geq S$, transfer to statement 275; if $i \neq 1$, transfer to statement 927.

*l. If $0.999 \leq x \leq 1.001$, STOP; otherwise transfer to statement 639. Do this in two ways:

    1. With a logical IF having two relations combined with an .AND..

    2. With a logical IF having only one test, using the absolute value function.

*m. XREAL and XIMAG are the real and imaginary parts of a complex number. Set SQUARE equal to 1 if XREAL and XIMAG are both less than 1 in absolute value; otherwise do nothing.

    n. Set CIRCLE equal to 1 if $\sqrt{XREAL^2 + XIMAG^2} \leq 1$; otherwise do nothing.

    o. Set DIAMND equal to 1 if the point with coordinates XREAL and XIMAG lies within a square of side $\sqrt{2}$ with its corners on the coordinate axes.

2. In the following exercises you are to write a complete program, including input and output.

*a. Read the value of ANNERN; print ANNERN and compute and print TAX according to the following table:

ANNERN (annual earnings)	TAX
Less than $2000	Zero
$2000 or more but less than $5000	2% of the amount over $2000
$5000 or more	$60 plus 5% of the amount over $5000

b. The current United States Withholding Tax on a weekly salary can be computed as follows: 14% of the difference between a man's gross pay and $13 times the number of dependents he claims. Do not assume that there will always be a tax: a man may not have earned more than his dependency allowance. Read the values of GROSS and DEPEND; compute and print TAX, along with GROSS and DEPEND.

*c. Y is to be computed as a function of X according to

$$Y = 16.7X + 9.2X^2 - 1.02X^3$$

There will be no data to read; compute and print both X and Y for X values from 1.0 to 9.9 in steps of 0.1. You may assume for this exercise that you are working with a decimal computer, so that adding 0.1 to X repeatedly will eventually give 9.9 *exactly*.

*d. Same as (c) but your computer is binary. Since the binary representation of decimal 0.1 is a nonterminating fraction, there is no guarantee that adding the binary representation of 0.1 to X repeatedly will give 9.9 *exactly*. Therefore, if you were to start X at 1.0, add 0.1 each time around, and each time test X against 9.9, the odds are you would not get exact equality and thus go on past 9.9. In fact, you might never get out of the loop. Write a program to solve this problem by letting an integer variable run from 10 to 99; convert to real, then divide by 10. Use the result as the independent variable.

e. Y is to be computed as a function of X according to the formula

$$Y = \sqrt{1 + X} + \frac{\cos 2X}{1 + \sqrt{X}}$$

for a number of equally spaced values of X. Three numbers are to be read from a card: XINIT, XINC, and XFIN. XINIT, we assume, is less than XFIN; XINC is positive. Y is to be computed and printed initially for X = XINIT. Then X is to be incremented by XINC, and Y is to be computed and printed for this new value of X, and so on, until Y has been computed for the largest value of X not exceeding XFIN. (The phrase "the largest value of X *not exceeding* XFIN" lets us ignore the problem presented in the last two exercises. However, this formulation does mean that if the data is set up with the intention of terminating the process with X exactly equal to XFIN it may not do so.)

CASE STUDY 7 — THE MASS-SPRING SYSTEM OR THE SERIES **RLC** CIRCUIT; FLOWCHARTS; THE ARITHMETIC **IF** STATEMENT

The second-order ordinary differential equation with constant coefficients is one of the most familiar in applied mathematics, for it describes a variety of common physical systems. For instance, suppose we have a particle of mass m oscillating along the x-axis with a force kx that tends to return it to the origin, with a frictional damping force $r(dx/dt)$ that opposes the motion, and under the influence of an external force $f(t)$, which is a function of time. The differential equation of the motion is

$$m \frac{d^2x}{dt^2} + r \frac{dx}{dt} + kx = f(t)$$

Or consider the series circuit containing a resistance R, a capacitance C, an inductance L, and a voltage source $f(t)$, shown in Figure 7.1. In this case the differential equation for the current is

$$L \frac{d^2i}{di^2} + R \frac{di}{dt} + \frac{i}{C} = \frac{df(t)}{dt}$$

In our discussion of the equation in this case study we use the electrical terminology as being perhaps slightly more familiar to many readers.

The solution of this equation is given in any text on differential equations. For our purpose here we simply take the applied voltage to be a constant dc voltage V and state the solution without proof.

Case 1. Oscillatory. If $R^2 - 4L/C < 0$, the solution is

$$i = \frac{V}{\omega_n L} e^{\alpha t} \sin \omega_n t$$

where

$$\omega_n = \sqrt{\frac{1}{LC} - \frac{R^2}{4L^2}}$$

$$\alpha = -\frac{R}{2L}$$

Case 2. Critically damped. If $R^2 - 4L/C = 0$, the solution is

$$i = \frac{V}{L} te^{st}$$

where

$$s = \frac{-R}{2L}$$

Case 3. Overdamped. If $R^2 - 4L/C > 0$, the solution is

$$i = K_1 e^{s_1 t} + K_2 e^{s_2 t}$$

where

$$K_1 = -\frac{V}{2L \sqrt{\frac{R^2}{4L^2} - \frac{1}{LC}}}$$

$$K_2 = -K_1$$

$$s_1 = \frac{-R}{2L} - \sqrt{\frac{R^2}{4L^2} - \frac{1}{LC}}$$

$$s_2 = \frac{-R}{2L} + \sqrt{\frac{R^2}{4L^2} - \frac{1}{LC}}$$

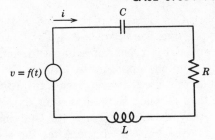

Figure 7.1. The circuit that is the basis in Case Study 7.

We wish to write a program that will read one card containing values of V, R, L, and C, then read a series of time cards, and compute the current for each, so that we can plot the response of the circuit. The program is to be designed to operate correctly, no matter which of the three cases is determined by the values of R, L, and C.

We use this familiar system equation as a vehicle for introducing flowcharts and the arithmetic IF statement.

A flowchart is a graphical representation of the operations performed in a program and the sequence in which they are performed. It is of great assistance in visualizing the interrelationships of the various parts of a complex program. The program for this case study is sufficiently complicated to make a flowchart of definite value.

A flowchart is made up of a set of boxes, the shapes of which indicate the nature of the operations described in them, and connecting lines and arrows that show the "flow of control" between the operations.

For our purpose here the notation, which is quite simple, contains the following symbols:

 A rectangle indicates any processing operation except a decision.

 A diamond indicates a decision. The lines leaving the box are labeled with the decision results that cause each path to be followed.

 A trapezoid indicates an input or output operation.

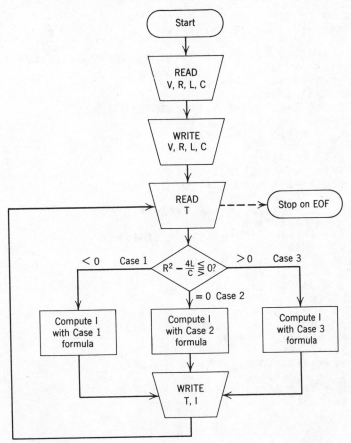

Figure 7.2. A flowchart of the method of solution in Case Study 7.

An oval indicates the beginning or ending point of the program.

A small circle indicates a connection between two points in a flowchart, where a connecting line would be too clumsy.

Arrows indicate the direction of flow through the flowchart; every line should have an arrow on it.

The flowchart of our program is shown in Figure 7.2. We start at the top with the "Start" box and follow the arrows. We read and write the four unchanging parameters, then read a value of time; the operating system will stop program execution on an end-of-file condition when there are no more cards. Now we come to a decision: which case is this? The decision box here has three exits, corresponding to the three cases and, by no coincidence, to the three paths out of an arithmetic IF statement. If the value of the expression $R^2 - 4L/C$ is negative. we have Case 1; if the value of the expression is zero, we have Case 2; and, if the value of the expression is positive, we have Case 3. The flowchart shows a transfer to one of three boxes to compute the current from the appropriate formula. Whichever formula was used, we now wish to print the result and then go back to read another time card.

The program shown in Figure 7.3 contains no new ideas for the first few lines. After the second READ we encounter the arithmetic IF statement. In contrast to the logical IF statement, the arithmetic IF contains an *arithmetic* expression in parentheses, rather than a logical expression, and the action to be taken as a result of evaluating the expression is predetermined in the definition of what the arithmetic IF does. It takes the general form

$$\text{IF } (e) \ s_1, s_2, s_3$$

where e is any arithmetic expression and s_1, s_2, and s_3 are statement numbers. The action is as follows: if the value of the expression e is negative when the statement is encountered in executing the program, statement s_1 is executed next; if the value of the expression is zero, statement s_2 is executed next;

```
C CASE STUDY 7
C THE MASS-SPRING SYSTEM OR THE SERIES RLC CIRCUIT
C
      REAL I, V, R, L, C, T, W, A, S1, S2, K1, K2, S, D
C READ THE UNCHANGING PARAMETERS
      READ (5, 1) V, R, L, C
    1 FORMAT (8F10.0)
C WRITE THE FIRST LINE WITH PARAMETERS
      WRITE (6, 2) V, R, L, C
    2 FORMAT (1H , 1P10E13.5)
C READ A TIME - PROGRAM REPEATS FROM HERE
    6 READ (5, 1) T
C DETERMINE WHICH CASE THIS IS
      IF (R**2 - 4.0*L/C) 20, 30, 40
C CASE 1 - OSCILLATORY
   20 W = SQRT(1.0/(L*C) - R**2/(4.0*L**2) )
      A = -R/(2.0*L)
      I = V/(W*L) * EXP(A*T) * SIN(W*T)
      GO TO 50
C CASE 2 - CRITICALLY DAMPED
   30 S = -R/(2.0*L)
      I = V/L*T*EXP(S*T)
      GO TO 50
C CASE 3 - OVERDAMPED
   40 D = SQRT(R**2/(4.0*L**2) - 1.0/(L*C))
      K1 = -V/(2.0*L*D)
      K2 = - K1
      S1 = -R/(2.0*L) - D
      S2 = -R/(2.0*L) & D
      I = K1 * EXP(S1*T) & K2 * EXP(S2*T)
   50 WRITE (6, 2) T, I
      GO TO 6
      END
```

Figure 7.3. A program to compute the performance of a mass-spring system of a series resonant circuit.

and, if the value of the expression is positive, statement s_3 is executed next. In our case we specify accordingly that if the expression R**2 — 4.0*L/C is negative, zero, or positive we shall want to go to statement 20, 30, or 40, respectively. At these points are found the statements for computing the current according to the correct formula.

For instance, if the expression is negative, we transfer to 20—the next statement, as it happens—where we find appropriate statements for computing the current in the oscillatory case; W and A are used as variable names for ω and α, which may not be good Greek but is commonly done. Naturally, the computation of the current could also have been done with just one arithmetic assignment

V	R	L	C
1.00000E 01	1.00000E 02	2.50000E-01	1.00000E-06
0. F		0. I	
2.00000E-04	7.48500E-03		
4.00000E-04	1.32588E-02		
6.00000E-04	1.65770E-02		
8.00000E-04	1.71249E-02		
1.00000E-03	1.50323E-02		
1.20000E-03	1.08198E-02		
1.40000E-03	5.28947E-03		
1.60000E-03	-6.18212E-04		
1.80000E-03	-5.97789E-03		
2.00000E-03	-1.00185E-02		
2.20000E-03	-1.22283E-02		
2.40000E-03	-1.24129E-02		
2.60000E-03	-1.06999E-02		
2.80000E-03	-7.49511E-03		
3.00000E-03	-3.39950E-03		
3.20000E-03	8.97022E-04		
3.40000E-03	4.72711E-03		
3.60000E-03	7.54549E-03		
3.80000E-03	9.00232E-03		
4.00000E-03	8.98124E-03		
4.20000E-03	7.59907E-03		
4.40000E-03	5.17017E-03		
4.60000E-03	2.14356E-03		
4.80000E-03	-9.75596E-04		
5.00000E-03	-3.70691E-03		
5.20000E-03	-5.66579E-03		
5.40000E-03	-6.61441E-03		
5.60000E-03	-6.48650E-03		
5.80000E-03	-5.38424E-03		
6.00000E-03	-3.54978E-03		
6.20000E-03	-1.31776E-03		
6.40000E-03	9.42592E-04		
6.60000E-03	2.88614E-03		
6.80000E-03	4.24236E-03		
7.00000E-03	4.85062E-03		
7.20000E-03	4.67614E-03		
7.40000E-03	3.80557E-03		
7.60000E-03	2.42452E-03		
7.80000E-03	7.81783E-04		
8.00000E-03	-8.53272E-04		
8.20000E-03	-2.23315E-03		
8.40000E-03	-3.16811E-03		
8.60000E-03	-3.55049E-03		
8.80000E-03	-3.36477E-03		
9.00000E-03	-2.68279E-03		
9.20000E-03	-1.64619E-03		
9.40000E-03	-4.39518E-04		
9.60000E-03	7.41071E-04		
9.80000E-03	1.71845E-03		
1.00000E-02	2.35995E-03		

Figure 7.4. The output of the program in Figure 7.3 for an oscillatory case.

V	R	L	C
1.00000E 01	1.00000E 03	2.50000E-01	1.00000E-06
0. F		0. I	
2.00000E-04	5.36256E-03		
4.00000E-04	7.18926E-03		
6.00000E-04	7.22866E-03		
8.00000E-04	6.46069E-03		
1.00000E-03	5.41341E-03		
1.20000E-03	4.35446E-03		
1.40000E-03	3.40536E-03		
1.60000E-03	2.60878E-03		
1.80000E-03	1.96731E-03		
2.00000E-03	1.46525E-03		
2.20000E-03	1.08041E-03		
2.40000E-03	7.90056E-04		
2.60000E-03	5.73723E-04		
2.80000E-03	4.14161E-04		
3.00000E-03	2.97450E-04		
3.20000E-03	2.12679E-04		
3.40000E-03	1.51473E-04		
3.60000E-03	1.07508E-04		
3.80000E-03	7.60686E-05		
4.00000E-03	5.36740E-05		
4.20000E-03	3.77777E-05		
4.40000E-03	2.65290E-05		
4.60000E-03	1.85913E-05		
4.80000E-03	1.30039E-05		
5.00000E-03	9.07999E-06		

Figure 7.5. The output of the program in Figure 7.3 for a critically damped case.

statement, but it would have been rather involved and our way costs nothing extra. After computing the value of I , we transfer to 50, the WRITE statement. It is now necessary to transfer around the intervening statements to avoid computing I according to the other formulas.

The critically damped case is a little easier to compute. It would have been quite easy to do this one with a single assignment statement, but we prefer this way as a preparation for the refinements that will be made in Case Study 8. After the completion of this computation we must again have a GO TO 50 statement to skip around what follows.

The overdamped case, as we have formulated it, requires a few more statements, but nothing of any real difficulty. It will be noted that we have introduced another intermediate variable to save some computing time and effort: the variable D gives the value of the radical, which is used in three subsequent places.

A GO TO 50 is not needed this time, for statement 50 is, in fact, the next one. (The GO TO would not have been illegal, just unnecessary; the program would operate correctly if the GO TO were inserted.)

The WRITE and the GO TO 6 involve no new ideas.

Figures 7.4, 7.5, and 7.6 show the output produced from runs of this program with data calling for each of the three cases.

V	R	L	C
1.00000E 01	2.00000E 03	2.50000E-01	1.00000E-06
0.	-0.		

2.00000E-04	3.88918E-03
4.00000E-04	4.36795E-03
6.00000E-04	4.12044E-03
8.00000E-04	3.74580E-03
1.00000E-03	3.37502E-03
1.20000E-03	3.03422E-03
1.40000E-03	2.72634E-03
1.60000E-03	2.44936E-03
1.80000E-03	2.20044E-03
2.00000E-03	1.97680E-03
2.20000E-03	1.77589E-03
2.40000E-03	1.59540E-03
2.60000E-03	1.43325E-03
2.80000E-03	1.28758E-03
3.00000E-03	1.15671E-03
3.20000E-03	1.03915E-03
3.40000E-03	9.33536E-04
3.60000E-03	8.38656E-04
3.80000E-03	7.53418E-04
4.00000E-03	6.76844E-04
4.20000E-03	6.08052E-04
4.40000E-03	5.46253E-04
4.60000E-03	4.90734E-04
4.80000E-03	4.40858E-04
5.00000E-03	3.96051E-04

Figure 7.6. The output of the program in Figure 7.3 for an over-damped case.

EXERCISES

*1. Rewrite your answers to the problems in Exercise 1 of Case Study 6, this time using the arithmetic IF statement instead of the logical IF.

*2. The value of the definite integral below is completely determined by the value of a. Write a program that reads a value of a and prints a and the value of the integral.

$$\int_0^\infty \frac{a\,dx}{a^2 + x^2} \begin{cases} = \dfrac{\pi}{2} & \text{for } a > 0 \\[2mm] = 0 & \text{for } a = 0 \\[2mm] = -\dfrac{\pi}{2} & \text{for } a < 0 \end{cases}$$

3. The value of the definite integral below is completely determined by the value of m. Write a program that reads a value of m and prints m and the value of the integral.

$$\int_0^\infty \frac{\sin x \cos mx}{m}\,dx \begin{cases} = 0 & \text{for } m^2 > 1 \\[2mm] = \dfrac{\pi}{4} & \text{for } m = 1 \text{ or } m = -1 \\[2mm] = \dfrac{\pi}{2} & \text{for } m^2 < 1 \end{cases}$$

*4. Rewrite the program in Figure 6.1, Case Study 6, to use an arithmetic IF to make the choice between the formulas.

5. Rewrite the program in Exercise 2a, Case Study 6, to use the arithmetic IF statement.

6. The equation

$$y = 200 - \sqrt{4 \cdot 10^4 - x^2}$$

is to be solved for x values from zero to 5.00, in steps of 0.01, to provide numerical data for milling a spherical surface. Write a program that will complete this task without reading any data. This can be done as follows. First give x the value zero in an initializing portion of the program which is not repeated. Compute y and print it; then execute the statement

$$X = X + 0.01$$

which, executed repeatedly, will eventually give a value of 5.00. Use an arithmetic IF statement to "break the loop"; that is, stop when X exceeds 5.00.

For small values of x the value of y will naturally be very close to zero. Include in the program a test of whether y is less than 0.001 and do not print the line if it is.

CASE STUDY 8 THE MASS-SPRING SYSTEM OR THE SERIES RLC CIRCUIT; THE COMPUTED GO TO STATEMENT

The program we developed in Case Study 7 has certain features that are somewhat undesirable in certain situations. In the first place it was necessary to decide which case we had for every time card, even though the choice was completely determined as soon as the card with the values of V, R, L, and C had been read. We should like to find some way to "store" the result of a decision made at the beginning and apply it repeatedly later on. The computed GO TO provides this capability.

Second, it is not really good practice to recompute the common factors for every time card when the factors never change. The quantities represented by K_1, K_2, s_1, s_2, s, ω, and α all depend solely on V, R, L, and C. They might as well be computed once and for all at the outset and then used when needed.

Let us first look into the computed GO TO which has the general form

$$\text{GO TO } (n_1, n_2, \ldots, n_m), i$$

In this statement i must be an integer variable written without a sign and $n_1, n_2, \ldots, n_m$ must be statement numbers of statements elsewhere in the program. If the value of the variable i at the time this statement is executed is j, control is transferred to the statement with the statement number n_j. For instance, in the program in Figure 8.1 we set the integer variable named CASE equal to 1, 2, or 3. Then several statements later we have the statement

$$\text{GO TO } (20, 30, 40), \text{CASE}$$

If the value of CASE is 1, we transfer to statement 20; if the value is 2, to statement 30; and if the value is 3, to statement 40.

In our situation we need a three-way branch, the same number provided by the arithmetic IF; we take advantage of the computed GO TO because it lets us make a decision just once, then apply that decision at many later times. However, the computed GO TO may also have *any* number of statement numbers, instead of just three, another useful feature. The value of the integer variable must be in the range of 1 to m, where m denotes how many statement numbers there are in parentheses. If it is not in this range, we cannot say, in general, what the program will do.

With this much background on the new programming feature introduced in the case study, we are ready to investigate the program in Figure 8.1. The REAL statement is the same as that of the earlier program in Figure 7.3, with the exception that the variable S does not appear here: S and A were, in fact, the same quantity, and we shall not compute both of them just to be consistent with the terminology of the formulation taken from a standard textbook. (Although a slight duplication for the sake of documentation would certainly be defensible; we have inserted a comment line to point out the change in notation.) An INTEGER

33

```
C CASE STUDY 8
C THE MASS-SPRING SYSTEM OR THE SERIES RLC CIRCUIT
C REVISED TO USE THE COMPUTED GO TO AND TO PRECOMPUTE
C CERTAIN EXPRESSIONS
C
      REAL I, V, R, L, C, T, W, A, S1, S2, K1, K2
      INTEGER CASE
  400 WRITE (6, 401)
  401 FORMAT (1H1)
C READ THE UNCHANGING PARAMETERS
      READ (5, 1) V, R, L, C
    1 FORMAT (8F10.0)
C WRITE THE FIRST LINE WITH PARAMETERS
      WRITE (6, 2) V, R, L, C
    2 FORMAT (1H , 1P10E13.5)
C DETERMINE WHICH CASE THIS IS
      IF (R**2 .LT. 4.0*L/C) CASE = 1
      IF (R**2 .EQ. 4.0*L/C) CASE = 2
      IF (R**2 .GT. 4.0*L/C) CASE = 3
C PRECOMPUTE UNCHANGING FACTORS
C NOTE THAT VARIABLE A IS USED IN PLACE OF BOTH A AND S, SINCE THE
C NUMERICAL VALUES ARE THE SAME.     SIMILARLY, THE ABSOLUTE
C VALUE OF W IS USED FOR W AND D.
      W = SQRT(ABS(1.0/(L*C) - R**2/(4.0*L**2)))
      A = -R/(2.0*L)
      K1 = -V/(2.0*L*W)
      K2 = -K1
      S1 = A - W
      S2 = A + W
C READ A TIME CARD - THIS PORTION OF PROGRAM IS REPEATED
    6 READ (5, 1) T
C CHOOSE FORMULA
      GO TO (20, 30, 40), CASE
C CASE 1 - OSCILLATORY
   20 I = V/(W*L) * EXP(A*T) * SIN(W*T)
      GO TO 50
C CASE 2 - CRITICALLY DAMPED
   30 I = V/L * T * EXP(A*T)
      GO TO 50
C CASE 3 - OVERDAMPED
   40 I = K1 * EXP(S1*T) + K2 * EXP(S2*T)
   50 WRITE (6, 2) T, I
      GO TO 6
      END
```

Figure 8.1.   A revised program to compute the performance of a mass-spring system or a series resonant circuit.

statement establishes the type of the variable CASE. Reading the fixed parameters and printing them in an initial line are also as before.

Next we have three logical IF statements that show which of the three cases is determined by the values of R, L, and C; CASE is set equal to 1, 2, or 3 by these statements.

Now there are seven statements that compute the values of all of the intermediate variables needed anywhere in the three formulas. This, of course, is a slight waste of time, for we cannot possibly need all of these factors for any one case. However, an important principle comes into play here: *these steps are in a part of the program that is not repeated.* A small waste of time here is wasted

only once. The same sort of waste in the repeated portion of the program would incur a much greater penalty, since the waste would be multiplied by the number of repetitions. It often happens that we trade some time at a point in a program that is seldom executed to remove steps from a part that is frequently executed. The net effect is, of course, to save time—occasionally a great deal of time. In our example it would be quite difficult to establish the savings with a stop watch because the entire program would probably be carried out in less than a second on a large computer, but we may as well learn good habits; the principles are of general validity; and on a small computer the difference would be important even here.

With the preliminaries out of the way, we are finally ready to get down to work. A time card is read, and we now ask, in effect, "What was it we decided earlier about which case this is?" The answer to this question has been "stored" for us as the value of the integer variable CASE, and we can call on the computed GO TO for the necessary branch. From here on things are like the earlier program, with the important difference that much less computation is required because everything possible has been computed in advance. The *logic*, however, is similar; that is, the "flow" of statement execution is similar.

The program was tested with the data used in the earlier version and gave identical results.

EXERCISES

*1. If N = 1, 2, or 8, transfer to statement 250; if N = 3 or 7, transfer to statement 251; if N = 4, 5, or 6, transfer to statement 249. You may assume that N is not less than 1 nor greater than 8.

2. Same as Exercise 1, but program a test to stop if N is less than 1 or greater than 8.

*3. Assume that a man's gross pay has been placed in GROSS by a earlier part of your program and that the integer variable CODE has been given a value of 1, 2, 3, or 4. If CODE = 1, place zero in BONDS; if CODE = 2, place 18.75 in BONDS; if CODE = 3, place 37.50 in BONDS; if CODE = 4, place 10% of GROSS in BONDS. In any case, place in NET the difference between GROSS and BONDS.

4. Assume that the integer variable LEG has been given a value by previous statements. Then compute the value of one of the first five Legendre polynomials as follows:

$$IF\ LEG = 0,\ P_0(X) = 1$$
$$1,\ P_1(X) = X$$
$$2,\ P_2(X) = \frac{3}{2}X^2 - \frac{1}{2}$$
$$3,\ P_3(X) = \frac{5}{2}X^3 - \frac{3}{2}X$$
$$4,\ P_4(X) = \frac{35}{8}X^4 - \frac{15}{4}X^2 + \frac{3}{8}$$

Your program should place in P the value it computes, whichever Legendre polynomial it is. Note that LEG ranges from zero to 4; it will be necessary to add 1 to it to get a value that may be used in a computed GO TO.

CASE STUDY 9 RESISTANCE TABLE; INPUT AND OUTPUT

In this case study we shall consider a particularly simple problem, mathematically, in order to concentrate on a new aspect of programming: input and output formats.

The program we shall write is required to read a small deck of cards. Each card contains two numbers: a B & S wire gauge number and the diameter of that gauge in thousandths of inches. For each card the program must print one line that will give gauge, diameter, and resistance per thousand feet at 20°C. A program that will do this job is by now an almost trivial task if we use the familiar input and output methods. Figure 9.1 shows a program and Figure 9.2, the output from it, when the deck of cards gave the data on gauges 1 to 40. Let us consider this output to see what additional facilities would be useful.

To begin with, it is obviously awkward to have the gauge numbers printed in floating point: nobody expects to have to look for No. 27 wire in the form 2.70000E 01. The first thing we can reasonably consider, therefore, is a way to read and print integers. Here, of course, the integers are only identifying numbers, but from time to time we have problems in which integer input is used in the calculation itself.

Let us change the specifications of the input card, too, while we are at it, to make the card preparation a little more reasonable. The new card form will have a two-digit integer in columns 1–2, punched without a decimal point. The wire diameter will be punched in columns 3–7 with a decimal point; five col-

umns were chosen because they are adequate for a number that ranges from 289.3 to 3.145 and never has more than four digits, plus a decimal point. Figure 9.3 shows a typical card, for gauge 8. We note that column 1 has been left blank and that the 8 is punched in column 2. If the 8 had been punched in column 1 and column 2 left blank, the number would have been interpreted as 80: blank columns in a data card are taken in most cases to be equivalent to zeros.

This card cannot be read with the standard FORMAT statement we have been using so far because it describes a card in which every input number is punched in 10 columns usually with a decimal point. We must now write a new FORMAT statement to tell the program that our data card this time has an integer in columns 1–2 and a floating point number in 3–7, punched with a decimal point. This is readily done; the required FORMAT is

FORMAT (I2, F5.0)

The I2 and F5.0 are called *field specifications;* they tell the program, in the case of input, how the data items are arranged on the card. For instance, the I2 says to expect that the first data value on the card will be an integer (no decimal point punched ever) and that it will occupy two columns. The F5.0 says to expect that the next five columns will contain a number that is to be a real number inside the computer. It may or may not have a decimal point punched; if not, interpret the number punched in the five col-

```
C CASE STUDY 9
C WIRE TABLE - INPUT AND OUTPUT METHODS
C
      REAL G, D, R
   12 READ (5, 1) G, D
    1 FORMAT (8F10.0)
      R = 8.1456E+3 / (0.7853916 * D * D)
      WRITE (6, 2) G, D, R
    2 FORMAT (1H , 1P10E13.5)
      GO TO 12
      END
```

Figure 9.1. A program to compute the resistance of a series of wire sizes.

umns as though there were zero digits after an assumed decimal point. If a decimal point *is* punched, as we shall do, then interpret the number as the decimal point dictates.

We have here a brief introduction to the general characteristics of the FORMAT statement and to two commonly used field specifications. We can summarize the way the FORMAT statement keeps "in step" with the READ statement and with the data values. The execution of a READ always initiates the reading of a new card; it is never possible to have one card read by two different READ statements, the second taking up where the first left off. Reading always begins with column 1. The *first* data field (group of columns) is the one described by the *first* field specification and the value is assigned to the *first* variable named in the READ; the *second* data field is associated with the *second* field specification in the FORMAT and the *second* variable named in the READ; and so on, for as many variables as there are in the READ.

One frequently used simplification in the writing of FORMAT statements is that if the same field specification is repeated we may write it once with a repetition number in front. We have seen this in all of the programs to date: instead of writing F10.0 eight times, we have been writing 8F10.0, which has exactly the same effect.

It will be noted, too, that in the simplified FORMATs we have been using there have been more field specifications than data fields. In such a case the extra field specifications are simply ignored. (If there are *fewer* field specifications than variables, "scanning" of the FORMAT begins again at the beginning. We shall not need this capability in this book, however. Case Study 17 takes up related matters.

Let us now consider the output side to see what variations may be called for there.

The first field specification for a line to be printed should always specify carriage control. The situation here is that on a line that is printed the first character, in fact, is *not* printed but used instead to control spacing of the paper, according to the following table:

Character	Action
blank	single space
0	double space
1	skip to top of next page
+	suppress spacing

The standard FORMAT statement, which we have used for all output so far, has started with a field specification that made the first character of each line a blank. This is the "1H," called a Hollerith

1.00000E 00	2.89300E 02	1.23920E-01
2.00000E 00	2.57600E 02	1.56295E-01
3.00000E 00	2.29400E 02	1.97083E-01
4.00000E 00	2.04300E 02	2.48485E-01
5.00000E 00	1.81900E 02	3.13452E-01
6.00000E 00	1.62000E 02	3.95191E-01
7.00000E 00	1.44300E 02	4.98086E-01
8.00000E 00	1.28500E 02	6.28103E-01
9.00000E 00	1.14400E 02	7.92474E-01
1.00000E 01	1.01900E 02	9.98823E-01
1.10000E 01	9.07400E 01	1.25962E 00
1.20000E 01	8.08100E 01	1.58821E 00
1.30000E 01	7.19600E 01	2.00288E 00
1.40000E 01	6.40800E 01	2.52576E 00
1.50000E 01	5.70700E 01	3.18435E 00
1.60000E 01	5.08200E 01	4.01576E 00
1.70000E 01	4.52600E 01	5.06300E 00
1.80000E 01	4.03000E 01	6.38597E 00
1.90000E 01	3.58900E 01	8.05174E 00
2.00000E 01	3.19600E 01	1.01537E 01
2.10000E 01	2.84600E 01	1.28046E 01
2.20000E 01	2.53500E 01	1.61392E 01
2.30000E 01	2.25700E 01	2.03598E 01
2.40000E 01	2.01000E 01	2.56711E 01
2.50000E 01	1.79000E 01	3.23691E 01
2.60000E 01	1.59400E 01	4.08188E 01
2.70000E 01	1.42000E 01	5.14352E 01
2.80000E 01	1.26400E 01	6.49146E 01
2.90000E 01	1.12600E 01	8.18013E 01
3.00000E 01	1.00300E 01	1.03094E 02
3.10000E 01	8.92800E 00	1.30115E 02
3.20000E 01	7.95000E 00	1.64098E 02
3.30000E 01	7.08000E 00	2.06905E 02
3.40000E 01	6.30500E 00	2.60896E 02
3.50000E 01	5.61500E 00	3.28956E 02
3.60000E 01	5.00000E 00	4.14855E 02
3.70000E 01	4.45300E 00	5.23036E 02
3.80000E 01	3.96500E 00	6.59706E 02
3.90000E 01	3.53100E 00	8.31843E 02
4.00000E 01	3.14500E 00	1.04857E 03

Figure 9.2. The output of the program in Figure 9.1.

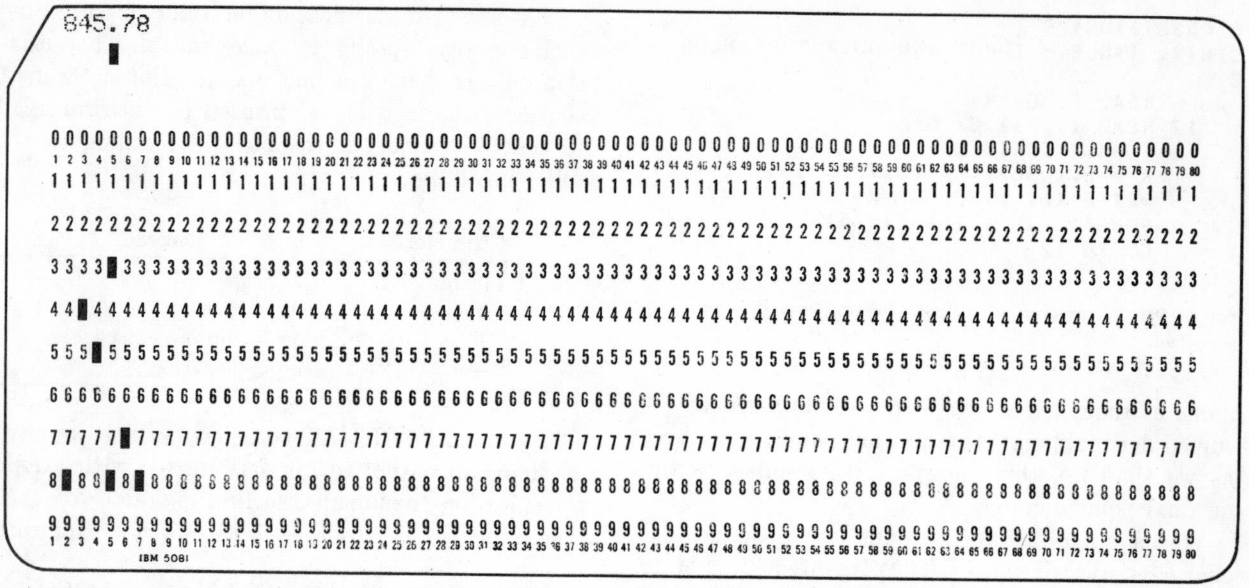

Figure 9.3. A sample of a card for reading by the program in Figure 9.4.

field specification.* The blank following the H is, of course, essential: this is the character that calls for single spacing. If we want to specify skipping to the top of the next page, which we shall often do at the start of a program, we use 1H1. We shall see later in this case study that the Hollerith field specification actually has a much broader application than just carriage control and may be used to do things like printing headings.

Next comes the field specification for the wire gauge. Assuming that we want the printing to start at the extreme left-hand side of the page, and there is no reason not to do so, we can use an I2 specification as we did for input. We know that the number can never be more than two digits, and there is no reason to allow blank spaces to the left of the number.

The specification for the wire diameter is a different matter. Here we must consider two things: the total amount of space to be allotted to the field and the handling of the decimal point. These are the functions of the two numbers in the field specification; if we indicate the specification symbolically as $Fw.d$, then w designates the total number of printing spaces that will be assigned to the result and d stands for the number of digits after the decimal point, which is always printed on output.

*So named in honor of Herman Hollerith, who invented a way of representing letters and digits on punched cards for use in the United States Census of 1890.

Looking at the decimal point situation first, with the data of this problem we need three places after the point for values like 3.145. The largest number, on the other hand, is 289.3, so that in some cases we shall have three digits before the decimal point. Thus we must allow at least seven positions for this number: possibly three before the point, the point itself, and the three places that will always come after the point. (When a number such as 289.3 is printed, it will be in the form 289.300.) We should allow some added space, however, to make the output simple to read; if we were to specify only seven places, the largest numbers would be crowded up against the wire size with no space (and if negative numbers were possible, a space would have to be allowed for the minus sign). If, by accident, we allow too few places for the number, digits at the left are simply deleted; no warning is given. Let us somewhat arbitrarily allow five extra positions for a total of 12. The field specification we want is F12.3. The F says that the internal number is the real type and that the output should be printed in a fixed format without an exponent; the 12 says 12 spaces total; the 3 says three digits after the decimal point.

We could handle the format for the resistance in the same way, using the F specification. If we wanted exactly the same format, we could indicate that two successive fields had the same format by writing

FORMAT (1H , I2, 2F12.3)

However, there is a minor problem in printing the resistance in this manner, and getting around the problem will give us an opportunity to see how the E specification works.

This one has the general form Ew.d, where w and d have the same meaning as before. E means that the internal number to be printed is of the real type and that it should be printed with an exponent. If we do not use a scale factor, discussed in the next paragraph, the standard form will be

$$\pm 0.dd \ldots dE \pm ee$$

where $E \pm ee$ is the exponent, and there will be as many digits between the decimal point and the letter E as specified by d in the field specification. Plus signs are not ordinarily printed, as we have seen.

This form of printing displays a number in the form of a multiplier between 0.1 and 1.0, times a power of 10. Many people prefer the form of a multiplier between 1.0 and 10.0, times a power of 10. This is easily called for in FORTRAN by the use of a *scale factor*, which takes the form sP and is written before the field specification. For our purposes here, with the E specification for output, the total effect is to place s digits before the decimal point and decrease the exponent by s. Thus, if with a field specification of E12.5, a certain number printed as

$$0.36289E \ 03$$

with 1PE12.5 it would print as

$$3.62894E \ 02$$

The same quantity is represented either way. Note that use of a scale factor does not change the number of digits after the decimal point, so that in this example the first form has five significant digits and the second form has six.

```
C CASE STUDY 9
C WIRE TABLE — INPUT AND OUTPUT METHODS
C REVISED TO USE NEW FIELD SPECIFICATIONS
C
      REAL D, R
      INTEGER G
   12 READ (5, 1) G, D
    1 FORMAT (I2, F5.0)
      R = 8.1456E+3 / (0.7853916 * D * D)
      WRITE (6, 2) G, D, R
    2 FORMAT (1H , I2, F12.3, 1PE14.3)
      GO TO 12
      END
```

Figure 9.4. A modified version of the program in Figure 9.1.

1	289.300	1.239E-01
2	257.600	1.563E-01
3	229.400	1.971E-01
4	204.300	2.485E-01
5	181.900	3.135E-01
6	162.000	3.952E-01
7	144.300	4.981E-01
8	128.500	6.281E-01
9	114.400	7.925E-01
10	101.900	9.988E-01
11	90.740	1.260E 00
12	80.810	1.588E 00
13	71.960	2.003E 00
14	64.080	2.526E 00
15	57.070	3.184E 00
16	50.820	4.016E 00
17	45.260	5.063E 00
18	40.300	6.386E 00
19	35.890	8.052E 00
20	31.960	1.015E 01
21	28.460	1.280E 01
22	25.350	1.614E 01
23	22.570	2.036E 01
24	20.100	2.567E 01
25	17.900	3.237E 01
26	15.940	4.082E 01
27	14.200	5.144E 01
28	12.610	6.522E 01
29	11.260	8.180E 01
30	10.030	1.031E 02
31	8.928	1.301E 02
32	7.950	1.641E 02
33	7.080	2.069E 02
34	6.305	2.609E 02
35	5.615	3.290E 02
36	5.000	4.149E 02
37	4.453	5.230E 02
38	3.965	6.597E 02
39	3.531	8.318E 02
40	3.145	1.049E 03

Figure 9.5. The output of the program in Figure 9.4.

What we would like in our example is to force the output to have just four significant digits. If we were to use an F12.3 field specification, the results might have as many as seven, as in a number like 1089.746. This is not really desirable because we should not like to suggest that the accuracy of the data justifies this kind of precision. One way to proceed is to use the E field specification, in which the number of significant digits is always the same, and arrange things so that we get four digits.

An appropriate field specification is readily devised. We want a total of four digits; we propose to use a scale factor of 1P, so that there will be one digit before the point; therefore we need three places after the point. The exponent takes four printing positions; then there are the three digits after the point, the point itself, and one digit before the point. This makes a total of nine positions. (Again, if the number could be negative we would have to allow for the minus sign.) Once more providing five extra spaces for readability, we arrive at the total printing width of 14. The field specification we need is therefore 1PE14.3.

With these changes in the FORMAT statements, the program becomes as shown in Figure 9.4. G has been identified as an integer, but nothing else in the program has been changed. The modified program was rerun with a modified data deck. The output is shown in Figure 9.5.

Let us now modify the program again to make it print identification and column headings. The output should begin at the top of a new page with the heading line

RESISTANCE PER THOUSAND FEET
OF COPPER WIRE GAUGES 1–40

The columns should be headed GAUGE, DIAMETER, and RESISTANCE. There should be a blank line between the page heading and the column headings and between the column headings and the first line of results. The results themselves should be single-spaced.

Getting the heading line will require a WRITE statement that refers to a FORMAT statement but lists no variables. All the heading information will be contained within the FORMAT itself in the form of a long Hollerith field. This may be seen in the program in Figure 9.6. The FORMAT begins with a Hollerith field 1H1 to force skipping to the

top of a new page before any printing begins. We then have a much longer Hollerith field:

55HRESISTANCE PER THOUSAND FEET
OF COPPER WIRE GAUGES 1–40

It is so long, in fact, that it cannot be contained in the allowable space, which ends in column 72. Accordingly, we specify a continuation card by punching a 1 in column 6 of the following card. As many as 19 continuation cards may be used.

The H designates a field of text and the 55 says that it will be 55 characters long. When counting characters, everything must be included: blanks, hyphens—everything.

The next WRITE refers to another FORMAT that contains Hollerith text only. This time the carriage control character is zero, to cause double spacing before printing. After the text, which gives the desired headings, we see two slashes. These cause one blank line to be inserted after the line. (If we had wanted *two* blank lines we would have written *three* slashes, etc.)

It would be possible in both cases to combine the carriage-control character with the text so that there would be only one Hollerith field in the FORMAT. This, however, is to be discouraged as confusing and error-prone.

The FORMAT for the output is almost the same as before, the only difference being that it is necessary somehow to specify shifting everything two spaces to the right. The purpose of the shift is to center the gauge numbers under the word GAUGE. The simplest way to do this would no doubt be to

```
C CASE STUDY 9
C WIRE TABLE - INPUT AND OUTPUT METHODS
C REVISED TO USE NEW FIELD SPECIFICATIONS AND PRINT HEADINGS
C
      REAL D, R
      INTEGER G
C WRITE HEADING LINE - NOTE ABSENCE OF VARIABLE IN WRITE
      WRITE (6, 30)
   30 FORMAT (1H1, 55HRESISTANCE PER THOUSAND FEET OF COPPER WIRE GAUGES
     1 1-40)
      WRITE (6, 31)
   31 FORMAT (1H0, 31HGAUGE     DIAMETER     RESISTANCE//)
   12 READ (5, 1) G, D
    1 FORMAT (I2, F5.0)
      R = 8.1456E&3 / (0.7853916 * D * D)
      WRITE (6, 32) G, D, R
   32 FORMAT (1H , 2X, I2, F12.3, 1PE14.3)
      GO TO 12
      END
```

Figure 9.6. A further modified version of the program in Figure 9.1.

RESISTANCE PER THOUSAND FEET OF COPPER WIRE GAUGES 1-40

GAUGE	DIAMETER	RESISTANCE
1	289.300	1.239E-01
2	257.600	1.563E-01
3	229.400	1.971E-01
4	204.300	2.485E-01
5	181.900	3.135E-01
6	162.000	3.952E-01
7	144.300	4.981E-01
8	128.500	6.281E-01
9	114.400	7.925E-01
10	101.900	9.988E-01
11	90.740	1.260E 00
12	80.810	1.588E 00
13	71.960	2.003E 00
14	64.080	2.526E 00
15	57.070	3.184E 00
16	50.820	4.016E 00
17	45.260	5.063E 00
18	40.300	6.386E 00
19	35.890	8.052E 00
20	31.960	1.015E 01
21	28.460	1.280E 01
22	25.350	1.614E 01
23	22.570	2.036E 01
24	20.100	2.567E 01
25	17.900	3.237E 01
26	15.940	4.082E 01
27	14.200	5.144E 01
28	12.610	6.522E 01
29	11.260	8.180E 01
30	10.030	1.031E 02
31	8.928	1.301E 02
32	7.950	1.641E 02
33	7.080	2.069E 02
34	6.305	2.609E 02
35	5.615	3.290E 02
36	5.000	4.149E 02
37	4.453	5.230E 02
38	3.965	6.597E 02
39	3.531	8.318E 02
40	3.145	1.049E 03

Figure 9.7. The output of the program in Figure 9.6.

change the I2 field specification to I4, which would provide two extra spaces at the left of the gauge number and would shift the rest of the line two spaces to the right also. Instead, we shall take advantage of the opportunity to see another useful field specification in action: the X. This has the simple function of skipping over as many spaces as the number written in front of it—2 in this case. It will often be found convenient to do this instead of providing extra positions in other field specifications. The X field specification may also be used with input, in which case it simply ignores as many card columns as specified. This is usually done to skip over blank columns, but in fact it makes no difference what—if anything—is punched in the columns skipped.

The output of this final modification of the program is shown in Figure 9.7. No change in the data was necessary.

We shall see in the following case study, and repeatedly thereafter, that a heading line with Hollerith text may also contain results, or, viewed alternatively, that a line of output may have identifications printed with the numerical values. Some examples appear in the exercises.

It should be noted, too, that the usage we have made of the slash in a FORMAT statement is also just one specialized example of its usefulness. More generally, the slash specifies that one "record" (card or line, roughly speaking) should be terminated and another begun. Suppose, for instance, that we have these statements:

WRITE (6, 34) A, B, C
34 FORMAT (1H , F10.3/2F12.5)

The action will be to print one number on the first line under control of F10.3, "terminate" that line, printing nothing more on it, and then print two values on the second line under control of F12.5. In the program in Figure 9.6 FORMAT statements 30 and 31 could have been written as one statement by using a slash between "1–40" and "1H0." Or we could have used *two* slashes between the two long Hollerith fields to get the required double spacing between the heading lines, which would have made the 1H0 unnecessary.

On input a slash means to ignore whatever remains on the current card, if anything, and move on to read from another card. We might write, for instance,

READ (5, 39) A, B, C
39 FORMAT (I4, F10.0/F10.0)

Values for A and B would be taken from one card and a value for C from a second. If anything were punched in columns 15–80 of the first card, it would be ignored.

This usage of the slash is explored further in the exercises; other examples appear in the answers at the back of the book.

EXERCISES

1. Show how the given data values would be printed under control of the field specifications stated.

   a. I5: 0, 1, 10, −587, 90062, 123456
   b. F7.2: 0.0, 1.0, 16.77, −586.21, 0.04, 12.34
   c. F5.0: 0.0, 1.0, 16.87, −12.32
   d. E10.3: 0.0, 0.00072, 601000., −473., −0.0123
   e. 1PE10.3: 0.0, 10.0, 0.000076, 6780000., −627., −0.000456

2. Given that the values of three variables are as follows,

$$M = 12$$

$$X = 407.8$$

$$Y = -32.9$$

show exactly what would be printed by

WRITE (6, 107) M, X, Y

with each of the following FORMAT statements, in which b stands for a blank:

  *a. 107 FORMAT (1Hb, 3HM = b, I3, 3HX = b, F6.1, 3HY = b, F6.1)
  b. 107 FORMAT
    (1H1, 17HREADINGbNUMBERb = b, I3/ 1H0, 11HPRESSUREb = b, F6.1, 7HTEMPb = b, F6.1)
  c. 107 FORMAT
    (1H1, 17HREADINGbNUMBERb = b, I3/ 1H0, 8HPRESSURE, 6X, 11HTEMPERATURE/1H0, F7.1, 9X, F6.1)

*3. Four numbers are punched on a card; they are new values of real variables named BOS, EWR, PHL, and DCA. Each number is punched in eight columns, the first beginning in column 1. Each number contains a decimal point. Write READ and FORMAT statements to read the card.

4. Same as Exercise 3, except that there is no decimal point. The numbers are to be treated as if they had two decimal places, that is, two places to the right of an assumed decimal point.

5. Same as Exercise 3, except that each number occupies 14 columns and is punched with a decimal point and an exponent.

*6. A card is punched in the following format:

Columns	Sample Format	Variable Name
1–3	$\pm$xx	LGA
4–6	xxx	JFK
7–20	$\pm$x.xxxxxxxE$\pm$ee	BAL
21–34	$\pm$x.xxxxxxxE$\pm$ee	TPA

LGA and JFK are integer variables; BAL and TPA are real. The small letters stand for any digits. Write statements to read such a card.

*7. Given a WRITE statement,

WRITE (6, 92) I, J, R, S

I and *J* are integer variables and *R* and *S* are real. For each of the following, write a FORMAT statement that could produce such a line or lines with the given WRITE statement.

  a. bbbb-16bb92017bbb16.82bb437.89
  b. bbbb-16bb92017bbb17.bb438.
  c. bbb-16bbb92017bb0.16824Eb02bb0.43789Eb03
  d. bbb-16bbb92017bb1.6824Eb01bb4.3789Eb02
  e. I = bbb-16bbJ = b92017bbR = bb16.8bbS = b437.9
  f. I = bbb-16
    J = b92017
    R = bb16.8
    S = b437.9

[Write only one FORMAT statement for (f).]

8. Same as Exercise 7, but given the WRITE statement

WRITE (6, 93) M, P, Q, R

M is an integer variable and the others are real.
 a. b9bb-33.b1.6E-04bb1.439024Eb06
 b. b9bb-32.62bb0.00016bb0.1439024Eb07
 c. b9bb-32.62bb0.16E-03bb1439024.

*9. The values of the real variables A, B, X, and Z are to be printed on one line. A and B are to be printed without exponents, X and Z with exponents. Twelve spaces should be allowed for A and B, and they should have four decimal places. Twenty spaces should be allowed for X and Z, and they should be printed with eight decimal places and no scale factor. Write appropriate statements.

10. Same as Exercise 13 except that a positive integer named K is to be printed in six spaces between A and B and the decimal point is to be moved one place to the right in X and Z.

CASE STUDY 10 FORTRAN REVIEW; NEWTON-RAPHSON METHOD; PROGRAM CHECKOUT

We are now in an ideal position to stop for review and consolidation of what we have learned about FORTRAN. Quite a bit has been covered, in fact—enough to permit interesting and useful work to be done with a computer. It seems appropriate to show most of the FORTRAN techniques in one case study which introduces no basically new language elements. This case study does, however, involve an important computational method that we may explore in brief before considering the program, and furthermore we shall use this case study as the vehicle for a look at the ways of finding mistakes in a program and guaranteeing that it is correct.

The Newton-Raphson method for finding a root of an equation is one of the most commonly used numerical methods, being valuable with or without a computer. Given a function of x, $F(x) = 0$, the method says that, subject to certain conditions, if x_i is an approximation to a root, a better approximation is given by

$$x_{i+1} = x_i - \frac{F(x_i)}{F'(x_i)}$$

where the prime denotes the derivative. This is called an *iteration formula*.

For instance, suppose we have the function $F(x) = x^2 - 25$. As a first approximation to the root, take $x_0 = 2$. Because $F'(x) = 2x$, a better approximation can be found from

$$x_1 = x_0 - \frac{x_0^2 - 25}{2x_0} = 2 - \frac{4 - 25}{4} = 7.25$$

Continuing in the same way, now substituting 7.25 into the formula, and so on, we get a succession of approximations:

$$x_0 = 2$$
$$x_1 = 7.25$$
$$x_2 = 5.35$$
$$x_3 = 5.0114$$
$$x_4 = 5.000001$$
$$x_5 = 5.0000000001$$

The approximation can be made as accurate as we please by continuing the process. One root of the original equation is indeed 5, the square root of the constant term.

The Newton-Raphson procedure is readily adapted to computer use. In fact, some variation of the scheme just sketched is commonly used for finding square roots. It can readily be adapted to finding roots of poly-nominal equations. For example, to find the roots of

$$x^3 - 1.473x^2 - 5.738x + 6.763 = 0$$

we would use the iteration formula

$$x_{i+1} = x_i - \frac{x_i^3 - 1.473x_i^2 - 5.738x_i + 6.763}{3x_i^2 - 2.946x_i - 5.738}$$

This is a simple enough formula to program; if many evaluations of high-order polynomials are required, there are more economical ways to evaluate the iteration formula than at first meet the eye.* This cubic has roots near —2.3, 1.1, and 2.7; the Newton-Raphson method must be applied three times, with appropriate starting values, to find all three roots. Alternatively, having found one root x_1 we can sometimes divide the original polynomial by $(x - x_1)$ to obtain a reduced polynomial, to which the Newton-Raphson method is applied again, although accuracy problems may develop. Complex roots can be found by using complex arithmetic and complex starting values.

For an example to program let us take a transcendental equation:

$$F(x) = \cosh x + \cos x - A$$

For A less than 2 this equation has no real roots; for $A = 2$ it has four (identical) roots at $x = 0$, and for A greater than 2 it has two roots equal in absolute value and opposite in sign. We shall try values in each category to provide some insight into the behavior of computers when carelessly programmed. We readily find that

$$F'(x) = \sinh x - \sin x$$

so the iteration formula is

$$x_{i+1} = x_i - \frac{\cosh x_i + \cos x_i - A}{\sinh x_i - \sin x_i}$$

We shall have to compute the hyperbolic sine and cosine from exponentials, for we have no built-in functions for evaluating them:

$$\cosh x = \frac{e^x + e^{-x}}{2}$$

$$\sinh x = \frac{e^x - e^{-x}}{2}$$

Figure 10.1 is a flowchart of the computational scheme. We begin by printing a heading line that identifies the program, then read a value for A and print it. Now the iteration process starts; we begin by arbitrarily setting $x = 1.0$. This becomes our "previous" approximation. In the iteration formula we wrote x_i, but the subscript is not needed here; all that is necessary is to have two different variables, one for the "old" value of X and another for the "new" value. We shall write XNEW for the new value and compute it from the iteration formula. Because it will be interesting to see the succession

* See, for example, McCracken and Dorn, *Numerical Methods and FORTRAN Programming*, Wiley, New York, 1964, pp. 77, 150.

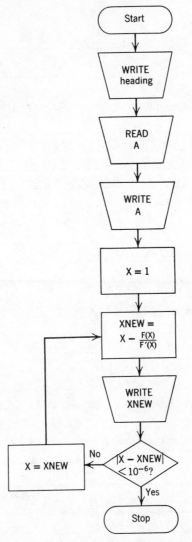

Figure 10.1. A flowchart of the method used in Case Study 10 to find the roots of a transcendental equation by the Newton-Raphson method.

of approximations, we now write out the value of XNEW. Ordinarily we would do this only at the conclusion, when the root had been found.

Next comes the test for convergence: are the last two approximations the same to within 10^{-6}? This question has to be asked in terms of the absolute value of the difference between X and XNEW, because we do not know, in general, whether the convergence will be from above or below. If the process has *not* converged we set X = XNEW, which makes the just-computed value the "old" value the next time around, and return to compute another XNEW. If the process has converged, we go back to read another value of A and repeat the entire computation.

```
C CASE STUDY 10
C NEWTON-RAPHSON METHOD, PROGRAM CHECKOUT
C
C THIS PROGRAM CONTAINS MANY DELIBERATE ERRORS
C
    1 REAL X, EX, EMX, XNEW, A
    2 WRITE (6,3)
    3 FORMAT (1H1, 31HROOT OF COSH X + COS X - A = 0)
    4 READ (5, 5) A
    5 FORMAT (8F10.0)
    6 WRITE (6, 7) A
    7 FORMAT (1HO, 3HA =, F10.5//)
C START ITERATIONS AT X = 1.0
    8 X = 1.0
    9 EX = EXP(X)
C GET EXP(-X) = 1.0/EXP(X)
   10 EMX = 1.0 / EX
C NEW APPROXIMATION
   11 XNEW = X + ((EX+EMX)/2. + COSF(X) - A)/((EX-EMX)/2. - SIN(X))
   12 WRITE (6, 13), XNEW
   13 FORMAT (1H , F10.6)
```

```
C CHECK FOR CONVERGENCE
   14 IF (ABS(X - XNEW) .LT. 1.E-6) GO TO 4
   15 XNEW = X
   16 GO TO 9
   17 END
```

Figure 10.2. A program to find the roots of an equation by the Newton-Raphson method. *This program contains many deliberate errors.*

The program shown in Figure 10.2 contains quite a number of deliberate errors. After noting a few features of the program we shall trace through the steps we might follow in discovering and correcting the various kinds of error. To simplify this discussion, all statements have been given statement numbers; several of the statements would not otherwise need them.

Statement 9 assigns the value of e^x to the variable EX. This will be used twice in the iteration formula, and it is wasteful to go through the function evaluation twice. Statement 10 assigns the value of e^{-x} to EMX, e^x and e^{-x} being reciprocals. Division is also faster than function evaluation. (Naturally, *in this program* the time savings could not be detected with a stopwatch, but the principles are valid.) The iteration formula of statement 11 is routine, as is the WRITE statement.

The IF statement uses the absolute-value function ABS to eliminate the sign of the difference between X and XNEW. Statements 15 and 16 are unexceptional.

There are a great many ways to make mistakes in a program. One of the easiest is in punching the program cards. If the programmer punches them himself, he may make mistakes because he is not experienced at operating the card punch. If they are punched by a cardpunch operator, there is the danger of misreading handwriting. Either way, much time can be saved in the long run by checking the deck carefully before trying to compile it.

One common way to prove the accuracy of punching is to *verify* the deck with a *verifier*. This machine has the same general appearance as a card punch but only examines the punching. A second operator strikes the same keys, we hope, as were struck in punching the cards, but all that happens is that the holes in the cards are matched to see if they correspond to the keys struck by the verifier operator. If not, a red light signals the discrepancy. If the error was made by the verifier operator, the red light can be cleared and the checking continued; if the hole in the card is incorrect, the card is notched and must be repunched.

A second way to check the accuracy of the punching, and one that is sometimes more readily available to the amateur programmer, is to *list* it with a *tabulator*. Figure 10.3 shows the listing of the deck punched from Figure 10.2. Study discloses three errors attributable to punching. In statement 6 the WRITE begins in column 6 instead of the required column 7; this would prevent compilation of the program. The actual punching error, of course, was in beginning the statement number one column too soon, but this placement of the statement number is not an error in FORTRAN and would, by itself, cause no trouble. In statement 11 there is a P where there should be a plus sign. This happens to be easy to do because of the way the cardpunch works. The same explanation applies to the mistake in punching statement number 16.

A study of this listing will no doubt also reveal to the careful reader many of the deliberate errors. This is generally true. Desk-checking the program listing before compiling is usually a good investment of the programmer's time as well as a saving in computer time. However, let us pretend that we have not seen these errors, so that we may learn how they can be detected later in the process.

Now we shall try to compile. All FORTRAN compilers include some degree of diagnostic checking for statements that are *syntactically* illegal; that is, statements that do not follow the rules for forming a statement, no matter what they might mean. FORTRAN cannot determine whether we wrote what we meant to, in general, but it can sometimes establish without question that a statement is illegal. The various FORTRAN systems vary widely in the degree of checking provided and in the amount of checking that is done at any one

```
C CASE STUDY 10
C NEWTON-RAPHSON METHOD, PROGRAM CHECKOUT
C
C THIS PROGRAM CONTAINS MANY DELIBERATE ERRORS
C
      1 REAL X, EX, EMX, XNEW, A
      2 WRITE (6, 3)
      3 FORMAT (1H1, 31HROOT OF COSH X + COS X - A = 0)
      4 READ (5, 5) A
      5 FORMAT (8F10.0)
     6 WRITE (6, 7) A
      7 FORMAT (1H0, 3HA =, F10.5//)
START ITERATIONS AT X = 1.0
      8 X = 1.0
      9 EX = EXP(X)
C GET EXP(-X) = 1.0/EXP(X)
      10 EMX = 1.0 / EX
C NEW APPROXIMATION
      11 XNEW = X + ((EX+EMX)/2. P COSF(X) - A)/((EX-EMX)2. - SIN(X)
      12 WRITE (6, 13), XNEW
      13 FORMAT (1H , F10.6)
C CHECK FOR CONVERGENCE
      14 IF (ABS(X - XNEW) .LT. 1.E-6) GO TO 4
      15 XNEW = X
      1L GO TO 9
      17 END
```

Figure 10.3. Tabulator listing of the deck punched from the program in Figure 10.2.

```
C CASE STUDY 10
C NEWTON-RAPHSON METHOD, PROGRAM CHECKOUT
C
C THIS PROGRAM CONTAINS MANY DELIBERATE ERRORS
C
      1 REAL X, EX, EMX, XNEW, A
      2 WRITE (6, 3)
      3 FORMAT (1H1, 31HROOT OF COSH X + COS X - A = 0)
      4 READ (5, 5) A
      5 FORMAT (8F10.0)
      6 WRITE (6, 7) A
      7 FORMAT (1H0, 3HA =, F10.5//)
START ITERATIONS AT X = 1.0

ERROR MESSAGE NUMBER     1

ERROR MESSAGE NUMBER     2

ERROR MESSAGE NUMBER     3
      8 X = 1.0
      9 EX = EXP(X)
C GET EXP(-X) = 1.0/EXP(X)
     10 EMX = 1.0 / EX
C NEW APPROXIMATION
     11 XNEW = X + ((EX+EMX)/2  + COSF(X) - A)/((EX-EMX)2. - SIN(X)

ERROR MESSAGE NUMBER     4
     12 WRITE (6, 13), XNEW

ERROR MESSAGE NUMBER     5
     13 FORMAT (1H , F10.6)
C CHECK FOR CONVERGENCE
     14 IF (ABS(X - XNEW) .LT. 1.E-6) GO TO 4
     15 XNEW = X
     16 GO TO 9
     17 END
```

Figure 10.4. Program listing produced by the compiler when the program in Figure 10.2 was run.

stage. Some, for instance, take no further action on a statement once any error has been found in it. A statement with many errors could thus require many compilation attempts to locate all the errors.

In our case the compiler detected what seemed to be six errors. Figure 10.4 shows the program listing, with its notes on where errors occurred, and Figure 10.5 gives the error messages produced. We see immediately that the first three messages are the result of the compiler's attempt to make some sense out of what was obviously meant to be a comment line but did not have the required C in column 1. The actual error messages are of little meaning. This will happen; the compiler can seldom guess what we meant to do when given a collection of characters that ought to follow certain rules but does not.

Error message number 4 tells us that in statement 11 there are not the same number of left and right parentheses. We are not told where the error is, of course, because there are many ways of mak-

ing the statement legal, having different meanings. In fact, we are missing a right parenthesis at the end of the statement.

Error 5 refers to a punctuation requirement in the FORTRAN language that is not strictly necessary. That is, the language could have been designed either to require a comma in the position noted or not; the compiler could have been designed to interpret the statement unambiguously either way. Most punctuation is distinctly *not* of this sort.

Error message 6 at first seems unlikely: the parentheses in statement 3 *do* balance! This is an example of the kind of error message that has to be decoded. As seen by the compiler, the parentheses do *not* balance; it looks to us as if they do. What has happened? The answer is that the count of 31 in the Hollerith field is one too long: it includes what was intended to be the closing right parenthesis of the statement.

The compiler used for this program consists of a number of phases or passes. If a disabling error is detected in the first phase, that phase is completed but no others are initiated. After we have corrected the errors found in the first phase, we try again. This time the error scan is continued, at which point another error is discovered. This time the compilation is again disabled. The error message (not reproduced) on statement 4 is, "types combined illegally." Inspection reveals that the constant 2 has been written without a decimal point; the decimal point was apparently omitted from the corrected card when the P was changed to $+$. Things like this are easy to do!

We correct it and try again to compile. The error message is again on statement 4: "operator missing before or after operand EMX." EMX appears twice in the statement; the reference is to the second appearance, where the division operator has been omitted.

Again. This time the program compiles, but we are told: "undefined control dictionary entries referenced COSF." This refers to the fact that in this system the mathematical functions are inserted into the object program after compilation, just before object program execution. In compilation COSF was assumed to be a function name, but no check was made to see that it existed in the "library" of standard functions. Then, at the loading of the ob-

ject program, a search was made to determine whether COSF was the name of a library function or of a function that we could provide at loading time by techniques to be explored in Case Study 22. Neither being the case, the error message was written and loading terminated. (The culprit here was that in an earlier version of FORTRAN all function names ended in *F*. Habit can be blamed for this one.)

With these corrections made, the program compiles. There may still be errors! (And there are.) There is a limit to the amount of checking that can be designed into a compiler economically, and there are certain types of error that cannot be detected in any case, as we now see.

When the program was executed, the computer wrote out several hundred feet of magnetic tape before the operator decided something was probably wrong and interrupted the program at the computer console. When the tape was printed, it was found to contain the heading line, the value $A = 4$, and the one value -4.743014 repeated approximately 50,000 times before the operator stopped the program.

What happened? Somehow the iteration system was not iterating correctly; the suspicion is that it never got a new value to work with. This was in-

DIAGNOSTIC   MESSAGES

1   SOURCE ERROR   9   LEVEL 3 — ASSEMBLY DELETED
ILLEGAL CHARACTER IN A NUMERIC FIELD.

2   SOURCE ERROR   7   LEVEL 1 — WARNING ONLY
INCORRECT EFN IN COLUMNS 1 TO 6.

3   SOURCE ERROR   169 LEVEL 2 — LOADING SUPPRESSED
   SYMBOL BEGINNING WITH   ITERAT TRUNCATED TO SIX CHARACTERS.

4   SOURCE ERROR   170 LEVEL 4 — COMPILATION TERMINATED ERROR SCAN CONTINUES
PARENTHESES DO NOT BALANCE.

5   SOURCE ERROR   328 LEVEL 1 — WARNING ONLY
REDUNDANT COMMA OR ILLEGAL PUNCTUATION

                    PHASE B DIAGNOSTIC MESSAGES

6   SOURCE ERROR   279 LEVEL 2 — LOADING SUPPRESSED
FORMAT STATEMENT   3   PARENS DO NOT BALANCE

Figure 10.5. The error listing that accompanies the program listing in Figure 10.4.

ROOT OF COSH X + COS X - A = 0

A =   4.00000

```
 -4.743014
 -5.658057
 -6.631848
 -7.624681
 -8.621946
 -9.620513
-10.619826
-11.619567
-12.619491
-13.619472
-14.619465
-15.619462
-16.619460
-17.619460
-18.619459
-19.619459
-20.619459
-21.619459
-22.619458
-23.619458
-24.619458
```

Figure 10.6. The output produced by a partly corrected version of the program in Figure 10.2.

deed the case: in statement 8 the variables are reversed from what we want. The compiler was quite unable to diagnose the error, for XNEW = X and X = XNEW are both completely legal statements.

With this statement corrected, we go back to the computer. This time the answers are different, but the process never converges. The first 20 approximations are shown in Figure 10.6. Now we have to wonder if the iteration formula itself is correct. Accordingly we carefully inspect the method of computing sinh x and cosh x, the formula for the derivative, and anything else we can think of to check. An excellent idea is to calculate what the first value of XNEW should be, using a book of tables and a desk calculator. This is revealing: the value should be about 6.7, instead of the —4.743014 we see. Sooner or later it will be noticed that in statement 11 the fraction is being *added* to X instead of subtracted from it.

This, of course, could have been discovered from the first run if an approximate value had been

```
C CASE STUDY 10
C NEWTON-RAPHSON METHOD, PROGRAM CHECKOUT
C
C THE ERRORS IN THE PROGRAM OF FIGURE 10.1 HAVE BEEN CORRECTED,
C AND AN ITERATION COUNTER HAS BEEN ADDED
C
      1 REAL X, EX, EMX, XNEW, A
        INTEGER N
      2 WRITE (6, 3)
      3 FORMAT (1H1, 30HROOT OF COSH X + COS X - A = 0)
      4 READ (5, 5) A
      5 FORMAT (8F10.0)
      6 WRITE (6, 7) A
      7 FORMAT (1H0, 3HA =, F10.5//)
C START ITERATIONS AT X = 1.0
      8 X = 1.0
C START ITERATION COUNTER AT 1
        N = 1
      9 EX = EXP(X)
C GET EXP(-X) = 1.0/EXP(X)
     10 EMX = 1.0 / EX
C NEW APPROXIMATION
     11 XNEW = X - ((EX+EMX)/2. +  COS(X) - A)/((EX-EMX)/2. - SIN(X))
     12 WRITE (6, 13) XNEW
     13 FORMAT (1H , F10.6)
C CHECK FOR CONVERGENCE
     14 IF (ABS(X - XNEW) .LT. 1.E-6) GO TO 4
C CHECK ITERATION COUNTER
        IF (N .GT. 20) GO TO 165
        N = N + 1
     15 X = XNEW
     16 GO TO 9
    165 WRITE (6, 166)
    166 FORMAT (1H , 34HFAILS TO CONVERGE IN 20 ITERATIONS)
        GO TO 4
     17 END
```

Figure 10.7. The final corrected version of the program in Figure 10.2.

computed for a check. It is always a very good idea to compute a test case by desk calculator. Even if the process does coverage to what seems to be a reasonable answer, there is usually no assurance of correctness unless a test case checks out.

The chances are getting better that the next time we may get a correct answer, but while we are making this change it would be an excellent idea to insert an *iteration counter* in the program to set a limit on the number of iterations. It is most disconcerting to leave a program to be run on the night shift and then discover in the morning that it ran two hours when we thought it might take two minutes at the longest. It is little comfort what the reason may have been; we shall see shortly a case in which the program was quite correct but the problem analysis was somewhat faulty. A program that never terminates is undesirable in any case.

The iteration counter is a simple matter of setting N equal to 1 after reading a value of A, adding 1 to it on each iteration, and in each iteration asking if it has yet exceeded some reasonable limit on the number of iterations. If the limit of 20 chosen here is exceeded, we print a notice to that effect and go back for another value of A.

The completed program is shown in Figure 10.7. Part of the output from it is given in Figure 10.8, where we see fairly rapid convergence for $A = 4$ and $A = 3$, a doubtful answer for $A = 2$, and nonsense for $A = 1$. Comparison with hand calculation shows that the first case is correct, and we are therefore willing to accept the second, but what about the others?

For $A = 2$ the graph of the function is tangent to the x-axis at the origin, where further analysis, which we shall not develop, shows that there is, in fact, a quadruple root. Study of the numerator in the iteration formula shows that it can become very close to zero while x is still some distance from zero.* The last two values for the root are indeed the same to within 10^{-6}, but both are some 20,000 times 10^{-6} from the true root. That's life.

For $A = 1$, when we knew in advance that there were no real roots, the system fell apart com-

---

* In the series expansions for cos x and cosh x note that when $x = 0.02$ only the $x^2/2!$ terms have any effect in a computer with roughly eight decimal digits in the fractional part of a floating point (real) number; $x^4/4!$ is lost entirely when added to a number near 1. Compare with the example on page 61. Thus cos 0.02 + cosh 0.02 = 2, to the limits of our available precision.

ROOT OF CCSH X + COS X - A = 0

A =   4.00000

```
   6.743014
   5.749290
   4.772434
   3.854463
   3.076748
   2.531097
   2.265857
   2.208072
   2.205626
   2.205622
   2.205622
```

A =   3.00000

```
   3.746581
   2.947440
   2.348734
   1.995278
   1.871750
   1.858076
   1.857921
   1.857921
```

A =   2.00000

```
   0.750149
   0.562647
   0.421954
   0.316498
   0.237376
   0.178039
   0.133538
   0.100160
   0.075157
   0.056415
   0.042472
   0.031974
   0.026502
   0.021703
   0.021703
```

A =   1.00000

```
  -2.246284
  -1.436134
  -0.070283
 641.662842
```

ERROR TRACE. CALLS IN REVERSE ORDER.

CALLING ROUTINE	IFN OR LINE NO.	ABSOLUTE LOCATION
FXPF	50	12314
10.X	6	03112

EXP(X), X GRT THAN 88.029692 NOT ALLOWED

SET RESULT = + OMEGA
640.662842

ERROR TRACE. CALLS IN REVERSE ORDER.

CALLING ROUTINE	IFN OR LINE NO.	ABSOLUTE LOCATION
FXPF	50	12314
10.X	6	03112

EXP(X), X GRT THAN 88.029692 NOT ALLOWED

SET RESULT = + OMEGA
639.662842

ERROR TRACE. CALLS IN REVERSE ORDER.

Figure 10.8. The output of the program in Figure 10.7.

pletely.* The error message, produced at execution time, says that we were trying to take the exponential of a number larger than 88; the result, if it were computed, would be too large to hold in one storage location in the computer used. The result was set equal to zero and the program continued. Numerically what happened was this. The process tried to get to $x = 0$. When x became small, the problem we noted above also applied to the denominator, except that here we were *subtracting* two approximate and nearly equal numbers. This almost always causes trouble; in this case it led to a huge value for the fraction. Then, with the exponential disabled, the process could never recover.

This example is meant to suggest that, useful as computers surely are, they are not giant brains. They blindly try to do what we tell them to, even if our instructions are not very intelligent.

A really thorough program for finding roots of equations, even simple-appearing functions such as polynomials, is a *very* complicated matter. The reader will often find it to his advantage to use "package" programs, developed for the purpose, in which great pains have been taken to avoid some of the more common difficulties. But the thoughtful reader even then will perhaps be inclined to be cautious.

Most programs either do not compile the first

* Which you can't depend on either! Things would be *much* simpler if we could be assured that the computer would always notify us when we ask a stupid question. Not so. See Case Study 12 for illustrations.

time they are tried or, if they do, they produce incorrect answers. Experienced programmers expect to have to spend time on program checkout and they plan accordingly. We conclude this brief introduction to the subject of program checkout with a few suggestions on how to go about it.

1. Checkout is usually facilitated if values of intermediate variables are available. Common practice inserts extra WRITE statements to get the values, then removes the cards and recompiles when checkout is complete.

2. Time spent in desk-checking a program will shorten the total time the programmer must spend on checkout and will also save computer time. This should be done both before and after the program cards are punched.

3. When you correct one mistake, try to avoid making two more.

4. Never assume that a program is correct just because the compiler detects no errors.

5. Accomplish as much as you can with each computer run. Even after locating one error, there may be others, and you probably have enough evidence to locate some of them. Resist the almost overwhelming temptation to rush back to the computer after finding one mistake. Make sure that your corrective action accounts for *all* the troubles. Stated otherwise: operate on the disease, not the symptoms.

6. The final test of a program is comparison with hand calculations wherever possible. When choosing test cases, try to select values that bring all parts of the program into operation.

CASE STUDY 11 THE SERIES RLC CIRCUIT; BASIC OPERATIONS WITH COMPLEX NUMBERS

In this case study we use a simple electric circuit as the vehicle for a study of how to program operations with complex numbers in FORTRAN. The electrical engineering content of the illustrative problem is strictly minimal so that the reader who wishes to learn how to use the FORTRAN complex operations capability will have no difficulty with the specific example.

On the other hand, the reader who anticipates no need for this material can postpone study of the entire case study with no loss of continuity. Only Case Study 24 makes any explicit use of FORTRAN complex numbers after this, and even then at least 99% of the emphasis is on other matters.

Let us begin by investigating how complex numbers are manipulated in FORTRAN. We recall that in introducing operations on real numbers we learned about constants, variables, arithmetic operations, mathematical functions, and input/output. We can follow much the same outline here.

A FORTRAN complex constant is composed of two real constants separated by commas and enclosed in parenthesis. Thus (2.0, 3.0) would be an acceptable representation of what we would write in ordinary mathematical notation as $2+3i$ or in electrical engineering notation as $2+j3$. A number that is pure real, but written in complex notation, simply has a zero imaginary part: such as (4.67, 0.0). A pure imaginary likewise has a zero real part: (0.0, 91.0). Any form permitted for a real constant may be used in writing complex constants, so that

all of the following, for example, are correct complex constants:

$$(0., 659.)$$
$$(1.57, 4.5E\text{-}10)$$
$$(.1905, 7.62E2)$$

A complex variable is simply one that has been named in a COMPLEX statement. There are no restrictions on naming. A FORTRAN complex variable consists of two numbers representing the real and imaginary parts; each of these numbers may take on any of the values permitted of real variables. For instance, we might have a program containing these statements:

```
COMPLEX Z, IMPED
Z = (2.0, 87.344)
IMPED = (0.0, 0.0056)
```

Z and IMPED would be set up as pairs of storage locations because of the COMPLEX statement. The second statement would make the real part of Z equal to 2.0 and the imaginary part equal to 87.344. Similarly, IMPED would be given real and imaginary parts of zero and 0.0056, respectively. It would then be possible to carry out operations using Z and IMPED in succeeding statements.

The four arithmetic operations and exponentiation are all represented by the same symbols as for the real and integer cases, although, of course, the various operations require separate actions on the real and

53

```
C CASE STUDY 11
C EXAMPLES OF OPERATIONS WITH COMPLEX NUMBERS
C
      REAL R, S
      COMPLEX X, Y, Z
      Z = (1.0, 2.0) + (3.0, 4.0)
      WRITE (6, 1) Z
    1 FORMAT (1H0, 36HLINE  1    (1.0, 2.0) + (3.0, 4.0) = , 2F8.2)
      Z = (2.0, 6.0) - (3.0, 3.0)
      WRITE (6, 2) Z
    2 FORMAT (1H0, 36HLINE  2    (2.0, 6.0) - (3.0, 3.0) = , 2F8.2)
      Z = (1.0, 3.0) * ( 2.0, 4.0)
      WRITE (6, 3) Z
    3 FORMAT (1H0, 36HLINE  3    (1.0, 3.0) * (2.0, 4.0) = , 2F8.2)
      Z = (5.0, 10.0) / (2.0, 1.0)
      WRITE (6, 4) Z
    4 FORMAT (1H0, 37HLINE  4    (5.0, 10.0) / (2.0, 1.0) = , 2F8.2)
      Z = (3.0, -2.0)**2
      WRITE (6, 5) Z
    5 FORMAT (1H0, 27HLINE  5    (3.0, -2.0)**2 = , 2F8.2)
      Z = (4.0, 0.0) * (5.0, 0.0)
      WRITE (6, 6) Z
    6 FORMAT (1H0, 36HLINE  6    (4.0, 0.0) * (5.0, 0.0) = , 2F8.2)
      Z = (0.0, 4.0) * (0.0, 5.0)
      WRITE (6, 7) Z
    7 FORMAT (1H0, 36HLINE  7    (0.0, 4.0) * (0.0, 5.0) = , 2F8.2)
      Z = (6.0, 0.0) + (-4.0, -2.0)
      WRITE (6, 8) Z
    8 FORMAT (1H0, 38HLINE  8    (6.0, 0.0) + (-4.0, -2.0) = , 2F8.2)
      Z = 6.0 + (-4.0, -2.0)
      WRITE (6, 9) Z
    9 FORMAT (1H0, 31HLINE  9    6.0 + (-4.0, -2.0) = , 2F8.2)
      Z = (5.0, 10.0) / 2.0
      WRITE (6, 10) Z
   10 FORMAT (1H0, 30HLINE 10    (5.0, 10.0) / 2.0 = , 2F8.2)
      X = (1.0, 2.0)
      Y = (3.0, 4.0)
      Z = X * Y
      WRITE (6, 11) Z
   11 FORMAT (1H0, 14HLINE 11    Z = , 2F8.2)
      R = CABS(Y)
      WRITE (6, 12) R
   12 FORMAT (1H0, 20HLINE 12    CABS(Y) = , F8.2)
      R = REAL(X)
      WRITE (6, 13) R
   13 FORMAT (1H0, 20HLINE 13    REAL(X) = , F8.2)
      R = AIMAG(X)
      WRITE (6, 14) R
   14 FORMAT (1H0, 21HLINE 14    AIMAG(X) = , F8.2)
      Z = CSQRT((5.0, -12.0))
      WRITE (6, 15) Z
   15 FORMAT (1H0, 32HLINE 15    CSQRT((5.0, -12.0)) = , 2F8.2)
      Z = CSQRT((-1.0, 0.0))
      WRITE (6, 16) Z
   16 FORMAT (1H0, 31HLINE 16    CSQRT((-1.0, 0.0)) = , 2F8.2)
      Z = CEXP((1.0, 2.0))
      WRITE (6, 17) Z
   17 FORMAT (1H0, 29HLINE 17    CEXP((1.0, 2.0)) = , 2F12.6)
      Z = CEXP((0.0, 3.14159265))
      WRITE (6, 18) Z
   18 FORMAT (1H0, 36HLINE 18    CEXP((0.0, 3.14159265)) = , 2F12.6)
      STOP
      END
```

Figure 11.1. A program to produce examples of operations with complex numbers.

imaginary parts. For reference, the definitions may be reviewed:

$$(a + bi) + (c + di) = (a + c) + (b + d)i$$

$$(a + bi) - (c + di) = (a - c) + (b - d)i$$

$$(a + bi) * (c + di) = (ac - bd) + (ad + bc)i$$

$$(a + bi) / (c + di) = \frac{(ac + bd)}{c^2 + d^2} + \frac{(bc - ad)}{c^2 + d^2} i$$

A meaning has not been shown for exponentiation because the method of raising a complex number to a power in the computer depends on the size of the exponent. For small powers an actual multiplication is used. For larger powers the number may be converted to polar form and use may be made of De Moivre's formula.* A complex quantity can be raised to an integer power only; it cannot be raised

* A complex number $a + bi$ can be converted to the form $\rho(\cos \theta + i \sin \theta)$ in which $\rho = \sqrt{a^2 + b^2}$ is called the absolute value and $\theta = \tan^{-1} b/a$ is the amplitude. Then $(a + bi)^n = \rho^n(\cos n\theta + i \sin n\theta)$.

to a power that is a real or complex number (using the FORTRAN operator **, that is; raising to a power is defined mathematically and can be done in FORTRAN with functions).

The four arithmetic operations are defined for pairs of complex quantities, based on the formulas just displayed; they may also be used when one of the numbers is complex and the other real. Thus, if Z is complex and R is real, we are permitted expressions of this sort:

$$2.0 + (3.0, 4.0)$$
$$6.5 * Z$$
$$Z / R$$

For future reference note that Tables A1.1 and A1.2 in Appendix 1 state explicitly which combinations of operators are permitted and which are not.

For further examples of FORTRAN operations on complex quantities consider the program of Figure 11.1 and the output of Figure 11.2. We see that three variables have been declared to be complex. Then there is a statement that carries out

```
LINE  1    (1.0, 2.0) + (3.0, 4.0) =        4.00     6.00
LINE  2    (2.0, 6.0) - (3.0, 3.0) =       -1.00     3.00
LINE  3    (1.0, 3.0) * (2.0, 4.0) =      -10.00    10.00
LINE  4    (5.0, 10.0) / (2.0, 1.0) =       4.00     3.00
LINE  5    (3.0, -2.0)**2 =        5.00   -12.00
LINE  6    (4.0, 0.0) * (5.0, 0.0) =       20.00     0.
LINE  7    (0.0, 4.0) * (0.0, 5.0) =      -20.00     0.
LINE  8    (6.0, 0.0) + (-4.0, -2.0) =      2.00    -2.00
LINE  9    6.0 + (-4.0, -2.0) =      2.00    -2.00
LINE 10    (5.0, 10.0) / 2.0 =      2.50     5.00
LINE 11    Z =     -5.00    10.00
LINE 12    CABS(Y) =       5.00
LINE 13    REAL(X) =       1.00
LINE 14    AIMAG(X) =       2.00
LINE 15    CSQRT((5.0, -12.0)) =       3.00    -2.00
LINE 16    CSQRT((-1.0, 0.0)) =       0.      1.00
LINE 17    CEXP((1.0, 2.0)) =    -1.131204    2.471727
LINE 18    CEXP((0.0, 3.14159265)) =    -1.000000    0.000000
```

Figure 11.2. Examples of operations with complex numbers; produced by the program in Figure 11.1.

complex addition on two complex constants. The WRITE statement mentions just the one complex variable, but because a FORTRAN complex number consists of two FORTRAN real numbers we must provide two field specifications in the FORMAT (and that is about all there is to input and output of complex numbers). The somewhat elaborate Hollerith field permits us to write an output line that is easy to discuss. Identified as LINE 1 in Figure 11.2, it shows the two numbers added as FORTRAN complex constants and the result simply as two numbers.

In similar fashion the next four lines illustrate the other three arithmetic operations and exponentiation. LINE 6 shows that if the two complex numbers are pure reals written in complex notation the result is correct. Applied to pure imaginaries, however, the result, of course, is different (LINE 7).

LINE 8 shows that just one of the two complex constants may, in fact, be a pure real. LINE 9 shows that one of the two may be an actual FORTRAN real, which will give the same complex result. LINE 10 does the same for division.

LINE 11 involves complex variables rather than just constants and shows that things work as we might expect.

LINE 12 exhibits the result of our first example of a function. This is the complex absolute value, which accepts a complex argument and returns a single real number as the function. The absolute value of a complex number $a + bi$ is simply $\sqrt{a^2 + b^2}$; it is also called its *magnitude*. This result can be viewed as a demonstration of the Pythagorean theorem for a 3-4-5 right triangle.

The next two lines illustrate the application of two frequently used functions. REAL accepts a complex argument and returns the real part as a single FORTRAN real number. AIMAG accepts a complex argument and returns the imaginary part as a single FORTRAN real number.

LINE 15 is an example of a function of a constant—this time the complex square root. Note the double parentheses, one for the function argument and one for the complex constant. Note also that this function returns only one of the two square roots, just as its real counterpart does; $-3 + 2i$ is also a square root of $5 - 12i$. Similarly, as LINE 16 shows, the complex square root function returns only one of the square roots of -1.

The complex exponential is demonstrated in LINE 17 and LINE 18. The second of these is a numerical verification of the famous equation due to Euler,

$$e^{\pi i} + 1 = 0$$

which combines in one simple equation the five most important numbers in mathematics.

For reference we may collect in one place the definitions of the common complex functions in terms of operations on their real and imaginary parts.

$$CABS(a + bi) = \sqrt{a^2 + b^2}$$

$$CEXP(a + bi) = e^a(\cos b + i \sin b)$$

$$CLOG(a + bi) = \tfrac{1}{2}\log(a^2 + b^2) + i\tan^{-1} b/a$$

$$CSIN(a + bi) = \sin a\,\frac{e^b + e^{-b}}{2} + i\cos a\,\frac{e^b - e^{-b}}{2}$$

$$CCOS(a + bi) = \cos a\,\frac{e^b + e^{-b}}{2} - i\sin a\,\frac{e^b - e^{-b}}{2}$$

With these rather long preliminaries out of the way, let us look into the problem that we shall use for a programming example. We have a series circuit containing a resistance R, an inductance L, a capacitance C, and a sinusoidal voltage V, with frequency f cps. This is the same circuit considered in Case Studies 7 and 8—almost: there the voltage was direct current. There, too, our goal was a picture of the instantaneous behavior of the circuit on a time scale much shorter than one complete cycle. Here we wish to know only the steady-state current flowing as the result of the steady application of the driving voltage V; we assume that all transients have died out. For such assumptions the equation we have to deal with is simply

$$I = \frac{V}{R + \left(2\pi f L - \dfrac{1}{2\pi f C}\right) i}$$

In this formulation V is a real number, but I will in general be complex.

Figure 11.3 shows a program that can be used to compute the current. We see I declared as the only complex variable. A card giving V, R, L, and C is read; we have then a WRITE statement which references a rather elaborate FORMAT to get a heading line, the parameters printed with identifications, and column headings. We next read a frequency card and are ready to compute the complex current.

The statement consists of a complex variable on the left, with a complex expression on the right;

```
C CASE STUDY 11
C CURRENT FLOW IN SERIES RLC CIRCUIT, USING COMPLEX VARIABLES
C
      COMPLEX I
      REAL V, R, F, L, C, ABSI, PHASE
C READ PARAMETERS
      READ (5, 1) V, R, L, C
    1 FORMAT (8F10.0)
C WRITE HEADING LINE AND PARAMETER VALUES
      WRITE (6, 2) V, R, L, C
    2 FORMAT (1H1, 18HSERIES RLC CIRCUIT//1H , 4HV = , 1PE18.5/
    1 1H , 4HR = , 1PE18.5/1H , 4HL = , 1PE18.5/1H , 4HC = ,
    2 1PE18.5////
    3 62H   FREQUENCY    I REAL      I IMAG       ABS          PHA
    4SE//)
C READ ANOTHER FREQUENCY CARD
   99 READ (5, 1) F
C COMPUTE COMPLEX CURRENT
      I = V / CMPLX(R, 6.2831853*F*L - 1.0/(6.2831853*F*C))
C GET MAGNITUDE OF CURRENT
      ABSI = CABS(I)
C GET PHASE OF COMPLEX CURRENT, CONVERT TO DEGREES
      PHASE = 57.29578 * ATAN2(AIMAG(I), REAL(I))
      WRITE (6, 4) F, I, ABSI, PHASE
    4 FORMAT (1H , 1P5E13.5)
      GO TO 99
      END
```

Figure 11.3. A program for finding the current flowing in a circuit containing resistance, inductance, and capacitance.

the entire expression is complex because one part of it is. The function CMPLX accepts two FORTRAN real expressions as arguments and delivers as the function value a FORTRAN complex number. The expressions in the arguments of the CMPLX function may be any real expressions. Here we have a single variable for the first argument and a slightly involved expression for the other. It is quite common for a single constant to appear as an argument of this function.

Dividing a real quantity (V) by a complex quantity (that delivered by the CMPLX function) produces a complex quantity which is assigned as the value of I.

One of the outputs we want is the magnitude of the complex current, which is readily found by use of the CABS function. Another desired output is the phase angle of the current, which requires the use of the ATAN2 function. This function accepts two arguments, the first being the ordinate of a point and the second its abscissa; it produces the angle between the x-axis and the line from the origin to the point. The arctangent function, mathematically, is multiple-valued; there are infinitely many angles having the same tangent. This function produces an angle between $-\pi$ and $+\pi$, in radians. We multiply by $180/\pi$ to convert to degrees.

Now we are ready to print the results. The WRITE statement list contains four variables, but the complex variable I has two parts; so the referenced FORMAT must contain five field specifications.

Figure 11.4 shows the output produced when the program was run with suitable data. We see that for frequencies below the resonant frequency of about 2250 cps the complex current lies in the first quadrant and therefore has a positive phase angle less than 90°. At resonance the current is pure real (lies on the x-axis) and has a phase angle of zero. Beyond resonance the current lies in the fourth quadrant so that the phase angle is less than 90° and negative.

EXERCISES

1. State the value stored by each of the following assignment statements, in which C1, C2, etc., are complex variables and R1, R2, etc., are real.

   a. C1 = 2.0*(1.0, 2.0)
b. C2 = (2.0, 0.0)(1.0, 2.0)
   c. C3 = (2.0, 3.0)**2
*d. C4 = CMPLX(1.0 + 4.0, 4.0**2)
   e. C5 = CMPLX(REAL((1.0, 2.0)),
            AIMAG((3.0, 4.0)))
   f. R1 = CABS((3.0, 4.0))
g. R2 = AIMAG((4.0, 0.0)(1.0, 3.0))*2.0
   h. R3 = AIMAG((4.0, 0.0)*(1.0, 3.0)*2.0)
   i. C6 = CEXP((0.0, 31415927)) + (4.0, 4.0)

SERIES RLC CIRCUIT

V	=	1.00000E 01
R	=	1.00000E 03
L	=	1.00000E-01
C	=	5.00000E-08

FREQUENCY	I REAL	I IMAG	ABS	PHASE
1.00000E 03	1.32857E-03	3.39420E-03	3.64495E-03	6.86235E 01
1.10000E 03	1.70901E-03	3.76422E-03	4.13402E-03	6.55813E 01
1.20000E 03	2.17170E-03	4.12319E-03	4.66015E-03	6.22241E 01
1.30000E 03	2.73036E-03	4.45519E-03	5.22528E-03	5.84980E 01
1.40000E 03	3.39763E-03	4.73629E-03	5.82892E-03	5.43458E 01
1.50000E 03	4.18160E-03	4.93257E-03	6.46653E-03	4.97103E 01
1.60000E 03	5.07999E-03	4.99936E-03	7.12741E-03	4.45416E 01
1.70000E 03	6.07220E-03	4.88369E-03	7.79243E-03	3.88087E 01
1.80000E 03	7.11087E-03	4.53257E-03	8.43260E-03	3.25140E 01
1.90000E 03	8.11786E-03	3.90883E-03	9.00992E-03	2.57113E 01
2.00000E 03	8.99146E-03	3.01135E-03	9.48233E-03	1.85163E 01
2.10000E 03	9.62899E-03	1.89010E-03	9.81274E-03	1.11055E 01
2.20000E 03	9.95849E-03	6.42944E-04	9.97922E-03	3.69402E 00
2.21000E 03	9.97331E-03	5.15942E-04	9.98665E-03	2.96140E 00
2.22000E 03	9.98484E-03	3.89022E-04	9.99242E-03	2.23118E 00
2.23000E 03	9.99311E-03	2.62302E-04	9.99656E-03	1.50357E 00
2.24000E 03	9.99815E-03	1.35903E-04	9.99908E-03	7.78762E-01
2.25000E 03	9.99999E-03	9.93926E-06	9.99999E-03	5.69478E-02
2.26000E 03	9.99867E-03	-1.15475E-04	9.99933E-03	-6.61682E-01
2.27000E 03	9.99423E-03	-2.40230E-04	9.99711E-03	-1.37694E 00
2.28000E 03	9.98672E-03	-3.64218E-04	9.99336E-03	-2.08867E 00
2.29000E 03	9.97619E-03	-4.87336E-04	9.98809E-03	-2.79667E 00
2.30000E 03	9.96271E-03	-6.09485E-04	9.98134E-03	-3.50080E 00
2.40000E 03	9.68049E-03	-1.75869E-03	9.83895E-03	-1.02968E 01
2.50000E 03	9.18662E-03	-2.73354E-03	9.58468E-03	-1.65707E 01
2.60000E 03	8.56476E-03	-3.50606E-03	9.25460E-03	-2.22622E 01
2.70000E 03	7.88742E-03	-4.08201E-03	8.88111E-03	-2.73631E 01
2.80000E 03	7.20736E-03	-4.48637E-03	8.48962E-03	-3.19011E 01
2.90000E 03	6.55778E-03	-4.75114E-03	8.09801E-03	-3.59235E 01
3.00000E 03	5.95646E-03	-4.90767E-03	7.71781E-03	-3.94859E 01
3.10000E 03	5.41065E-03	-4.98311E-03	7.35571E-03	-4.26445E 01
3.20000E 03	4.92113E-03	-4.99938E-03	7.01507E-03	-4.54519E 01
3.30000E 03	4.48510E-03	-4.97342E-03	6.69709E-03	-4.79554E 01
3.40000E 03	4.09805E-03	-4.91798E-03	6.40160E-03	-5.01962E 01
3.50000E 03	3.75486E-03	-4.84248E-03	6.12769E-03	-5.22100E 01

Figure 11.4. The output of the program in Figure 11.3.

j.   R4 = SQRT(REAL(C6)**2 + AIMAG(C6)**2)/CABS(C6)

*2. Write a program, using complex variables throughout, to solve two simultaneous equations in two unknowns.

$$(2 + 3i)x + (4 - 2i)y = (5 - 3i)$$
$$(4 + i)x + (-2 + 3i)y = (2 + 13i)$$

The solution is $x = 1 + i$, $y = 2 - i$.

3. Write a program, using complex variables throughout, to solve the quadratic equation

$$ax^2 + bx + c = 0$$

where a, b, and c are complex. Use the formula

$$x = \frac{-b \pm \sqrt{b^2 - 4ac}}{2a}$$

Run the program with a data card that specifies

$$a = 1 + i$$
$$b = 2 + 3i$$
$$c = -7 + i$$

Run again, this time with a card that gives

$$a = 2$$
$$b = 3$$
$$c = -12$$

(Read them as complex numbers, i.e., as $2 + 0i$, etc., and use the program without change.)

*4. Write a program, using complex variables throughout, to compute and print z and e^z for a succession of values along a vertical line through (1, 0), such as $1 - 5i$, $1 - 4i$, $1 - 3i$, . . . , $1 + 4i$, $1 + 5i$. This will demonstrate how the complex exponential function maps a vertical line.

5. Write a program, using complex variables throughout, to compute and print z and e^z for a succession of values along a 45° line through the origin: $-5 - 5i$, $-4 - 4i$, . . . , $4 + 4i$, $5 + 5i$.

6. Write a program, using complex variables throughout, to compute and print z and e^z for 20 points equally spaced along the arc of a circle with the center at the origin and radius 1.

7. Write a program, using complex variables throughout, that reads a value of z, then computes and prints $a = \cos^2 z + \sin^2 z$. Make up a series of data cards and run the program as a demonstration that $\cos^2 z + \sin^2 z = 1$ holds with complex variables.

8. Write a program, using complex variables throughout, to read z and c, then compute and print $A = z^c = \exp(c \log z)$. Use the program to demonstrate that $(-i)^i = e^{\pi/2}$.

9. Write a program to demonstrate that $\log 3 - \pi i$, $\log 3 + \pi i$, $\log 3 + 3\pi i$, and $\log 3 + 5\pi i$ are all solutions of $e^z + 3 = 0$.

10. Write a program to demonstrate that the transformation

$$w = \frac{z - 1}{z + 1}$$

transforms a point anywhere in the right half of the z-plane into the unit circle in the w-plane and that the transformation

$$w = \frac{i - z}{i + z}$$

transforms a point anywhere in the upper half of the z-plane into the unit circle in the w-plane.

*11. The "transfer function" of a certain servomechanism is given by

$$T = \frac{K(1 + j0.4\omega)(1 + j0.2\omega)}{(1 + j2.5\omega)(1 + j1.43\omega)(1 + j0.02\omega)^2}$$

where j has been written for i, following usual electrical engineering practice; K is a real number to be read from a card. From the same card you should also read values

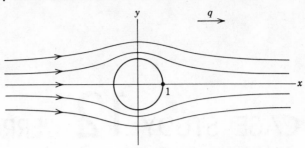

Figure 11.5. The flow problem considered in Exercise 14.

for FIRST, LAST, and INC, all real numbers. Find the value of T for $\omega =$ FIRST. Then repeatedly multiply ω by INC and unless ω is now greater than LAST re-evaluate the formula.

12. Same as Exercise 11, except that the transfer function is

$$T = \frac{100}{j\omega(1 + 0.25j\omega)(1 + 0.0625j\omega)}$$

13. Same as Exercise 11, except that the transfer function is

$$T = \frac{1260}{j\omega(1 + 0.25j\omega)(1 + 0.001j\omega) + \dfrac{20(j\omega)^4}{(1 + 0.5j\omega)(1 + 0.4j\omega)}}$$

14. A long circular cylinder of unit radius is placed in a large body of fluid flowing with a uniform velocity, with its axis perpendicular to the direction of flow, as shown in Figure 11.5. If the velocity of the fluid is A far from the cylinder, the velocity in the region of the cylinder (and indeed everywhere) can be shown* to be given by

$$q = A\left(1 - \frac{1}{\bar{z}^2}\right)$$

where $\bar{z}$ is the complex conjugate of z. Write a program that reads cards or generates values within the program to provide points for plotting the velocity in the region of the cylinder.

* See Ruel V. Churchill, *Introduction to Complex Variables and Applications*, McGraw-Hill, New York, 1948, p. 167.

CASE STUDY 12 ERRORS; DOUBLE PRECISION

In this case study we shall consider a variety of examples of the effect of various kinds of error in computation. This sketch does no more than hint at the subject of error analysis, which is a basic aspect of the study of numerical methods but outside the main scope of this book. The survey indicates some of the problems and perhaps will instill an attitude of caution; the serious user of computers will want to pursue the matter in further study.

We shall consider the assistance in relieving some of the problems that can often (but not always) be offered by the use of FORTRAN double precision operations.

Let us consider in simple examples how it can happen that a result even if printed with many digits of precision may be inaccurate.

The first problem is that the input data may not be accurate, and under unfavorable circumstances this *inherent error* may be amplified by the mathematical formulation of the method of solution or by the details of the arithmetic processes. Consider the simple system of two equations in two unknowns:

$$2.34x + 5.46y = 1.56$$

$$1.58x + 3.71y = 1.03$$

An exact solution is readily found to be $x = 3$, $y = -1$. It might appear to the unsuspecting computer user that an accurate answer has been found, especially if it is presented on a page of a computer output as $x = 3.0000000$ and $y = -1.0000000$; but suppose the numerical values were experimental measurements which are inherently approximations. How sensitive might the solution be to small errors in the data? For instance, let us change the coefficient 3.71 to 3.712, a difference of only 0.05%. Reworking, we find that the solution is now $x = 2.813$, $y = -0.92$. The change in y is 8%, some 400 times greater than the change in the data value. We conclude that unless the coefficients are, in fact, far more accurate than experimental procedures ordinarily justify, we must assume that the solution is largely meaningless, regardless of the number of digits printed by the "giant brain."

The difficulty in that example is that the two lines represented by the two equations are nearly parallel or, stated in algebraic terms, the determinant of the system is nearly zero. This factor "amplifies" errors in the coefficients, quite apart from the method of solution.

That is hardly the only problem. Major complications are introduced also by the fact that in a digital computer we must deal with finite representations of values. The number of digits varies from one machine to the next but is in the general range of 6 to 10 decimal digits, 8 being the most common. Eight digits of precision is more than we ordinarily try to maintain in pencil-and-paper work and a great deal better than slide-rule accuracy, but problems can still occur. For instance, what will FORTRAN tell us is the value of $3 \times (\frac{1}{3})$? You say, "But three times one-third is *obviously* one?" Obvious to whom? If we call for this bit of computation, using FORTRAN real quantities, the fraction $\frac{1}{3}$ must be approximated, for it has no terminating decimal fraction representation nor has it a terminating binary representation if the machine is binary. In a ternary (base 3) computer the representation would be the simple ternary fraction 0.1, but ternary computers are hard to find. In an eight-digit decimal machine, $\frac{1}{3}$ is represented as 0.33333333. Three times that is 0.99999999. If a calculation were set up to require an exact match between this result and 1.00000000, naturally the computer would have to answer "not equal."

For the next three examples let us assume that we are working with a computer in which a FORTRAN real number is represented by four digits rather than the more typical eight. The conclusions we shall be able to draw will be valid, and the arithmetic will be easier to follow. We understand that in FORTRAN a real number is represented as a fraction times a power of 10 (or 2, in a binary machine, or 16 in a hexadecimal machine). We shall not show the exponent in most cases, but this form of representation should be understood. The important thing to keep in mind is that we can *never* have more than four digits in the result of an arithmetic operation. If an operation produces more than four, the extra ones must be dropped, perhaps after rounding: some FORTRAN systems round results, others simply drop the extras, which is usually called *chopping*. We shall round in these examples; chopping would make the examples even more dramatic.

For the simplest sort of illustration of the impact of a limited number of digits for representation consider this arithmetic sequence:

$$.4000 + 1234. - 1233.$$

To begin with, we cannot even say what is meant by it until we agree on an order of evaluation—which is a bit of a shock to people who have been taught (by silence, usually) that addition is both commutative and associative. But, look: if we try to do the addition first, the result will be 1234.4, which is too many digits and therefore is rounded back to 1234.; the .4000 has been dropped entirely and can never re-enter the computation. Now we subtract to get 1.000 as the final result. On the other hand, if the subtraction had been done first, the result would be 1.000, to which we could add .4000 to get 1.400 as the final result.

We can see the effect of rounding by studying another system of two equations in two unknowns.

$$1.000x + 592.0y = 437.0$$

$$592.0x + 4308.y = 2251.$$

Taking these in the order written, we apply the method of Gauss elimination, which requires that we multiply the first equation by 592.0 and subtract from the second. The coefficient of x in the second equation then becomes zero, which was the point of the operation; we then need to subtract from 4308. the product of 592.0 and 592.0, which is 350464. But, of course, we cannot maintain such a large number in our four-digit computer, so we round it off and write the number as $3505. \cdot 10^2$. Similarly, the product of 592.0 and 437.0 is 258704.00, which becomes $2587. \cdot 10^2$. Now, when we subtract $3505. \cdot 10^2$ from 4308., it is necessary to ignore the last two digits in 4308. to "line up the decimal points." The same thing happens when $2587. \cdot 10^2$ is subtracted from 2251. The net result of subtracting 592.0 times the first equation from the second equation is therefore

$$-3462. \cdot 10^2 y = 2564. \cdot 10^2$$

Doing the division, we obtain $y = .7406$.

This result can be substituted into the first equation to get x:

$$x = (437.0 - 592.0 \cdot .7406)/1.000$$

Again we find a problem in the multiplication, the result of which is 438.4352 and requires rounding to 438.4. Accordingly, we find that $x = -1.4000$.

This seems reasonable enough. Now, however, let us try something that at first glance must seem utterly pointless: let us solve the system again, but this time with the two equations interchanged. This surely cannot make any difference!

The process this time is to multiply the first equation by 1.000/592.0 (= 0.001689) and subtract from the second. We find ourselves subtracting 7.277 from 592.0; the 7.277 becomes 7.3 and the result of subtraction is 584.7. For the modified constant terms we get 433.2; division gives $y = .7409$, which is not too greatly different from the earlier result. But now we substitute this value into the first equation:

$$x = (2251. - 4308. \cdot .7409)/592.0$$

The product here gives 3191.7972, which must be rounded to 3192. Completing the arithmetic we find that $x = -1.590$, about 12% different from the earlier result.

The culprit here, clearly, is the necessity for dropping digits because of the limited length of numbers in our hypothetical computer. If we carry out the same procedures with the same data, but carry eight digits in all results, the values given for the two

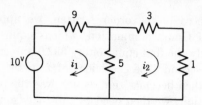

Figure 12.1. The network used to illustrate roundoff problems.

solutions are

$$x = -1.5888863$$

$$y = 0.74085961$$

for the first case and

$$x = -1.5888905$$

$$y = 0.74085961$$

In other words, in the second solutions we carried double the precision of the first. The roundoff problem has clearly been reduced a great deal, but the two solutions are still not identical.

Double precision is sometimes the answer to the kinds of problems we are presenting, and we must learn accordingly how to use the FORTRAN double-precision capability in which we get numbers represented with something on the order of 16 decimal digits instead of something around eight.

First, however, let us look into one more example that is particularly frustrating in a perverse sort of way.

Consider the dc network shown in Figure 12.1. We find that the two currents are given by the solution to

$$14i_1 - 5i_2 = 10$$

$$-5i_1 + 9i_2 = 0$$

The solution is found, with no particular problems, to be

$$i_1 = 90/101 = .8911$$

$$i_2 = 50/101 = .4950$$

(We continue to use four-digit arithmetic throughout. The solution just displayed was found by determinants, but the method of Gauss elimination gives the exact same result if the equations are reversed, and an answer that is only slightly different if solved in the order written.)

Now let us do something that ought to get us into trouble: let us formulate this problem so that it will require the solution of three equations in three unknowns by adding a third loop current as shown in Figure 12.2. There are, in fact, only two independent loops in the network; we are free to choose any two

of the three we please, but according to any text we must not write these three equations:

$$14i_1 - 5i_2 + 9i_3 = 10$$

$$-5i_1 + 9i_2 + 4i_3 = 0$$

$$9i_1 + 4i_2 + 13i_3 = 10$$

The reason for the prohibition is that we have created a system of *linearly dependent* equations, which means that one of them can be found from the other two by a process of multiplying by constants and adding. Indeed: the third equation, for instance, is simply the sum of the first and second. Such a system has infinitely many solutions: we are free to pick a value at random for one of the variables, then solve for the other two. For example, if we pick $i_3 = 0$, we get the same values for i_1 and i_2 that we found from solving the 2 x 2 system. We could also pick $i_2 = -91$ and find values for i_1 and i_3 that would satisfy the system of three equations and which, furthermore, would give the same physical information about the network. That is, once we have found three values for i_1, i_2, and i_3, we can get the actual current circulating in the left loop by adding i_1 and i_3 and the current in the right loop by adding i_2 and i_3. (Try it with the values given below.)

This is foolishness, of course: we know better than to formulate such a problem this way; but, being persistent, what would happen if we did? After all, with the extremely complicated networks that sometimes appear in practice, it may not always be easy to tell how many independent loops there are. Would it not be nice to be able to depend on the computer to warn us that an error had been made?

The question is a tempting one, but the answer is frustrating. It does indeed seem, from what we all learned in high school algebra, that the computer ought to give us the bad news; but it fails to do so. Let us examine both sides of this highly instructive situation.

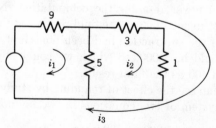

Figure 12.2. A three-loop version of the network in Figure 12.1. This formulation produces a system of linearly dependent equations.

We might think that the computer "ought" to warn of an error because we would expect that the Gauss elimination process would develp a zero divisor, and most computers give a warning of this occurrence. Consider the system of equations and follow the steps of the elimination method.

We are to multiply the first equation by $-5/14$ and subtract from the second. *Doing this with rational fractions rather than converting to decimal fractions*, the second and third equations become

$$0i_1 + \frac{101}{14} i_2 + \frac{101}{14} i_3 = \frac{50}{14}$$

$$0i_1 + \frac{101}{14} i_2 + \frac{101}{14} i_3 = \frac{50}{14}$$

If we now continue the elimination process to the next step, we of course get

$$0i_3 = 0$$

This, in fact, is satisfied by *any* finite number i_3, which is what we knew already. The computer ought to send up a warning flag on trying to do the division. This will be the outcome, no matter how the equations are arranged before starting the elimination process.

And what will the computer do? Answer: usually it will find some solution or other, printed with lots of decimal places if we ask for them. This solution will be correct in the sense that it does satisfy the system. There will be no warning whatsover that anything unusual has happened, namely, that there are an infinite number of other solutions. Let us see how this can happen.

Taking the equations in the order written, our first step is to multiply the first equation by $-.3571$ and subtract from the second. We round the result of every arithmetic operation to four digits; the second equation becomes

$$7.214i_2 + 7.214i_3 = 3.571$$

The third equation becomes

$$7.215i_2 + 7.210i_3 = 3.570$$

Multiplying the new second equation by 1.000 and subtracting from the new third equation, we get

$$.0060i_3 = -.0010$$

(The i_2 term does not actually go to zero, exactly, but, because we never use it, it does not matter.) Now we have $i_3 = -.1667$. We can substitute this value of i_3 into the "new" second equation to get

$i_2 = .6618$, and these two values substituted into the first equation give $i_1 = 1.058$.

Note that $i_1 + i_3 = .8910$ and $i_2 + i_3 = .4941$, giving currents in the two loops very close to the solution found with the two-equation system.

If we now reverse equations 1 and 3 and solve, again, using the procedure just outlined, the solution is

$$i_1 = .4911$$

$$i_2 = .09519$$

$$i_3 = .4000$$

Quite different! Yet this solution also satisfies the system of equations quite closely, and once again the loop currents are quite close to the values found before:

$$i_1 + i_3 = .8911 \quad \text{and} \quad i_2 + i_3 = .4952$$

Finally, if we interchange equations 1 and 2, we find that equation 1 should be multiplied by -2.800 and subtracted from equation 2 and equation 1 should be multiplied by -1.800 and subtracted from equation 3. Both of these multipliers are exact because the fractions involved have terminating decimal representations. This time, finally, we get a situation in which a zero divisor would turn up in the elimination process to signal the situation.

It is perhaps obvious that only by relatively rare chance will the numbers come out so that a zero divisor will appear to give warning, considering that we are seldom, as here, concerned with coefficients that are simple whole numbers.

The source of difficulty in much of what we have seen so far has been, in part, the limited number of digits in the numbers. It is now time to look into FORTRAN double precision to get some idea of how we might attack this aspect of the problem.

As we have with the other kinds of quantities—integer, real, and complex—we can organize our study into the investigation of constants, variables, operations, functions, and input/output.

A double-precision constant is written in exponent form with a D instead of an E. The following are acceptable double-precision constants:

1.5D0	
5.0D4	$(= 5.0 \cdot 10^4)$
5.0D-4	$(= 5.0 \cdot 10^{-4})$
1.2345678923456D0	

In many FORTRAN systems a long number such as the last may be written without an exponent;

the compiler can recognize from the length that it must be double precision.

A double-precision variable is one that is named in a DOUBLE PRECISION statement, which is the fourth type of statement we have seen (the others: INTEGER, REAL, and COMPLEX). There are no restrictions on the naming of a double-precision variable.

A FORTRAN double-precision quantity is represented and used more or less like a FORTRAN real quantity, but it has more digits. Often there are actually twice as many digits, as suggested by the name double precision, but not always. It would probably be safe to say *at least* twice as many digits; one major computer (the IBM System/360) has about $2\frac{1}{3}$ times as many digits in a double-precision floating point number as in a single-precision number.

Arithmetic expressions are formed in double precision according to the same rules and using the same operation symbols $(+, -, *, /, **)$ as in real. It is permissible to combine real and double-precision operands in the same expression; the result is always double-precision. Thus simple constants need not be written in double-precision form; double-precision constants are needed only for numbers that cannot be represented with sufficient accuracy in single precision. Appendix 1 shows the permissible combinations of various types of quantities in an expression. Note that the prohibition against mixing integer and real extends to integer and double precision.

Double-precision versions of most of the common mathematical functions are provided, as may be seen in an inspection of Appendix 2. Note that with one exception they all begin with the letter *D* and that they require a double-precision argument and return a double-precision function value. The one exception is the function SNGL ("single") that accepts a double-precision argument and returns as a real number the most significant half of the argument. This is useful at the end of a computation, for instance, when double precision has been used throughout to protect against rounding error, but the final result is known not to have more than single-precision reliability and it is desired to print only the most significant half of the result.

This same action can also be accomplished with a simple assignment statement in which the right-hand side is a double-precision expression and the variable on the left is real. As a matter of fact, *any* combination of integer, real, and double-precision

right- and left-hand sides is permitted. In any such case all arithmetic is done in the mode of the expression on the right, and the result is converted to the mode of the variable on the left. (This does not alter the prohibition against mixed mode *expressions*. The expression on the right must follow the usual rules, according to which it is of one type or the other.)

A few examples may make matters clearer. Suppose that R1, R2, etc., are all real variables, that D1, D2, etc., are all double-precision variables, and that I1, I2, etc., are all integer variables. Then all of the following are acceptable arithmetic assignment statements:

D1 = D2*D3 + (D4 — 8765.65438764D0)/D5

D1 = 4.0*D2 — D3 / 1.1D0

D1 = R1 + D1 + R2

R1 = (D1*D2 — D3*D4) / (D1*D5 — D3*D6)

D1 = R1 + 2.0

D1 = (I1 — 8)*I2

I1 = R1 + D1

D1 = D2**3

D1 = 3.7 * DEXP(D2/4.0)

D1 = 3.7D0 * DEXP(D2/4.0)
　　　　　　— DSIN(D3) / DCOS(D3)

It may be noted that in the second example there is a point to making 1.1D0 a double-precision constant: In a binary computer 1.1 has no exact representation of any length, but the double-precision constant would be closer. On the other hand, 4.0, being a whole number, is represented exactly in almost all floating point systems, so that there is no loss of accuracy in using it as a real (single-length) constant.

Input and output of double-precision quantities is handled with a new field specification, D. D-conversion is the same as E-conversion, except that (1) the list variable associated with such a field specification should be double precision, (2) there can be more digits (17 is representative), and (3) D is used for the exponent indicator rather than E.

So far we have not used E-conversion for input, so a note is in order. When E-conversion or D-conversion is used for input, the data value must be punched on the card in exponent form. The exponent may be punched in any of the simplified forms (E + 02, E 02, E02, E + 2, E2, + 2 all being acceptable forms for the exponent *plus two*), but however the exponent is punched it *must* be at the extreme right-hand side of the field. This is because blanks are treated as zeros on input. Thus, if one blank is left to the right of the exponent E-2, the number will be entered as though the exponent were E-20. The same comments apply to double precision, with D for E.

For a first example of a FORTRAN program that uses double-precision operations let us turn to a problem that cannot be solved accurately otherwise. Consider the following system of equations:

$$3c_1 + 1011c_2 + 1{,}000{,}101c_3 = 999{,}159$$
$$1011c_1 + 1{,}000{,}101c_2 + 1{,}000{,}001{,}001c_3 = 999{,}024{,}153$$
$$1{,}000{,}101c_1 + 1{,}000{,}001{,}001c_2 + 1{,}000{,}000{,}010{,}001c_3 = 999{,}023{,}011{,}323$$

This system has the exact solution

$$c_1 = 23$$
$$c_2 = -1$$
$$c_3 = 1$$

The example was obviously contrived, but it is nonetheless representative of realistic problems that can develop in practice. Let us begin by trying to solve it with single precision, using the program of Figure 12.3.

Note first that the FORMAT statement has only

```
C CASE STUDY 12
C THREE SIMULTANEOUS EQUATIONS
C
C THIS VERSION USES REAL (SINGLE PRECISION) VARIABLES
C
      REAL A11, A12, A13, A21, A22, A23, A31, A32, A33, B1, B2, B3
      REAL K, C1, C2, C3
      WRITE (6, 1)
    1 FORMAT (1H1, 48HSIMULTANEOUS EQUATIONS, SINGLE PRECISION VERSION)
      READ (5,2) A11, A12, A13, B1, A21, A22, A23, B2, A31, A32, A33, B3
    2 FORMAT (F20.0)
      K = A21 / A11
      A22 = A22 - K * A12
      A23 = A23 - K * A13
      B2 = B2 - K * B1
      K = A31 / A11
      A32 = A32 - K * A12
      A33 = A33 - K * A13
      B3 = B3 - K * B1
      K = A32 / A22
      A33 = A33 - K * A23
      B3 = B3 - K * B2
      C3 = B3 / A33
      C2 = (B2 - A23 * C3) / A22
      C1 = (B1 - A12 * C2 - A13 * C3) / A11
      WRITE (6, 12) C1, C2, C3
   12 FORMAT (1H0,4HC1 =, 1PE20.8/1H0,4HC2 =, E20.8/1H0,4HC3 =, E20.8)
      STOP
      END
```

Figure 12.3. A single-precision program to solve a system of three simultaneous equations.

one field specification. What happens when the READ or WRITE has more variables than there are field specifications in the FORMAT is this: when the closing parenthesis of the FORMAT is reached, the "scanning" of the field specifications begins again at the beginning. Furthermore, a new record (card or line) is begun. Thus we can punch one number on each card but call for reading all of the cards with one READ statement.

Several of the numbers contain more significance than can be contained in a FORTRAN real number. A number such as 999,023,011,323, for instance, might be rounded to a form of the type

$$9.9902301 \cdot 10^{11}$$

in which the last four digits have, in effect, been discarded.

We shall solve this system by the method of Gauss elimination, which we can describe more conveniently by introducing the following notation

$$a_{11}c_1 + a_{12}c_2 + a_{13}c_3 = b_1$$

$$a_{21}c_1 + a_{22}c_2 + a_{23}c_3 = b_2$$

$$a_{31}c_1 + a_{32}c_2 + a_{33}c_3 = b_3$$

The task is first to eliminate a_{21} by multiplying the first equation by a_{21}/a_{11} and subtracting from the

SIMULTANEOUS EQUATIONS, SINGLE PRECISION VERSION

```
C1 =      2.18828124E 01
C2 =     -7.94426394E-01
C3 =      9.99795532E-01
```

Figure 12.4. The output of the program in Figure 12.3.

second, then to eliminate a_{31} by multiplying the first equation by a_{31}/a_{11} and subtracting from the third, and then to eliminate a_{32} by multiplying the (modified) second equation by a_{32}/a_{22} and subtracting from the (modified) third. Actually we shall not reduce a_{21}, a_{31}, and a_{32} to zero; once the required operations have been carried out on the other elements, those eliminated are never used again, so we simply ignore them.

Figure 12.4 shows the output. The solution bears some resemblance to the known exact answer, but it is surely not very accurate.

Figure 12.5 is the same program, modified to use double precision throughout. This time the field specification for input is D20.0. There is no double precision equivalent of the F field specification, so we must punch the data with a D exponent. The data cards were punched and sequenced as indicated in Figure 12.6, in which the zero in the exponent is in column 20 in each case. We see that no

```
C CASE STUDY 12
C THREE SIMULTANEOUS EQUATIONS
C
C THIS VERSION USES DOUBLE PRECISION VARIABLES
C
      DOUBLE PRECISION A11, A12, A13, A21, A22, A23, A31, A32, A33
      DOUBLE PRECISION B1, B2, B3, K, C1, C2, C3
      WRITE (6, 1)
    1 FORMAT (1H1, 48HSIMULTANEOUS EQUATIONS, DOUBLE PRECISION VERSION)
      READ (5,2) A11, A12, A13, B1, A21, A22, A23, B2, A31, A32, A33, B3
    2 FORMAT (D20.0)
      K = A21 / A11
      A22 = A22 - K * A12
      A23 = A23 - K * A13
      B2 = B2 - K * B1
      K = A31 / A11
      A32 = A32 - K * A12
      A33 = A33 - K * A13
      B3 = B3 - K * B1
      K = A32 / A22
      A33 = A33 - K * A23
      B3 = B3 - K * B2
      C3 = B3 / A33
      C2 = (B2 - A23 * C3) / A22
      C1 = (B1 - A12 * C2 - A13 * C3) / A11
      WRITE (6, 12) C1, C2, C3
   12 FORMAT (1H0,4HC1 =,1PD23.15/1H0,4HC2 =,D23.15/1H0,4HC3 =,D23.15)
      STOP
      END
```

Figure 12.5. A double-precision version of the program in Figure 12.3.

```
           3D+0
        1011D+0
     1000101D+0
      999159D+0
        1011D+0
     1000101D+0
   1-00-01001D+0
   999024153D+0
     1000101D+0
   1000-01001D+0
 1000000010001D+0
 999023011323D+0
```

Figure 12.6. A listing of the data cards used with the program in Figure 12.5.

decimal point has been punched; the second zero in D20.0 means that the numbers have zero places to the right of an *assumed* decimal point.

The output is shown in Figure 12.7, in which we see that the results are exact to 16 digits. Thus a problem caused by the finite length of representation of approximate quantities and by roundoff has been solved nicely by the use of double precision.

We must hasten to point out, however, that 16 or 17 digits do not automatically guarantee that problems will never arise, as many a naïve user has learned to his sorrow. For a final example let us investigate an application in which double precision helps, but only to a degree; the basic difficulty overwhelms even double precision after a point. The application is of some mild interest in itself.

The Taylor series for the sine

$$\sin x = x - \frac{x^3}{3!} + \frac{x^5}{5!} - \frac{x^7}{7!} + \cdots$$

is usually described as valid for any finite angle, and the truncation error committed by stopping the summation after a finite number of terms is said to be less in absolute value than the first term neglected. These statements would be true *if there were some way to keep an infinite number of digits in each arithmetic result.* In actual fact it is useless for large values of x. We shall determine why this happens in an attempt to understand the need for double-precision quantities even when the final results are much less accurate than the number of digits carried in program variables.

We shall write a program to evaluate the series

SIMULTANEOUS EQUATIONS, DOUBLE PRECISION VERSION

C1 = 2.3000000000000000D 01

C2 = -1.0000000000000000D 00

C3 = 1.0000000000000000D 00

Figure 12.7. The output of the program in Figure 12.5.

directly; that is, we shall start with the first term and continue to compute terms until one is reached that is less in absolute value than, say 10^{-17}. The sum of the series then ought to be within 10^{-17} of the correct value of the sine.

The program requires a strategem to avoid producing intermediate results too large to be represented in the computer. The largest angle we shall consider is about 70 radians; if we were to try to raise 70 to the large powers that will be required, we should greatly exceed the sizes permitted of variables in most systems. Therefore we take another approach—that of computing each term from the preceding term. The recursion relation is not complicated. Having the first term x, we can get the next term by multiplying by $-x^2$ and dividing by 2×3. Having the second term, we can get the third by multiplying by $-x^2$ and dividing by 4×5. In short, given the preceding term, we can always get the next one by multiplying by $-x^2$ and dividing by the product of the next two integers.

A flowchart is shown in Figure 12.8. We are to read cards, each containing an angle in degrees. The angle is first converted to radians by dividing by $180/\pi$; the result is called X. Now we need to get the recursion process started. We shall continually be adding a new term to a sum that will eventually become the sine when enough terms have been computed. To get started we set this sum equal to X; the first term computed by the recursion relation will be $-x^3/3!$. Thus the preceding term is initially X. To get the successive integers (mathematical integers, not FORTRAN integers!) we start a variable named DENOM at 3.0. To save recomputing x^2 repeatedly we compute it once before entering the loop, giving it the name XSQ.

Getting a new term is a simple matter of multiplying the preceding term by $-$XSQ and dividing by the product of DENOM and DENOM-1.0. This new term replaces the preceding term and is added to the sum. We are now ready to find out whether enough terms have been computed. We ask whether the absolute value of the term just computed is less than 10^{-17}. If it is, we are ready to print the result and go back to read another card. If not, the value of DENOM must be incremented by 2.0 before returning to compute another term.

Besides printing the value of the sine computed by this method, it might be interesting to use the double-precision sine routine supplied with the FORTRAN system to compute it also. This is done just before printing.

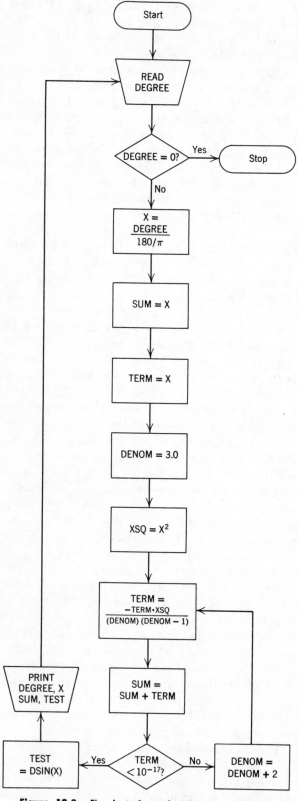

Figure 12.8. Flowchart of a method for computing a sine.

The program shown in Figure 12.9 illustrates many of the aspects of double-precision programming. In the READ statement we ask for a double-precision value, which is associated with a D field specification in the FORMAT statement. The field width of 23 allows space for a sign, one digit before the decimal, the decimal point, 16 digits after the decimal, and a four-position exponent.

The statement that converts from degrees to radians shows something that we ordinarily do not see: a division of one constant by another, which should be made with a desk calculator beforehand—but desk calculators that do 17-digit arithmetic are not readily available. If this program were to be run a great many times, it might be worthwhile to make a preliminary run to get the conversion constant.

Down to statement 24 we follow the flowchart quite closely. The only point to note is the occasional mixing of real and double-precision quantities in an arithmetic expression.

At **24** we use the function SNGL to get the more significant half of DEGREE into the form for a real variable. The idea here is that we shall actually be computing the sine for angles that are a whole number of degrees. Therefore we should rather print them in single-length form than as whole numbers, but we cannot write DEGREE in the WRITE statement and then use an F field specification in the FORMAT statement: double-precision variable names in the WRITE list must be associated only with D field specifications. To use the F field specification we must convert to real form. (This is a valid example of the use of the SNGL function, but it may be noted that we could also have written SDEGR = DEGREE to get the same conversion from double to real. Other examples can be found in which the SNGL function is strictly required or is at least a significant convenience.)

The results shown in Figure 12.10 are for 30° plus multiples of 360°. The exact result in every case should therefore be ½. The entry for 30° is as close as we should reasonably expect: 15 correct digits. At 390° we still have 14 good digits; at 750° the number is down to 11; at 1110° there are only 9; at 1470° only 6; at 1830° only 3. At 2190° there are *no* correct digits, and for 2550° the computer tells us that the sine is −130! Larger values likewise lead to complete nonsense.

Let us consider the first entry in which the method fails completely (2550°) to see if we can

```
C CASE STUDY 12
C DOUBLE PRECISION SINE ROUTINE
C
          DOUBLE PRECISION X, TERM, SUM, XSQ, DENOM, TEST, DEGREE
  506 READ (5, 504) DEGREE
  504 FORMAT (D23.16)
C TEST FOR ZERO END-OF-DECK SENTINEL
          IF (DEGREE .EQ. 0.0D0) STOP
C CONVERT FROM DEGREES TO RADIANS
          X = DEGREE / (180.0/3.14159265358897932)
C SET UP INITIAL VALUES
          SUM = X
          TERM = X
          DENOM = 3.0
          XSQ = X**2
C COMPUTE NEW TERM FROM PREVIOUS TERM
   25 TERM = -TERM * XSQ / (DENOM * (DENOM - 1.0))
C GET SUM OF TERMS SO FAR
          SUM = SUM & TERM
C TEST FOR CONVERGENCE
          IF (DABS(TERM) .LT. 1.0D-17) GO TO 24
          DENOM = DENOM & 2.0
          GO TO 25
C CONVERT FROM DOUBLE TO SINGLE PRECISION
   24 SDEGR = SNGL(DEGREE)
C GET VALUE FROM SUPPLIED DP SINE ROUTINE FOR COMPARISON
          TEST = DSIN(X)
          WRITE (6, 89) SDEGR, X, SUM, TEST
   89 FORMAT (F8.0, 1P3D26.16)
          GO TO 506
          END
```

Figure 12.9. A program for computing a sine, using double-precision techniques.

determine what happened. Using a separate program, not shown, the values of the individual terms and sums were printed. The first term is just the value of X:

$$44.505895925855405$$

The second term, to 17 digits, is

$$-14,692.692643741607$$

When these two terms were added, only 17 digits can be kept. The result is

$$-14,648.186747815750$$

The three rightmost digits of the first term had to be shifted off before adding and have been permanently lost. The third term is

$$1,455,145.7461934667$$

When the sum of the first two terms is added to this term, once again we can keep only 17 digits, so that the preceding sum must be shifted right two places before adding. That means that five digits of the first term have been permanently lost in this sum:

$$1,440,497.5594456506$$

SDEGR	X	SUM	TEST
30.	5.2359877559829884D-01	5.0000000000000004D-01	5.0000000000000004D-01
390.	6.8067840827778849D 00	5.0000000000001790D-01	4.9999999999999840D-01
750.	1.3089969389957470D 01	5.0000000000241549D-01	4.9999999999999969D-01
1110.	1.9373154697137056D 01	5.0000000623000525D-01	4.9999999999999969D-01
1470.	2.5656340004316645D 01	4.9999930135221125D-01	5.0000000000000089D-01
1830.	3.1939525311496229D 01	5.0041391231164414D-01	4.9999999999999840D-01
2190.	3.8222710618675815D 01	7.0621763237687443D-01	4.9999999999999840D-01
2550.	4.4505895925855397D 01	1.7804534635696152D 01	4.9999999999993361D-01
2910.	5.0789081233034983D 01	-2.4851218126611132D 04	4.9999999999999840D-01
3270.	5.7072266540214578D 01	-1.2429565075273363D 07	4.9999999999999840D-01
3630.	6.3355451847394155D 01	-2.5305244077708817D 10	4.9999999999999840D-01
3990.	6.9638637154573741D 01	-2.4793429800504405D 12	4.9999999999993361D-01

Figure 12.10. The output of the program in Figure 12.9.

The next few terms are

$$-68,626,571.055772073$$
$$1,887,969,175.3547401$$
$$-33,996,742,846.947168$$
$$431,665,965,172.74578$$
$$-4,071,585,970,550.4682$$
$$29,650,348,430,733.331$$
$$-171,727,082,331,700.35$$

By now all but four of the original digits of the first terms have been lost; they can never re-enter the computation after the terms start decreasing, when the sum should, of course, be reduced to a number less than 1. The largest term in the series is

$$1.2617630349119573 \cdot 10^{18}$$

After adding this to the sum of the preceding terms, *all* of the digits of the first term have left the scene, along with some of the digits of other terms.

A big part of our accuracy problem is clearly that we lose digits in the small numbers when they are added to the very much larger terms.

This is not the only problem. This example was run on a computer in which double-precision variables are represented with the equivalent of about 17 decimal digits. Consider a term such as this one, which is the largest in the series for sin 3270°.

$$3.2246921509364463 \cdot 10^{23}$$

Writing this out, we have

$$322,469,215,093,644,630,000,000$$

The zeros at the right, of course, are not significant; they merely locate the decimal point. In other words, this value is an approximation to some rational fraction; the zeros stand for digits that we cannot keep in the computing system. Putting it another way, this approximation could differ from the true value by as much as 5,000,000. This kind of error makes it hopeless to expect any significance at all in computing a final value that is properly never greater than 1.

We may note briefly what happened to the sine function supplied with the system. In all cases there are at least 14 good digits. The slightly decreased accuracy for larger angles is caused by losses of significance in the conversion from degrees to radians and in the reduction of the argument to an angle less than $\pi/2$. The latter is the first action in a practical sine routine, after which only a relatively few terms of the series are required and the roundoff problems are minor.

And how well would this program have fared if it were written to use single precision? The computation would become useless after about 800°, whereas the double-precision version here was of some value out through about 1400°. Naturally, sines are not computed this way at all, but the point is still valid: double-precision made possible a computation in a range in which single precision broke down.

What conclusions may be drawn from all of this? We suggest three.

1. Errors are a constant problem. No results should be accorded more validity than a careful analysis of the data, the method of solution, and the roundoff error justify. Never be fooled by eight-digit printouts!

2. Double precision sometimes permits results to be obtained in situations in which single precision is useless.

3. But not always! Eschew the mentality that says, "I'm not sure if there are any problems with roundoff, but I'll just use double precision and then I *know* I'll be safe."

EXERCISES

1. Consider the simultaneous equations

$$ax + by = c$$
$$dx + ey = f$$

and the solution by Cramer's rule

$$x = \frac{ce - bf}{ae - bd}$$
$$y = \frac{af - cd}{ae - bd}$$

Show that if $ae - bd$ is small the accuracy of the solution may be poor, even if the coefficients have no inherent errors. Illustrate by showing that the solution of the system

$$0.2038x + 0.1218y = 0.2014$$
$$0.4071x + 0.2436y = 0.4038$$

obtained with four-digit floating point arithmetic is $x = -1.714$, $y = 4.286$, whereas the exact solution, which can be obtained with eight-digit floating arithmetic, is $x = -2.000$, $y = 5.000$. If the coefficients themselves are inexact, as they almost always are, the "solution" of this system can be totally meaningless.

2. The following problem, suggested by Richard V. Andree, demonstrates effectively that roundoff is not the only problem in numerical computation. Consider the system

$$x + 5.0y = 17.0$$

$$1.5x + 7.501y = 25.503$$

Show that if enough digits are carried to make all roundoff "errors" zero the system will have a unique solution, $x = 2$, $y = 3$. Now show that if the constant term in the second equation is changed to 25.501, a modification of one part in about 12,000, a greatly different solution will be obtained.

If the coefficients and constant terms were experimental results, with a corresponding doubt about their exact values, the "solution" would be meaningless.

3. Are mixtures of the following data types permitted? D1, D2, etc., represent double precision variables, R1, R2, etc., represent real variables, and I1, I2, etc., represent integer variables.

*a. D2 = D1*4.0
b. D2 = D1*4
c. D2 = D1**4
*d. IF (D1 .EQ. 0) R2 = I1*I2/I3
*e. R1 = I1 + 6
f. D1 = I1 + 5
g. I1 + R1**2
*h. I1 = D1**3
i. D2 = 2.0/D1 + R1 − I1

4. Consider the following system of equations:

$$140679x + 556685y = 146710$$

$$81152x + 321129y = 84631$$

Write two programs to solve this system, using Cramer's rule (see Exercise 1), in one program using real variables and in the other, double precision. (It is assumed that your computer has real numbers with about 6 to 8 decimal digits of significance.) Why should there be a difference in the results when the data can be represented exactly with single precision (real)?

5. Write a program, using double precision throughout, to compute the values of D1 and D2 from

$$D1 = (A + \sin B - 10^9)*C$$

$$D2 = (A - 10^9 + \sin B)*C$$

The data card read by the program should enter the following values:

A: 1.0D9
B: 2.0D0
C: 1.0D7

Why the difference in the two results? Are not the two formulas equivalent, since addition is commutative?

CASE STUDY 13 THE DESIGN OF A BINARY ADDER; OPERATIONS WITH LOGICAL VARIABLES

Everything we have done so far has involved computations with quantities that were numbers of one sort or another. Now we turn to another type of quantity, the FORTRAN *logical*, which is one that is permitted to take on only the values "true" and "false." We shall see in small examples that FORTRAN logical operations are useful in programs that are primarily concerned with "ordinary" numerical computations, and in the case study we shall see an example of the kind of computation that is best done with logical operations as the heart of the programming technique.

This case study may be omitted without serious loss of continuity.

A *logical constant* is either of the following:

.TRUE.

.FALSE.

A *logical variable* is a variable that has been declared in a LOGICAL type statement. As with double precision and complex variables, there is no initial-letter naming convention, for a variable is never assumed to be the logical type unless it has appeared in a LOGICAL statement. A logical variable can take on only the values .TRUE. and .FALSE..

A *logical assignment statement* has the form

$$a = b$$

in which a is a logical variable and b is a logical expression.

A logical expression, as we have already seen many times in connection with the logical IF statement, may be a relational expression, such as I .EQ. 20. Thus, if L1, L2, etc., are logical variables, we may write logical assignment statements like these:

L1 = .TRUE.

L2 = .FALSE.

L3 = A .GT. 25.0

L4 = I .EQ. 0

L5 = L6

The first two are the logical equivalent of statements of the form

variable = constant

L3 would be set to .TRUE. if the value of the real variable A were greater than 25.0, and to .FALSE. if A were less than or equal to 25.0. Likewise, L4 would be set to .TRUE. if the value of I were in fact zero and to .FALSE., otherwise. L5 would be set to the same truth value as L6 currently has.

We have seen that a *relational expression* is one that compares integer, real, or double-precision values by using the relational operators .LT., .LE., .EQ., .NE., .GT., and .GE.. A *logical expression* combines logical values and/or relational expressions by using the *logical operators* .AND., .OR., and .NOT..

Thus we can write logical assignment statements like these:

L1 = D .LT. EPS .OR. ITER .GT. 20

L2 = D .GE. EPS .AND. ITER .LE. 20

L3 = BIG .GT. TOLER .OR. SWITCH

L1 would be given the value .TRUE., if *either or both* of the relations were true, and .FALSE., otherwise. L2 would be given the value .TRUE. if and only if *both* relations were true, and .FALSE. otherwise. L3 would be given the value .TRUE. if the relation were satisfied, if the logical variable SWITCH were true, or both.

The logical operator .NOT. reverses the truth value of the expression it operates on. For instance, consider this statement:

L = A .GT. B .AND. .NOT. SWITCH

If SWITCH is true, .NOT. SWITCH is false, and if SWITCH is false, .NOT. SWITCH is true. Thus L would be given the value .TRUE. if the value of A were greater than the value of B *and* the value of SWITCH were .FALSE..

In the absence of parentheses the hierarchy of logical operators is .NOT., .AND., and .OR.. For instance, take this expression, in which all variables are logical:

L = A .AND. B .OR. C .AND. .NOT. D .AND. E
The .NOT. D is performed first, then the three .AND.s, then the .OR..

The .AND. .NOT. and .OR. .NOT. combinations are the only ones in which two operators may appear side by side.

Parentheses may be used to dictate a meaning other than that implied in the hierarchy of operators. For instance, we may write

L = A .AND. (B .OR. C) .AND. .NOT.

(D .AND. E)

Most FORTRAN systems provide for input and output of logical values by use of the L field specification. This is of the form *Lw*, where *w* specifies the number of columns or printing positions. On input the first letter in the card field must be T or F, which may be followed by any other characters. Thus it is possible, if we wish, to use the field specification L5 and punch TRUE or FALSE. On output a T or F is printed at the right side of the *w* printing positions.

One common use of logical variables is to "save" the result of a decision to avoid having to repeat the test on which it is based. Consider the following simple example.

In Chapter 2 we defined a function of x

$$y = 0.5x + 0.95 \quad \text{if } x \leq 2.1$$

$$y = 0.7x + 0.53 \quad \text{if } x > 2.1$$

This computation can be written in a compact and easily understood form:

LOGICAL FIRST

FIRST = X .LE. 2.1

IF (FIRST) Y = 0.5*X + 0.95

IF (.NOT. FIRST) Y = 0.7*X + 0.53

Obviously it can also be written in many other ways, but this way is about as easy to follow as

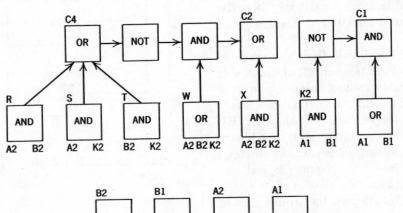

Figure 13.1. Arrangement of input binary digits (B2, B1, A2, and A1) and the logical operations leading to the sum digits (C4, C2, and C1) in one logical design of a binary adder.

any and about as fast. The speed is based on the fact that, as it happens, a test of a logical variable can be done very quickly by the computer.

Suppose that we have values of X and Y, which represent a point in Cartesian coordinates. We wish to set QUAD equal to the number of the quadrant in which the point lies.

```
LOGICAL XPOS, YPOS

INTEGER QUAD

XPOS = X .GT. 0.0

YPOS = Y .GT. 0.0

IF (XPOS .AND. YPOS) QUAD = 1

IF (.NOT. XPOS .AND. YPOS) QUAD = 2

IF (.NOT. XPOS .AND. .NOT. YPOS) QUAD = 3

IF (XPOS .AND. .NOT. YPOS) QUAD = 4
```

In this case the logical variables make it unnecessary either to repeat the determination of the signs of X and Y or to go into a complicated series of GO TO statements. (We have not worried here about points on the axes. With suitable agreements about the conventions needed, this aspect could easily be included.)

Another example of the usefulness of logical variables would be a calculation in which a certain factor would be included or omitted from a formula. The program could be set up to read a value of a logical variable, then used to determine the course of the calculation.

For our example of a complete program built around logical variables we turn to the computer itself to investigate in a very small way how one computer can be used to assist in the design of another.

Digital computers are built from thousands of individual logical elements, each of which is able to take on one of just two states, depending on some logical function of its inputs. In this case study we begin with four logical variables named B2, B1, A2, and A1, which are to represent two binary numbers, B and A, of two digits each. From these, using formulas representing computer logical elements, we are to produce the three digits C4, C2, and C1 of the sum. We visualize the arrangement of the digits as

$$
\begin{array}{r@{\;}c@{\;}c}
   & A2 & A1 \\
+ & B2 & B1 \\
\hline
C4 & C2 & C1
\end{array}
$$

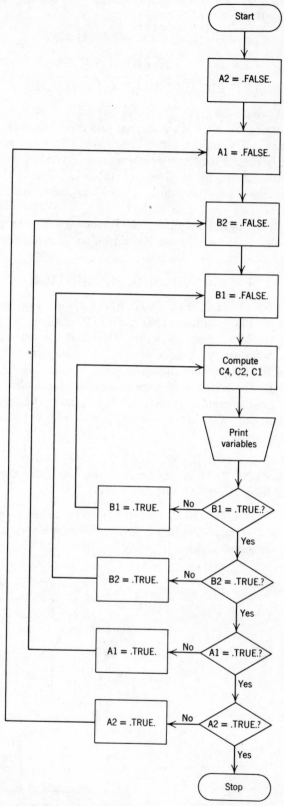

Figure 13.2. Flowchart of a method of testing the logical design of a binary adder, using FORTRAN logical variables.

```
C CASE STUDY 13
C THE LOGICAL DESIGN OF A BINARY ADDER
C
      LOGICAL A1, A2, B1, B2, K2, C1, C2, C4, R, S, T, W, X
      INTEGER A1OUT, A2OUT, B1OUT, B2OUT, C1OUT, C2OUT, C4OUT
C STATEMENTS 1-4 ARE PART OF THE SCHEME FOR OBTAINING ALL
C COMBINATIONS OF VALUES FOR THE FOUR INPUT VARIABLES
1     A2 = .FALSE.
2     A1 = .FALSE.
3     B2 = .FALSE.
4     B1 = .FALSE.
C THE NEXT 9 STATEMENTS ARE ALL LOGICAL ASSIGNMENT STATEMENTS,
C COMPUTING THE VALUES OF THE SUM DIGITS
5     K2 = A1 .AND. B1
      C1 = .NOT. K2 .AND. (A1 .OR. B1)
      X = A2 .AND. B2 .AND. K2
      W = A2 .OR. B2 .OR. K2
      T = B2 .AND. K2
      S = A2 .AND. K2
      R = A2 .AND. B2
      C4 = R .OR. S .OR. T
      C2 = X .OR. (W .AND. .NOT. C4)
C THE NEXT GROUP OF STATEMENTS CONVERTS FROM LOGICAL VALUES TO
C ZEROS AND ONES, ZERO STANDING FOR FALSE AND ONE FOR TRUE
      A2OUT = 0
      A1OUT = 0
      B2OUT = 0
      B1OUT = 0
      C4OUT = 0
      C2OUT = 0
      C1OUT = 0
      IF (A2) A2OUT = 1
      IF (A1) A1OUT = 1
      IF (B2) B2OUT = 1
      IF (B1) B1OUT = 1
      IF (C4) C4OUT = 1
      IF (C2) C2OUT = 1
      IF (C1) C1OUT = 1
C THE FORMAT STATEMENTS THAT ARE REFERENCED BY THE WRITE STATEMENTS
C CONTAIN HOLLERITH FIELD SPECIFICATIONS, EXPLAINED IN THE TEXT
      WRITE (6, 200) A2OUT, A1OUT
      WRITE (6, 201) B2OUT, B1OUT
      WRITE (6, 202) C4OUT, C2OUT, C1OUT
  200 FORMAT (3HC  , 2I2)
  201 FORMAT (3H   , 2I2)
  202 FORMAT (1H , 3I2)
C THE REMAINING STATEMENTS ARE A PART OF THE SCHEME FOR OBTAINING ALL
C COMBINATIONS OF VALUES OF THE FOUR INPUT VARIABLES
      IF (B1) GO TO 6
      B1 = .TRUE.
      GO TO 5
6     IF(B2) GO TO 7
      B2 = .TRUE.
      GO TO 4
7     IF (A1) GO TO 8
      A1 = .TRUE.
      GO TO 3
8     IF(A2) STOP
      A2 =.TRUE.
      GO TO 2
      END
```

Figure 13.3. A program for testing the logical design of a binary adder.

As an example,

$$\begin{array}{r} 0\ 1 \\ +\ 1\ 1 \\ \hline 1\ 0\ 0 \end{array}$$

We shall be working with logical variables which have truth values only; a digit of one will be represented by .TRUE., zero by .FALSE.. We shall generate all 16 possible combinations of binary digits for a complete test of the adder circuit.

The circuit is shown in schematic form in Figure 13.1. The four digits are represented by boxes at the bottom. All the other boxes stand for logical elements, the nature of which is stated in the box. Boxes representing values to which a name is given in the program have the name shown at the top left. In the middle row inputs are shown below the boxes; in the top row inputs are marked by arrows.

This diagram presents the logical design of the adder in graphic form. For instance, the units digit of the sum (C1) is seen to be 1 if either A1 *or* B1 is 1 *and* it is *not* true that both A1 *and* B1 are 1. The second digit (C2) is 1 if either all three of K2 (the carry from position 1), A2, and B2 are 1 or any one of them, but not two, is 1. C4 is 1 if any two or more of K2, A2, and B2 are 1.

The logical assignment statements to evaluate the various expressions here will be shown shortly. First, we should look at the flowchart of Figure 13.2, which shows how we shall work through the 16 combinations of values of B2, B1, A2, and A1. This scheme uses a nest of loops. The result is that B1 alternates between true and false, B2 is false twice, then true twice, then false twice, and so on.

The type declarations and the first four logical assignment statements in the program of Figure 13.3 present no difficulties. In the statement for C1 we see the need for parentheses. The intention, as written, is to evaluate the expression in the following sequence:

1. .NOT. K2
2. A1 .OR. B1
3. .AND. of the above.

Without the parentheses, the sequence would be

1. .NOT. K2
2. .NOT. K2 .AND. A1
3. .OR. of the above.

The two expressions (with and without parentheses) do not mean the same thing, just as $-A*(B+C)$ and $-A*B+C$ do not mean the same thing.

```
  0 0
  0 0
0 0 0

  0 0
  0 1
0 0 1

  0 0
  1 0
0 1 0

  0 0
  1 1
0 1 1

  0 1
  0 0
0 0 1

  0 1
  0 1
0 1 0

  0 1
  1 0
0 1 1

  0 1
  1 1
1 0 0

  1 0
  0 0
0 1 0

  1 0
  0 1
0 1 1

  1 0
  1 0
1 0 0

  1 0
  1 1
1 0 1

  1 1
  0 0
0 1 1

  1 1
  0 1
1 0 0

  1 1
  1 0
1 0 1

  1 1
  1 1
1 1 0
```

Figure 13.4. The output of the program in Figure 13.3. The 16 triples can be read as exercises in binary addition.

In the next statement X will be set to .TRUE. only if all three of the variables are true.

In the statement in which C2 is computed the parentheses make the intention clear but are not actually required; the sequence of evaluation and the meaning would be the same without them.

After computing C2, we are ready to print the results. This could be done with the L field specifications, giving T's and F's. We prefer, however, to make the printout more attractive by making it look like what it represents: binary additions. To this end we convert each logical variable to an integer variable, in which zero stands for false and 1 stands for true. This translation is accomplished by first setting all seven of the output variables to zero and then setting to 1 any that correspond to logical values of .TRUE..

Three WRITE statements are used to print the augend, the addend, and the sum on three separate lines. In FORMAT statement 200 we see a Hollerith field specification 3H0bb, in which the b's stand for blanks. The zero is the carriage control character, which will cause double spacing. The two blanks are inserted into the line, with the net effect of starting the printing on that line two spaces to the right of the point at which it would otherwise start. FORMAT statement 201 begins with a Hollerith field specification 3Hbbb; the first blank specifies single spacing, and the other two blanks are inserted in the line as before. In statement 202 we have a 1Hb for spacing, followed by the 3I2 for the three integers to be printed; we do not want the two blank spaces this time, which were used before to make the other numbers line up with this one.

The testing that comes after the output closely follows the logic of the flowchart.

The output from the program is shown in Figure 13.4. We see that all 16 cases, viewed as exercises in binary addition, give correct results, and we conclude that the adder design (and the program) is correct.

We hasten to point out that this is by no means the only way to set up a binary adder. As a matter of fact, this one is a copy of a system that may be used to demonstrate to elementary school children how computers work. Seventeen children, arranged as in Figure 13.1, but spread out so that each logical element can see his inputs, and with each child holding a card defining his logical operation, will function beautifully.

EXERCISES

1. Have the following pairs of logical expressions the same truth values?

*a. L1 .AND. L2 .OR. L3
   (L1 .AND. L2) .OR. L3

b. L1 .AND. L2 .OR. L3
   L1 .AND. (L2 .OR. L3)

*c. L1 .AND. L2 .OR. L3 .AND. L4
   (L1 .AND. L2) .OR. (L3 .AND. L4)

d. .NOT. L1 .AND. L2
   .NOT. (L1 .AND. L2)

*e. .NOT. (L1 .OR. L2)
   .NOT. L1 .AND. .NOT. L2

*f. L1 .AND. L2 .AND. L3 .AND. 1 .EQ. 2
   L4 .AND. .NOT. L4

2. Modify the program of Figure 13.3 so that C1, C2, and C4 can be computed without the use of intermediate variables, using the following formulas:

C1 = A1 .AND. .NOT. B1 .OR. .NOT. A1 .AND. B1
C2 = (A2 .AND. .NOT. B2 .OR. .NOT. A2 .AND. B2)
     .AND. (.NOT. A1 .OR. .NOT. B1)
     .OR. A1 .AND. B1 .AND. (.NOT. A2 .AND.
     .NOT. B2 .OR. A2 .AND. B2)
C4 = A1 .AND. B1 .AND. B2 .OR. A1 .AND. B1 .AND.
     A2 .OR. A2 .AND. B2

3. Case Study 7 can be done entirely without logical variables. Let the values of A1, A2, etc., be zero or 1 instead of false or true. Then write statements of the following sort:

IF (K2 .EQ. 0 .AND. (A1 .EQ. 1
                      .OR. B1 .EQ. 1)) C1 = 1

Modify the program along these lines and run it to see if it gives identical results.

CASE STUDY 14 LINEAR INTERPOLATION IN A FILTER TRANSMISSION TABLE; SUBSCRIPTED VARIABLES

Table 14.1 gives the percent transmission of various wavelengths of light through a Kodak Wratten No. 47 filter. This is a projection filter that transmits blue, a little green, a little ultraviolet, and considerable infrared. The transmission curve is presented in graphical form in Figure 14.1.

We wish to write a program that will permit us to interpolate in this table for wavelength values other than those in the table. We should like the search to be fairly rapid, and we do not want to waste time or storage by entering all the zero values.

Such an assignment would ordinarily fall to one small part of a larger program. Here we shall make a complete program of it by using the task as a convenient way of seeing how useful subscripted variables are and how to take advantage of them.

Let us begin by deciding which entries from Table 14.1 we need to use. Observe that if we enter zero values for 0.24 and 0.34 any value between the two will be correctly interpolated as zero. These two will contain all the information in the first six lines of Table 14.1. Likewise, zeros for 0.54 and 0.70 will contain all the information in another nine lines. Table 14.2 is therefore the one in which we want to interpolate.

The result of this condensation is a time and space saving, but it also means that we have a table with unevenly spaced values of wavelength. This will create no problems if we write the program to take account of the spacing, as we shall.

TABLE 14.1

Wavelength	Percent Transmission
0.24	0.0
0.26	0.0
0.28	0.0
0.30	0.0
0.32	0.0
0.34	0.0
0.36	1.5
0.38	10.0
0.40	25.1
0.42	49.0
0.44	48.0
0.46	43.7
0.48	30.3
0.50	12.5
0.52	1.3
0.54	0.0
0.56	0.0
0.58	0.0
0.60	0.0
0.62	0.0
0.64	0.0
0.66	0.0
0.68	0.0
0.70	0.0
0.72	2.5
0.74	11.0
0.76	31.0
0.78	55.0
0.80	72.0
0.82	83.0
0.84	89.0
0.86	91.0
0.88	91.0
0.90	91.0

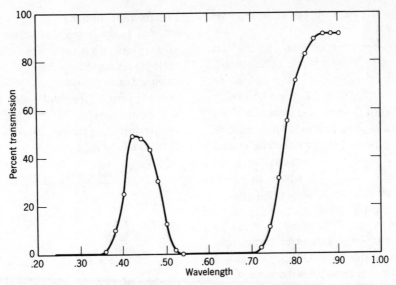

Figure 14.1. The transmission curve of a filter used as the basis of Case Study 14.

The program we shall write must first be able to read the table we have created. Then it will be required to accept an unspecified number of additional data cards, each containing a wavelength. For each the program is to interpolate in the stored table to find the corresponding transmission, then print both the wavelength and the transmission.

The idea of linear interpolation is simply to assume that two adjacent points are connected by a straight line. When a wavelength falls between two table entries, we use simple geometry to compute the corresponding transmission, as suggested in Figure 14.2. If a wavelength is exactly the same as a table value, we do not need to interpolate but may merely print out the corresponding transmission from the table. As a final requirement on the program, we must test to be sure that a wavelength is never less than the smallest table value nor greater than the largest. We do not propose to extrapolate from the table; a wavelength outside the range

TABLE 14.2

Wavelength	Percent Transmission
0.24	0.0
0.34	0.0
0.36	1.5
0.38	10.0
0.40	25.1
0.42	49.0
0.44	48.0
0.46	43.7
0.48	30.3
0.50	12.5
0.52	1.3
0.54	0.0
0.70	0.0
0.72	2.5
0.74	11.0
0.76	31.0
0.78	55.0
0.80	72.0
0.82	83.0
0.84	89.0
0.86	91.0
0.88	91.0
0.90	91.0

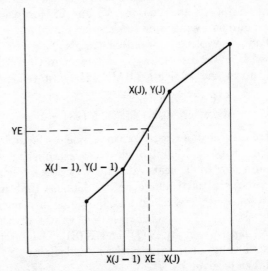

Figure 14.2. The geometrical basis of the method of linear interpolation, once the given x-value (XE) has been found to be bracketed by X(J — 1) and X(J).

represented here would probably be an erroneous data value.

Now we come to the programming issue that these requirements raise. How are we to represent the table values in the program and how are we to search through the table when a wavelength is entered for interpolation? With the programming methods we know up to now, it would be necessary to give a different name to every wavelength and every transmission in the table—46 names in all—and write an extremely awkward program in which every value would have to be named explicitly in a very long sequence of statements.

Such an approach is most unacceptable. It is wasteful of programming time, it is error-prone, and the resulting program is inflexible because it cannot easily be modified to handle a table with some other number of entries. Clearly we need a better approach, and the use of subscripted variables provides the tool we need.

Suppose we agree that X is the general name for the wavelength in the table and that Y is the general name for the transmission. (It is hoped that using the familiar x-y coordinate names rather than something like WAVE and TRANS will make it easier to keep the geometrical situation more clearly in mind.) Now, however, instead of standing for one value, a name such as X stands for 23 different values, the fact of which we inform FORTRAN by writing the type statement as

$$\text{REAL X(23), Y(23)}$$

FORTRAN is thus ordered to set up not one storage location for X and one for Y, but 23 for each. These are now said to be *subscripted variables*.

The subscripting information can also be given, if we wish, by writing the type statement in the ordinary form and adding a DIMENSION statement:

$$\text{REAL X, Y}$$
$$\text{DIMENSION X(23), Y(23)}$$

There is no strong reason to choose one or the other.

With these subscripted variables established as standing for 23 elements each, we may now see how to call for a particular element. Suppose that we have somehow established that the value of a variable named XE lies between the values of the wavelength given by X(7) and X(8). The interpolated value of the transmission, called, say, YE, will then be given by

$$\text{YE} = \text{Y(7)} + (\text{Y(8)} - \text{Y(7)})/$$
$$(\text{X(8)} - \text{X(7)}) * (\text{XE} - \text{X(7)})$$

The formula follows the scheme of Figure 14.2. We see that to specify a particular element in an array we may enclose a *subscript* in parentheses after the variable name. The subscript in this case is a single unsigned integer constant, which is one of the allowable forms of a *subscript expression*. The other forms permitted in FORTRAN are given in the following table, in which I stands for an integer variable and L and L′ are integer constants:

General Form	Example
I	J12
L	8
I ± L	K + 21
L * I	2 * LIMIT
L * I ± L′	3 * LAST − 4

The use of variables in subscript expressions is the basis for the great importance of subscripted variables in programming. For now we can write statements that refer to a "generalized" element of an array; the value of the variable in the subscript expressions determines the actual element. For instance, we can arrange our interpolation program to search through the table for two X values that "bracket" a given value, with the identification of the bracketing elements provided by the subscript expression values. It will turn out in the program we shall write, for example, that the X value just smaller than the given value will be the one called X(J − 1), where the search process will have given a value to J; the X value just larger will be X(J), where the J is the same as in X(J − 1). The same J can be used to select elements from the Y array, thus letting us write a statement such as

$$\text{YE} = \text{Y(J−1)} + (\text{Y(J)} - \text{Y(J−1)})/$$
$$(\text{X(J)} - \text{X(J−1)}) * (\text{XE} - \text{X(J−1)})$$

This statement now performs the interpolation calculation *no matter which* two values of wavelength and transmission are required.

Before going on to write the linear interpolation program, let us illustrate the usefulness of subscripted variables in a simpler application. Suppose that we have a real array named S that contains 20 elements and that we are required to form the sum of the squares of those 20 elements and place it in SUMSQ. In the usual mathematical subscript notation the requirement is to compute

$$\text{SUMSQ} = \sum_{i=1}^{20} s_i{}^2$$

```
      INTEGER I
      REAL SUMSQ, S(20)
      SUMSQ = 0.0
      I = 1
  189 SUMSQ = SUMSQ + S(I)**2
      I = I & 1
      IF (I .LE. 20) GO TO 189
      STOP
      END
```

Figure 14.3. A program illustrating operations with subscripted variables.

The computation can be done with the program in Figure 14.3, in which it is assumed that values are given to the 20 elements in a segment of the program that is not shown here.

The variable S is identified as a subscripted variable by the addition of the 20 in parentheses in the REAL statement. We first set SUMSQ, which is a nonsubscripted variable, equal to zero. This is done so that we may write the statement at 189 and be assured that the first time it is executed SUMSQ will be zero. Then I is made 1, so that when statement 189 is first executed we get the *first* element from the array of S values. Then 1 is added to I and a test is made to determine whether all of the values have been processed.

The first time statement 189 is executed the net effect is to add s_1^2 to zero and place the sum in SUMSQ. The second time s_2^2 is added, etc., etc.; when the statement has been executed with I equal to 20, the sums of the squares of all 20 elements will have been added to SUMSQ. Note that when the IF shows that I = 20 we must still go back once again because I is incremented *before the test*.

Let us now return to the interpolation problem and begin by stating more precisely the form of the input deck.

We shall require that the X- and Y-values be punched one pair to a card. The X-value is to be punched in columns 1–10 and the corresponding Y-value in columns 11–20, both in form suitable for reading with an F10.0 field specification. The deck is assumed to be in ascending sequence on the X-values, and there may be no more than 200 pairs. We have no need for so many in this problem, of course, but we are interested in the question of flexibility; the program will carry out the same general type of processing that we need on *any* table with no more than 200 entries. This flexibility of program use is, of course, one of the major attractions of subscripted variables.

Figure 14.4 is a flowchart of the work of the

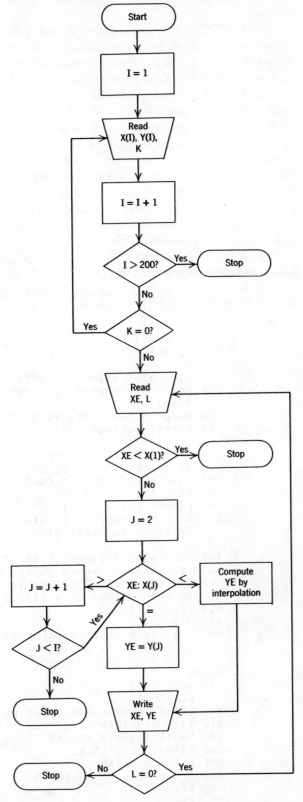

Figure 14.4. Flowchart of a program for linear interpolation.

program of Figure 14.5. Let us examine the two together.

The DIMENSION statement says X and Y are subscripted variables, each with storage space for 200 elements. The first assignment statement sets I to 1 to start storing X–Y pairs at the start of the arrays. The READ statement will read and store values in element locations according to the value of I, which is incremented each time around the reading loop. Each time through the loop we check to make sure that I has not exceeded 200; if it has, the data is invalid, for we have no more space in the arrays. Note that a variable name in a READ statement may be subscripted; the latest value of the subscript expression determines which element will receive the value from the card.

It is assumed that column 21 will be blank on all table-value cards except the last. A blank column in a numeric field on a data card is taken to be zero, and the second IF statement will cause a return to the READ statement on all X–Y cards before the last. In other words, we continue to read until the last card is detected, thus increasing the value of I by 1 each time. The pairs are stored in successive location in the two arrays. On detection of the last card, with its nonzero punch in column 21, the IF statement will not transfer back to the READ, and we go on to the READ at 43 where the processing begins.

In previous examples we have not needed to signal the end of the data with an explicit *sentinel*, as the nonzero punch in column 21 is usually called. We have simply let the attempt to read the end-of-file card signal the end. That is impossible here because what follows the last X–Y pair is *not* the end-of-file card, but more data: what comes next are the cards containing X-values for which we want to find interpolated Y-values.

These further data cards each contain just one number in columns 1–10. No sentinel will be re-

```
C CASE STUDY 14
C LINEAR INTERPOLATION
C
      DIMENSION X(200), Y(200)
C FOLLOWING STATEMENTS READ THE CARDS DEFINING THE CURVE
      I = 1
   56 READ (5,57) X(I), Y(I), K
   57 FORMAT (2F10.0, I1)
      I = I + 1
      IF (I .GT. 200) STOP
      IF (K .EQ. 0) GO TO 56
C THIS STATEMENT READS A CARD CONTAINING AN X VALUE
   43 READ (5, 70) XE, L
   70 FORMAT (F10.0, I1)
C CHECK FOR X VALUE LESS THAN SMALLEST X ON CURVE
      IF (XE .LT. X(1)) STOP
      J = 2
C THE FOLLOWING ARITHMETIC IF STATEMENT COMPARES THE GIVEN X VALUE
C WITH SUCCESSIVE VALUES ON THE CURVE, LOOKING FOR AN X VALUE ON THE
C CURVE THAT IS GREATER THAN THE GIVEN VALUE
   69 IF (XE - X(J)) 112, 111, 110
C NOT FOUND YET
  110 J = J + 1
C CHECK WHETHER GIVEN X VALUE IS LARGER THAN LARGEST X ON CURVE
      IF (J .LT. I) GO TO 69
      STOP
C EQUAL
  111 YE = Y(J)
      GO TO 200
C TWO CURVE VALUES BRACKET GIVEN X -- INTERPOLATE
  112 YE = Y(J-1) + (Y(J)-Y(J-1))/(X(J)-X(J-1))*(XE-X(J-1))
  200 WRITE (6,201) XE, YE
  201 FORMAT (1P2E15.6)
C CHECK FOR SENTINEL
  202 IF (L .EQ. 0) GO TO 43
      STOP
      END
```

Figure 14.5.   A program for linear interpolation.

quired this time, for we can now use the usual method of attempting to read the end-of-file card.

When we read an X card we first check to see that the value, named *XE*, is not less than the first element of the X array; if it is, the data card is in error. Assuming that this test will pass, we set J to 2 to start a search through the X-values, looking for one that is greater than or equal to XE.

In the arithmetic IF statement at 69 we go to statement 112 if XE is less than $X(J)$, which would mean that XE is bracketed by $X(J-1)$ and $X(J)$. We go to 111 if XE is equal to $X(J)$, in which case the value we want is simply $Y(J)$. A transfer to 110 will occur if XE is greater than $X(J)$, which means that we have not yet found two X-values that bracket XE. In this last case we need only increment J, check to see that we have not exhausted the list of X values without finding one larger than XE, and go back to the IF statement if the search should continue.

Once we have found that some $X(J)$ is larger than XE, we will have arrived at the situation in Figure 14.2. We have already considered the assignment statement, using subscripted variables, which computes the interpolated value in this case.

By whatever method YE was found, we now print XE and YE and then return to read another XE card.

The program was run with 23 X–Y cards containing the values in Table 14.2 and a sampling of XE cards. Figure 14.6 shows the output produced.

We close this case study by consolidating what we have learned about subscripted variables and making the statement of the techniques a bit more precise. At the same time we shall see that subscripted variables are not restricted to the one-dimensional type we have seen so far.

A subscripted variable in FORTRAN may have one, two, or three subscripts, and it then represents a one-, two-, or three-dimensional array. (When used in this connection, one dimensional refers to the number of *subscripts*, not to the number of *ele-*

ments: a one-dimensional array can have many elements, and it would be permissible for a three-dimensional array to have only one element.)

The first element of a one-dimensional array is element number 1, the second is element number 2, etc., up to the number of elements in the array. In mathematical notation we might write x_1, x_2, x_3, ..., x_{19}, x_{20}; in FORTRAN subscript notation we write $X(1)$, $X(2)$, $X(3)$, ..., $X(19)$, $X(20)$. We must always number elements consecutively starting with 1.

A two-dimensional array is composed of horizontal rows and vertical columns. The first of the two subscripts then refers to the *row number*, running from 1 to the maximum number of rows, and the second to the *column number*, running from 1 to the maximum number of columns. For instance, an array of two rows and three columns might be shown in mathematical notation as

$$a_{1,1} \quad a_{1,2} \quad a_{1,3}$$

$$a_{2,1} \quad a_{2,2} \quad a_{2,3}$$

In FORTRAN subscript notation the elements would be written $A(1,1)$, $A(1,2)$, $A(1,3)$, $A(2,1)$, $A(2,2)$, $A(2,3)$. We note that the subscripts are separated by commas, as they are in three-dimensional variables.

A three-dimensional array may be viewed as composed of planes, each of which contains rows and columns. The interpretation, however, depends somewhat on the purpose of the computation; other interpretations are possible.

Some versions of FORTRAN allow more than three subscripts, seven being the number permitted in several systems.

The name of a subscripted variable is formed in the same way as a nonsubscripted variable name. If the name is not mentioned in a type statement, the array is assumed to consist entirely of integer or entirely of real elements, depending on the initial letter. If the name is mentioned in a DOUBLE PRECISION, COMPLEX, or LOGICAL statement, the elements are assumed to be entirely of the one kind. In the case of double-precision and complex variables the extra computer storage required is automatically taken into account; no special handling of the subscripts is required. To emphasize: the elements of any given array must *all* be of one kind. An array, for instance, cannot consist partly of real and partly of integer elements.

When subscripted variables are used in a pro-

```
4.300000E-01    4.850000E 01
7.450000E-01    1.600000E 01
4.770000E-01    3.231000E 01
6.000000E-01    0.
2.400000E-01    0.
2.800000E-01    0.
8.700000E-01    9.100000E 01
9.000000E-01    9.100000E 01
```

Figure 14.6. The output of the program in Figure 14.5, run with data relating to the transmission graph in Figure 14.1.

gram, certain information about them must be supplied to the FORTRAN compiler:

1. Which variables are subscripted?
2. How many subscripts are there for each subscripted variable?
3. What is the maximum size of each subscript?

These questions are answered by the DIMENSION statement. Every subscripted variable in a program must be mentioned in a DIMENSION or type statement, and this statement must appear before the first occurrence of the variable in the program. A common practice is to give the dimension information for all subscripted variables in DIMENSION statements at the beginning of the program. One DIMENSION statement may mention any number of variables, and there may be any number of DIMENSION statements.

The DIMENSION statement takes the form

$$\text{DIMENSION V, V, V, . . .}$$

where the V's stand for variable names followed by parentheses enclosing one, two, or three unsigned integer constants that give the maximum size of each subscript. When FORTRAN processes a DIMENSION statement, it sets aside enough storage locations to contain arrays of the sizes specified by the information in the statement. Thus if a program contains the statement

$$\text{DIMENSION X(20), A (3,10), K(2,2,5)}$$

the FORTRAN compiler will assign 20 storage locations to the one-dimensional array named X; 30 (3×10) to the two-dimensional array A; and 20 ($2 \times 2 \times 5$) to the three-dimensional array K. If any of these variables had been named in a DOUBLE-PRECISION or COMPLEX statement, twice as much storage would have been assigned to each. The extra storage is automatically taken into account in all usage of the subscripts; the programmer need give no thought to the extra storage.

It is the programmer's responsibility to write the program so that no subscript is ever larger than the maximum size he has specified in the DIMENSION statement. Furthermore, subscripts must never be smaller than 1; zero and negative subscripts are not permitted. If these restrictions are violated, the source program will in some cases be rejected by the compiler. In other cases, in which the illegal subscripts are developed only at execution time, the program will be compiled but will in all probability give incorrect results.

The DIMENSION statement is said to be *nonexecutable;* that is, it provides information only to the FORTRAN compiler and does not result in the creation of any instructions in the object program. The FORMAT statement is also nonexecutable, as are the type statements and several others. A nonexecutable statement may appear almost anywhere in the source program, even between two assignment statements. As already noted, however, the dimension information for each subscripted variable must be given before the first appearance of that variable in the program. Furthermore, a DIMENSION statement must not be the first statement in the range of a DO statement (see Case Study 15).

EXERCISES

Note. Include an appropriate DIMENSION or type statement in each program segment for these exercises, realizing, of course, that in a complete program each subscripted variable is mentioned only once in a DIMENSION statement, usually at the beginning of the program.

*1. Suppose that the coordinates of a point in space are given by the three elements of a one-dimensional array named X. (Note the different usages of the word dimension: the elements of a one-dimensional array are being used as the coordinates of a point in three-dimensional space!) Write a statement to compute the distance of the point from the origin, which is given by the square root of the sum of the squares of the coordinates.

2. If the coordinates of a point in space are x_1, x_2, and x_3, the direction cosines of the line from the origin to the point are given by

$$CA = \frac{x_1}{\sqrt{x_1{}^2 + x_2{}^2 + x_3{}^2}}$$

$$CB = \frac{x_2}{\sqrt{x_1{}^2 + x_2{}^2 + x_3{}^2}}$$

$$CC = \frac{x_3}{\sqrt{x_1{}^2 + x_2{}^2 + x_3{}^2}}$$

Write statements to compute these three numbers, assuming that the coordinates are the elements of a one-dimensional array named X.

*3. Given two arrays named A and B, both two-dimensional, write statements to compute the elements of another two-dimensional array named C from the following equations. The maximum value of all subscripts is 2.

$$c_{11} = a_{11}b_{11} + a_{12}b_{21}$$
$$c_{12} = a_{11}b_{12} + a_{12}b_{22}$$
$$c_{21} = a_{21}b_{11} + a_{22}b_{21}$$
$$c_{22} = a_{21}b_{12} + a_{22}b_{22}$$

Readers familiar with matrix notation will recognize the multiplication of two 2×2 matrices.

4. Given a two-dimensional array named R, the elements of which are to be viewed as the elements of a 3×3 determinant, write a statement to compute the value of the determinant by any method you know. The value of the determinant should be named DET.

*5. Two one-dimensional arrays named A and B each contain 30 elements. Compute

$$D = \left(\sum_{i=1}^{30} (A_i - B_i)^2 \right)^{1/2}$$

(a distance function in 30-space). Write a program segment to perform the calculation.

*6. If we have a list of tabular values represented by a one-dimensional array, then the *first differences* of the list are formed by subtracting each element except the last from the element immediately following it. Suppose we have a one-dimensional array named X that contains 50 elements. Compute the 49 elements of another array named DX from

$$DX(I) = X(I + 1) - X(I) \qquad I = 1, 2, \ldots, 49$$

Write a program segment to perform this calculation.

7. Suppose we have a one-dimensional array named Y that contains 32 elements; these are to be regarded as the 32 ordinates of an experimental curve at equally spaced abscissas. Assuming that a value has already been given to H, compute the integral of the curve represented approximately by the Y values from

$$\text{TRAP} = \frac{H}{2} (Y_1 + 2Y_2 + 2Y_3 + \cdots + 2Y_{31} + Y_{32})$$

8. A two-dimensional array named AMATR contains 10 rows and 10 columns. A one-dimensional array named DIAG contains 10 elements. Write a program segment to compute the elements of DIAG from the formula

$$\text{DIAG}(I) = \text{AMATR}(I, I) \qquad I = 1, 2, \ldots, 10$$

*9. Given a one-dimensional array named Y, with 50 elements, and numbers U and I, write a statement to compute the value of S from the following equation, written in ordinary mathematical subscript notation.

$$S = y_i + u \frac{y_{i+1} - y_{i-1}}{2} + \frac{u^2}{2} (y_{i+1} - 2y_i + y_{i-1})$$

This is called *Stirling's interpolation formula* (through second differences), which may be described as follows: we have three points of a curve: (x_{i-1}, y_{i-1}), (x_i, y_i), and (x_{i+1}, y_{i+1}), such that $x_{i+1} - x_i = x_i - x_{i-1} = h$, and a value of x. We write $u = (x - x_i)/h$. Then the formula stated gives the interpolated value of y corresponding to x, found by passing a quadratic through the three given points.

10. Using the assumptions of Exercise 9, write a statement to compute the value of T from the following equation:

$$T = y_i + u(y_{i-1} - y_i)$$
$$+ \frac{u(u-1)(y_{i+2} - y_{i+1} - y_i + y_{i-1})}{4}$$
$$+ \frac{(u - \frac{1}{2}) u(u-1)(y_{i+2} - 3y_{i+1} + 3y_i - y_{i-1})}{6}$$

This is called *Bessel's interpolation formula* (through third differences). With the notation used in Exercise 9, it finds a value of y corresponding to x by passing a cubic through the four given points. The arrangement of the differences is somewhat different from that in Stirling's formula, however.

*11. Given two one-dimensional arrays named A and B, of seven elements each, suppose that the seven elements of A are punched on one card and the seven elements of B are punched on another card. Each element value is punched in 10 columns in a form suitable for reading with an F10.0 field specification. Write a program to read the cards, then compute and print the value ANORM from

$$\text{ANORM} = \sqrt{\sum_{i=1}^{7} a_i b_i}$$

Use a 1PE20.7 field specification for ANORM.

A *norm* may be thought of as a generalization of the concept of distance.

12. Using the assumptions of Exercise 11, write a program to read the data cards and then carry out the following procedure. If every $a_i > b_i$, for $i = 1, 2, \ldots, 7$, print an integer 1; if this condition is not satisfied, print a zero.

*13. Rewrite the program segment for Exercise 5 to use double precision variables for the arrays.

14. Rewrite the program segment for Exercise 10 to use double precision variables for the arrays.

*15. COMPLX is a one-dimensional array containing 30 complex numbers. Write statements to form the sum of the absolute values of the 30 elements, using the CABS function.

16. With assumptions as in Exercise 15, write statements to form the sum of all the imaginary parts of the 30 elements of COMPLX, but include in the sum only those imaginary parts that are positive.

*17. A one-dimensional array named TRUTH contains 40 truth values. Write a program to place in TRUE the count of the number of elements of TRUTH that are .TRUE., and in FALSE the count of the number of elements that are .FALSE.. Note that TRUE is a legitimate variable name; .TRUE., with periods, is a logical constant.

18. A one-dimensional array named COMPLX contains 30 elements. A two-dimensional array of logical variables named QUAD contains 30 rows and 4 columns. For all 30 values of I, if the point represented by COMLX(I) lies in quadrant J, make *QUAD*(I, J) .TRUE. and the other three elements in row I of QUAD .FALSE..

CASE STUDY 15 TOWER GUY WIRES; THE DO STATEMENT

The FORTRAN DO statement provides for the repeated execution of a segment of a program, with automatic modification of an integer variable between repetitions. We shall see in almost all case studies from now on that such a statement has great usefulness, particularly in connection with subscripted variables. In this case study we introduce the statement in a widely applicable technique, not, however, involving subscripted variables.

Let us suppose that in a design job for guy wires for a tower an engineer discovers that he is frequently using a desk calculator to find the value represented by X in Figure 15.1, given H, D, and θ. The formula is

$$X = \sqrt{D^2 + H^2 + 2HD \sin \theta}$$

Let us first take H = 300 and D = 400 and compute X for all values of θ from -20 to $+20°$ in steps of $1°$. The computational task is trivial: we know how to evaluate the formula, and it is no chore either to read the values of the angle or to generate them. The point of this case study is that there is a much simpler way to generate the angle values than we have previously had available to us.

The DO statement that we wish to study calls for a group of statements to be executed repeatedly, with an integer variable taking on a succession of values. In the program in Figure 15.2 we first have a few statements to give values to H and D and write heading lines and then the statement

DO 30 I = 1, 41

This has the meaning: "carry out the state-

ments down to and including the one numbered 30, with the variable I taking on successive values 1 to 41 in steps of 1." Consider the effect. The first time the group of repeated statements is executed I will have the value 1 and so the statement

DEGREE = I —21

will produce the value of -20 as an integer, convert it from real form because DEGREE is real, and store that value. The value is then used in the next statement to compute X, which is finally written. The WRITE statement is the one with the statement number 30, mentioned in the DO, so after writing a line we carry out the entire group of three statements again, except that this time the variable I has the value 2 and DEGREE becomes -19.0. The entire process is done 41 times, with I taking on the successive values 1 to 41. The output is shown in Figure 15.3.

This technique of producing a sequence of values for a variable with the DO statement can be applied to other ranges of values. For instance, suppose we wanted all angles from -20 to $+20$ in steps of $0.5°$. We could get the values needed by replacing the statements

DO 30 I = 1, 41

DEGREE = I — 21

with the statements

DO 30 I = 1, 81

DEGREE = 1 — 41

DEGREE = DEGREE / 2.0

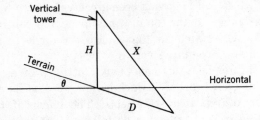

Figure 15.1. Sketch of the parameters in a tower guy-wire computation.

To summarize, I takes on the integer values of 1 to 81; subtracting 41 produces values that range from −40 to +40, which are converted to real in the first assignment statement; the second assignment statement produces values that range from −20 to +20 in steps of 0.5. It should be kept in mind that the DO statement requires the *index*, the variable that takes on successive values, to be an *integer* variable. Furthermore, the starting value must not be less than 1. If we were able, we might like to write something like

$$\text{DO 30 DEGREE} = -40.0, 40.0$$

but this is not permitted in FORTRAN. Further, it may be worth reminding ourselves that integer arithmetic always truncates to an integer. Accordingly, we cannot get the variation in steps of 0.5 by writing

$$\text{DEGREE} = (I - 41) / 2$$

because after the integer division the remainder would simply be dropped.

On the other hand, there is a function available that makes the conversion from integer to real form at any point in a statement we may wish. Its name, FLOAT, refers to the fact that real variables are sometimes called floating point variables because of the way they are stored in the computer. We

```
H = 300     D = 400

DEGREE     X

-20.0     409.8
-19.0     414.6
-18.0     419.3
-17.0     424.1
-16.0     428.8
-15.0     433.5
-14.0     438.1
-13.0     442.7
-12.0     447.3
-11.0     451.9
-10.0     456.4
 -9.0     460.9
 -8.0     465.4
 -7.0     469.8
 -6.0     474.3
 -5.0     478.6
 -4.0     483.0
 -3.0     487.3
 -2.0     491.6
 -1.0     495.8
  0.      500.0
  1.0     504.2
  2.0     508.3
  3.0     512.4
  4.0     516.5
  5.0     520.5
  6.0     524.5
  7.0     528.4
  8.0     532.4
  9.0     536.2
 10.0     540.1
 11.0     543.9
 12.0     547.6
 13.0     551.4
 14.0     555.0
 15.0     558.7
 16.0     562.3
 17.0     565.8
 18.0     569.4
 19.0     572.8
 20.0     576.3
```

Figure 15.3. The output of the program in Figure 15.2.

```
C CASE STUDY 15
C GUY WIRE COMPUTATION - FIRST EXAMPLE OF A DO STATEMENT
C
      H = 300.0
      D = 400.0
      WRITE (6, 20)
   20 FORMAT (1H1, 18HH = 300     D = 400///1H , 11HDEGREE     X//)
      DO 30 I = 1, 41
      DEGREE = I - 21
      X = SQRT(D**2 + H**2 + 2.0*H*D*SIN(DEGREE/57.2957795))
   30 WRITE (6, 25) DEGREE, X
   25 FORMAT (1H , F5.1, F8.1)
      STOP
      END
```

Figure 15.2. Use of a DO statement for a guy-wire computation.

may write

DO 30 I = 1, 81

DEGREE = FLOAT(I — 41) / 2.0

The FLOAT function converts the integer value I — 41 to real form, after which it may immediately be used in any way permitted a real value.

So far we have seen the DO statement in the form in which the increment between repetitions is 1. There is another form of the statement in which we may specify this increment to be any positive integer value we please. Let us illustrate by supposing that the program is to be modified to let H and D take on a variety of values instead of remaining fixed. We shall produce a table in which H ranges from 100 to 400 in steps of 50 and D ranges from a starting value that is 50 greater than the current value of H, up to a maximum of 500, in steps of 50. The angle is to range from —20 to +20 in steps of 1°, and the 41 values of X for each combination of values of H and D are to be printed on a separate page.

In the program of Figure 15.4 we see how simply this may be done. The first DO statement illustrates the new form in a fairly obvious way: when a third *indexing parameter* is present, it specifies the increment to be applied between repetitions. The specified range of 100 to 400 in steps of 50 is thus taken care of in one statement. After converting the integer value to real, we set the variable M to 50 more than whatever LH is and then use M as one of the indexing parameters in the following DO statement. This form is permitted: each of the two or three indexing parameters may be an integer constant or an integer variable, written without a sign.

With H and D having been given values, we write these values at the top of a new page and continue to the DO statement that causes the angle to run through the required values.

All three DO statements have statement 30 as the end of their "ranges," which is how we describe the group of repeated statements. The effect of this *nesting* of DO ranges is as follows. The first DO says to carry out its range with a succession (as it happens) of seven values. Within *each* of the seven executions the second DO is encountered, which says to carry out its range some number of times. The third DO is within the range of each of the others, and its range is carried out 41 times for both. The range of the third (outer) DO is thus carried out 1435 times. A sampling of the output is given in Figure 15.5.

The overlapping of ranges of nested DO's can be even more flexible than indicated here, where all ranges end with the same statement. We might, for instance, wish to arrange three DO statements as shown in Figure 15.6. The "outer" DO, the larger one, contains two "inner" DO's; the two inner DO's do not overlap, and the ending statement of the second inner DO is not the same as that of the outer DO. Such arrangements are quite permissible, and, as we shall see in many later case studies, very useful. The one thing we cannot do is to have an inner DO with a range that extends beyond the end of the range of the outer DO that contains it, as described in Figure 15.7. This prohibition is quite reasonable; if such a situation were permitted, it would be most awkward to define what we wanted the outer DO to do.

A similar remark applies to one final prohibition: it is not permitted to execute a transfer into the

```
C CASE STUDY 15
C SECOND EXAMPLE OF A DO STATEMENT
C
      DO 30 LH = 100, 400, 50
      H = LH
      M = LH + 50
      DO 30 LD = M, 500, 50
      D = LD
      WRITE (6, 20) LH, LD
   20 FORMAT (1H1, 4HH = , I3, 4X, 4HD = , I3///1H , 11HDEGREE      X//)
      DO 30 I = 1, 41
      DEGREE = I - 21
      X = SQRT(D**2 + H**2 + 2.0*H*D*SIN(DEGREE/57.2957795))
   30 WRITE (6, 25) DEGREE, X
   25 FORMAT (1H , F5.1, F8.1)
      STOP
      END
```

Figure 15.4. A modified version of the program in Figure 15.2 which uses three DO statements.

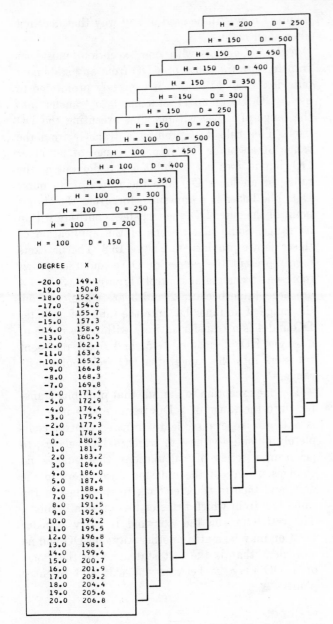

H = 200   D = 250
H = 150   D = 500
H = 150   D = 450
H = 150   D = 400
H = 150   D = 350
H = 150   D = 300
H = 150   D = 250
H = 150   D = 200
H = 100   D = 500
H = 100   D = 450
H = 100   D = 400
H = 100   D = 350
H = 100   D = 300
H = 100   D = 250
H = 100   D = 200

H = 100   D = 150

DEGREE	X
-20.0	149.1
-19.0	150.8
-18.0	152.4
-17.0	154.0
-16.0	155.7
-15.0	157.3
-14.0	158.9
-13.0	160.5
-12.0	162.1
-11.0	163.6
-10.0	165.2
-9.0	166.8
-8.0	168.3
-7.0	169.8
-6.0	171.4
-5.0	172.9
-4.0	174.4
-3.0	175.9
-2.0	177.3
-1.0	178.8
0.	180.3
1.0	181.7
2.0	183.2
3.0	184.6
4.0	186.0
5.0	187.4
6.0	188.8
7.0	190.1
8.0	191.5
9.0	192.9
10.0	194.2
11.0	195.5
12.0	196.8
13.0	198.1
14.0	199.4
15.0	200.7
16.0	201.9
17.0	203.2
18.0	204.4
19.0	205.6
20.0	206.8

Figure 15.5. A sampling of the output produced by the program in Figure 15.4.

range of a DO from outside its range. If this were permitted, the DO would not "know" that it should be called into action.

Before continuing, in the next case study, to a consideration of the application of DO statements to subscripted variables, let us summarize what we have learned so far and add a few notes about related matters.

The DO statement may be written in either of these forms:

$$DO \; n \; i = m_1, m_2$$

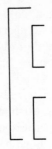

Figure 15.6. Schematic representation of a legal nesting of DO's.

or

$$DO \; n \; i = m_1, m_2, m_3$$

Here n must be a statement number, i must be a nonsubscripted integer variable written without a sign, and m_1, m_2, and m_3 must each be either an unsigned integer constant or a nonsubscripted unsigned integer variable. The value m_1 must be positive and greater than zero, as must m_2 and m_3. The value of m_2 must be greater than or equal to m_1.

The action of the DO statement is as follows. The statements following the DO, up to and including the one with the statement number n are executed repeatedly. They are executed first with $i = m_1$; before each succeeding repetition i is increased by m_3; repeated execution continues until the statements have been executed with i equal to the largest value that does not exceed m_2.

The variable i is called the *index* of the DO; m_1, m_2, and m_3 are called the *indexing parameters*; the group of statements from the DO down to and including the one numbered n is called the *range* of the DO.

There are two ways by which control can transfer outside the range of a DO. The *normal exit* occurs when the DO is *satisfied*, that is, at the completion of the number of executions of the range specified by the indexing parameters. When this happens, control passes to the statement following that named in the DO. The second method by which control can get outside the range of a DO is in a GO TO or

Figure 15.7. An illegal nesting of DO's.

IF statement; this can happen when we wish to specify the *maximum* number of executions of the range in the parameters but also wish to set up tests in the range to determine the *actual* number of executions.

When control is transferred outside the range of a DO *before* the DO is satisfied, the index i is available for any purpose permitted for an integer variable. This can be quite valuable. After a normal exit, however, i is *not* available, at least not in most versions of FORTRAN, but this is no serious inconvenience in practice.

We may close this section with a statement of a few rules governing the use of the DO statement. We shall see the application of some of these rules later on, but it will be found that they are not actually restrictive and that the DO remains one of the most heavily used statements in the FORTRAN language.

Rule 1. The first statement in the range of a DO must be one that can be executed. This excludes the DIMENSION and FORMAT statements and the various type statements. These *specification statements* provide information to the compiler about the program rather than causing any computation to take place. We shall see later that several other statements are also nonexecutable.

Rule 2. It is permissible for the range of one DO (which we call the *outer* DO) to contain another DO (which we call the *inner* DO). When this occurs, it is required that all statements in the range of the inner DO also be in the range of the outer DO. This does not prohibit the ranges of two or more DO's from ending with the same statement but it does prohibit a situation in which the range of an inner DO extends past the end of the range of an outer DO.

Rule 3. The last statement in the range of a DO must not be a GO TO in any form, nor an arithmetic IF, RETURN (discussed in Case Study 21), STOP, PAUSE, or DO statement. Neither can it be a logical IF that contains any of the others. These statements, with the exception of the RETURN, may be used freely anywhere else in the range. The CONTINUE statement, described later, is provided for situations that would otherwise violate this rule.

Rule 4. No statement within the range of a DO may redefine or otherwise alter any of the indexing parameters of that DO; that is, it is not permitted within the range of a DO to change the values of i, m_1, m_2, or m_3. As noted before, these numbers may still be used in any way that does not alter their values.

Rule 5. Control, with one exception, must not transfer into the range of a DO from any statement outside its range. Thus it is expressly prohibited to use a GO TO or an IF statement to transfer into the range of a DO without first executing the DO itself. This rule *does* prohibit a transfer from the range of an outer DO into the range of an inner DO, but it *does not* prohibit a transfer out of the range of an inner DO into the range of an outer DO. The latter is permissible, for, from the standpoint of the outer DO, the transfer is entirely within its range. Some illustrations of the application of this rule are provided in Figure 15.8. The brackets here represent the ranges of DO's and the arrows represent transfers of control. Tranfers 2, 3, and 4 are acceptable because 2 and 3 are transfers from the range of an inner DO into the range of an outer DO and 4 is a transfer entirely within the range of a single DO. Transfers 1, 5, and 6 all represent transfers into the range of a DO from outside its range.

The one exception to the rule that prohibits transfers into the range of a DO from outside its range is this: it is permissible to transfer control completely outside the nest to which a DO belongs, to perform a series of calculations that makes no changes in any of the indexing parameters in the nest, and then to transfer back to the range of the same DO from which transfer was originally made. The restriction on the exit and re-entry transfer location may be stated another way: no DO and no statement that is the last statement in the range of a DO may lie between the exit and re-entry points.

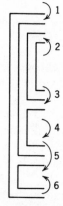

Figure 15.8. Examples of permissible nests of DO's and some correct and incorrect transfers of control. Transfers 2, 3, and 4 are acceptable; 1, 5, and 6 are not.

CONTINUE is a dummy statement that causes no action when the object program is executed. It merely satisfies the rule that the last statement in the range of a DO must not be one that can cause transfer of control. It is also used to provide a statement to which an IF can transfer when the computations in the range of a DO have been completed. This is necessary because a transfer within the range of a DO is not permitted to return to the DO itself, unless, of course, we really intend to start the execution of the DO from the beginning again. An example of the use of the CONTINUE appears in Figure 16.3, in the next case study.

EXERCISES

*1. The formula

$$Y = 41.298\sqrt{1 + x^2} + x^{1/3}\,e^x$$

is to be evaluated for

$$x = 1.00, 1.01, 1.02, \ldots, 3.00$$

Each xy pair is to be printed on a line, with 1PE20.7 field specifications. Write a program using a DO loop to do this.

2. The formula

$$z = \frac{e^{ax} - e^{-ax}}{2}\sin(x + b) + a\log\frac{b + x}{2}$$

is to be evaluated for all combinations of

$$x: 1.0(0.1)2.0$$
$$a: 0.10(0.05)0.80$$
$$b: 1.0(1.0)10.0$$

where x: 1.0(0.1)2.0 means $x = 1.0$, 1.1, 1.2, . . . , 2.0, and so on. For each combination of x, a, and b (there are 1650 combinations) a line giving x, a, b, and z is to be written. Write a program containing three DO loops to do this.

3. Modify the program of Figure 6.1 so that the values of S from 20 to 200 in steps of 5 are generated by a DO statement, making it unnecessary to read any data.

4. Modify the program of Figure 7.3 so that it uses a DO statement to generate the time values in the oscillatory case shown in Figure 7.4.

CASE STUDY 16 THE GAUSS-SEIDEL METHOD FOR SIMULTANEOUS EQUATIONS; THE DO STATEMENT WITH SUBSCRIPTED VARIABLES

In this case study we shall see the DO statement used in conjunction with subscripted variables, which is the most common application. We shall begin with simple examples, eventually arriving at a frequently used numerical technique, the Gauss-Seidel iteration method for solving a certain class of simultaneous linear algebraic equations.

For a starter suppose that we have three one-dimensional arrays named A, B, and C, each having 20 elements. We wish to form in C the sum of A and B, in the sense that any element c_i of C should be the sum $a_i + b_i$, $i = 1, 2, \ldots, 20$. If we view A, B, and C as vectors, the problem is to add two vectors, element by element. The program segment shown in Figure 16.1 does the task quite simply. With the DIMENSION statement to indicate that the variables are subscripted, we come to a DO statement that causes I to run through the values 1 to 20 in steps of 1. The range of the DO this time is the one statement named in the DO so that the effect is to carry out statement 16 exactly 20 times. This performs the vector addition.

For another example suppose that we wish to form the scalar product of A and B, defined by

$$\text{PROD} = \sum_{i=1}^{20} a_i b_i$$

The task is hardly any more difficult than the vector addition. We do have to make sure that PROD, the variable that will contain the scalar product at the end of the execution of the program, contains zero to start; we must never assume anything about the contents of a variable to which we have not assigned a value. Setting PROD to zero is thus the first action in the program of Figure 16.2. This done, we move to the DO statement which again causes I to run through the values 1 to 20. The first time the one-statement range is executed the product A(1)* B(1) is added to the previous contents of PROD—which are zero. The second time, the product A(2)*B(2) is added to the previous contents of PROD—which are now A(1)*B(1). The process, when completed, produces the sum of products of all of the corresponding pairs of values, which is the scalar product.

The next example introduces a situation that requires two variables for subscripting, even though the arrays are still one-dimensional, in a particularly instructive illustration.

We shall suppose that the input to a program consists of a series of experimentally measured values. Each point in the experiment consists of an x-value and a y-value, corresponding to the abscissa and ordinate of a point on a graph. The data points were gathered and entered into the computer in random order; that is, we know that the first x-value goes with the first y-value, the second x-value goes with the second y-value, and so on, but we cannot assume that the first x is

```
DIMENSION A(20), B(20), C(20)
DO 16 I = 1, 20
16 C(I) = A(I) + B(I)
```

Figure 16.1. A program segment using a DO statement for vector addition.

the smallest of the x-values. For the purposes of the calculations that are to be done later in the program it is necessary to rearrange the data points in storage so that the first x-value *is* the smallest and that the second x-value is the next larger and so on. In other words, we must order the data points into ascending sequence on the x-values.

We shall assume that the x-values as they were originally read (i.e., in scrambled order) are the elements of an array named X and that there are 25 of them. The y-values are the 25 elements of another array called Y.

The FORTRAN program to rearrange these data points into ascending sequence of the x-values involves a nest of two DO loops. We shall show the development of the program by displaying a simplified version of the inner loop before writing the full program. This simplified loop will place the smallest x-value in the first position of the x-array. This can be done by the following process. First compare the first and second x-values in the original array. If $X(1)$ is smaller than or equal to $X(2)$, leave them as they are; but if $X(1)$ is larger than $X(2)$, interchange these two values within the array. Having inspected the first and second elements and interchanged them if necessary to get the smaller in $X(1)$, we compare $X(1)$ and $X(3)$ and interchange them if $X(1)$ is larger. What is in $X(1)$ as we make this second comparison may well be the value that was originally in $X(2)$, but that does not matter. Similarly, we compare $X(1)$ and $X(4)$, $X(1)$ and $X(5)$, etc., each time interchanging if necessary to get the smaller in $X(1)$. This process guarantees that the smallest x-value in the entire array will finally be in $X(1)$, wherever it may have been to begin with. Remembering that to each element of X there corresponds an element of Y, we naturally carry out the same interchange operations on the Y-array as we do on the X-array, but there is no testing of the Y-array values.

```
DIMENSION A(35), B(35)
PROD = 0.0
DO 17 I = 1, 35
17 PROD = PROD + A(I) * B(I)
```

Figure 16.2. A program segment using a DO statement to produce a scalar product.

To interchange two values from the array in storage, we follow a three-step process: (1) move the first value to a temporary location which we call TEMP; (2) move the second value to the location originally occupied by the first; (3) move the first value, now in TEMP, to the location originally occupied by the second.

A program to carry out all of the steps of this process is shown in Figure 16.3. We are assuming that the data values have been read in by an earlier part of the program, and we are not showing the statements that complete the rearrangement or use of the data values.

The program illustrates a number of points worth noting. We see another example of a DO loop in which the index does not start with 1. We see an example of the use of the CONTINUE statement. This is required, for if the IF statement shows that $X(1)$ is already less than or equal to the other x-value with which we are comparing it a transfer of control must be made to skip around the six statements that interchange elements in the X- and Y-arrays. What we want to do in this case is simply to repeat the whole process with the index J increased by 1. As we have already noted, however, it is not possible to transfer control back to the DO. This would result in starting the DO loop again with J equal to 2—which is not what we want. Therefore we transfer control to the CONTINUE, which has been identified in the DO statement as the end of the range.

In reading this program, it is well to recall the meaning of an arithmetic assignment statement: the value of the variable on the left side of the equal sign is replaced by the value of the expression on the right. Thus a statement such as

$$X(1) = X(J)$$

means that the number identified by the variable name $X(J)$ is to be moved to the location for the

```
DIMENSION X(25), Y(25)
DO 12 J = 2, 25
IF (X(1) .LE. X(J)) GO TO 12
TEMP = X(1)
X(1) = X(J)
X(J) = TEMP
TEMP = Y(1)
Y(1) = Y(J)
Y(J) = TEMP
12 CONTINUE
```

Figure 16.3. A program segment to place the smallest value of X in an array of X's in the first position of the array and to place the corresponding Y in the first position of an array of Y's.

```
      DIMENSION X(25), Y(25)
      DO 12 I = 1, 24
      IP1 = I + 1
      DO 12 J = IP1, 25
      IF (X(I) .LE. X(J)) GO TO 12
      TEMP = X(I)
      X(I) = X(J)
      X(J) = TEMP
      TEMP = Y(I)
      Y(I) = Y(J)
      Y(J) = TEMP
   12 CONTINUE
```

Figure 16.4. A program segment to rearrange a set of X–Y data points into ascending sequence on the X values.

number identified by the name X(1). The value in the location for X(J) is unchanged.

In the data as described earlier there may or may not be two equal x-values. As the program has been written, it does not matter; if they are equal, there is no point in exchanging them, and we simply transfer control down to the CONTINUE and go around the loop again.

When this loop has been completed (when the DO is satisfied), we are guaranteed that the data points will have been rearranged so that the smallest x-value is in X(1) and the corresponding y-value is in Y(1). What we would like to do next is to get the next larger x-value in X(2). This can be done by comparing X(2) with X(3) and following, interchanging when necessary. After that we would like to get the next larger element in X(3). We would similarly like to get the successively larger values in X(4), X(5), and so on.

It appears that what we need to do for the complete program is to make variables of all subscripts that appear as 1's in Figure 16.3. The subscript will then select the element to be compared with all following elements (interchanging if necessary). This subscript, which we shall call I, will start at 1 and run through 24. The subscript that appears as J in the program of Figure 16.3 will still be J, but it will have to start at one more than the value of the I subscript and run through 25. All of this is easily done with another DO statement which controls the I subscript. This is the outer DO. There is one complication, however. The inner DO must specify that the J subscript will start at one more than I. Look-

ing back at the definition of the DO statement, we see that each indexing parameter must be either an integer constant or a single integer variable. It is *not* permitted to write a statement such as

$$DO\ 12\ J = I + 1, 25$$

To avoid this restriction we simply insert a statement that computes the value of a variable IP1, which is always one more than the value of I. The complete program to arrange the data points in sequence is shown in Figure 16.4.

This technique is by no means the most efficient sorting method known. It is presented here only as an interesting application of the DO statement.

To illustrate a slightly different type of DO loop let us now make some further assumptions about the purpose of the program just written. Suppose that the data points which have been read and arranged into ascending sequence on the x-values lie on a curve and that we are required to find the area under the curve; that is, to find the definite integral of the curve represented approximately by these points. If we now make the further assumption that the distance between successive x-values is equal to a constant value h, the approximate integral given by the trapezoidal rule is

$$AREA = \frac{h}{2}(y_1 + 2y_2 + 2y_3 + \cdots + 2y_{23} + 2y_{24} + y_{25})$$

A DO loop may conveniently be used to form the sum of the y-values with subscripts of 2 through 24. Having done so, we can multiply this sum by 2, add the first and last y-values, and multiply by $h/2$. This program segment, shown in Figure 16.5, would follow the CONTINUE statement of the segment shown in Figure 16.4. It includes the computation of h on the assumption that it is the interval between any two x-values. If, in fact, the x-values are not equally spaced, the program naturally gives an incorrect result. If it were required that the program be able to handle unequally spaced x-values, the numerical integration method would have to be modified.

Simpson's rule, with the subscripting scheme we

```
      DIMENSION X(25), Y(25)
      SUM = 0.0
      DO 20 I = 2, 24
   20 SUM = SUM + Y(I)
      AREA = (X(2) - X(1))/2.0 * (Y(1) + 2.0*SUM + Y(25))
```

Figure 16.5. A program segment for numerical integration by the trapezoidal rule.

```
      DIMENSION X(25), Y(25)
      ODD = 0.0
      EVEN = 0.0
      DO 47 I = 2, 24, 2
   47 EVEN = EVEN + Y(I)
      DO 48 I = 3, 23, 2
   48 ODD = ODD + Y(I)
      AREA = (X(2) - X(1))/3.0 * (Y(1) + 4.0*EVEN + 2.0*ODD + Y(25))
```

Figure 16.6. A program segment for numerical integration by Simpson's rule.

are using, is

$$AREA = \frac{h}{3} (y_1 + 4y_2 + 2y_3 + 4y_4 + 2y_5 + \cdots$$
$$+ 2y_{23} + 4y_{24} + y_{25})$$

A program to evaluate this formula will be a little more complex because of the alternating coefficients of 2 and 4. One fairly obvious way to handle the problem is to set up two DO loops and to accumulate separately the sums of the y's corresponding to the two coefficients. See Figure 16.6.

The computation may also be done with only one DO loop, which saves a little time in the running of the object program, if we proceed as follows. Suppose we set up an index that runs from 2 to 22 in steps of 2. Such an index always references an element that should be multiplied by 4; one plus the index references an element that should be multiplied by 2. The Y(24) element must be added in separately; this is caused by the fact that there are more odd elements than even. See Figure 16.7.

Flexibility in the manner of writing subscripts is very useful here. It is not possible to form the sum of the y-values with one DO loop unless some such subscripting arrangement is used.

It may be noted that we have written these integration formulas with the subscripts starting at 1, whereas it is conventional to write them with the subscripts starting at zero. This was done to make it easier to describe the problem, for we recall that a subscript must always be positive and nonzero.

It is possible to program formulas that have zero and negative subscripts, but it is somewhat more effort than it is worth to us at this point. It may be noted that some FORTRAN systems, or languages

similar to FORTRAN, do permit zero and negative subscripts.

All of the examples we have seen so far have involved one-dimensional arrays. Let us now turn to an application of the DO statement in which a two-dimensional array appears.

Suppose that we have a system of three simultaneous linear algebraic equations, such as

$$12x_1 + 7x_2 - 3x_3 = 16$$
$$-3x_1 + 9x_2 + 5x_3 = 21$$
$$1x_1 + 7x_2 + 13x_3 = -8$$

To work with such a system in FORTRAN we would most naturally put the coefficients into a two-dimensional array of three rows and three columns and the constant terms in a one-dimensional array of three elements. Alternatively, we would put all of these numbers into a two-dimensional array of three rows and four columns, the last column containing the constants. Let us do the first.

It happens that the method we shall employ for solving the system of equations does not apply to all systems. When it does apply, it is often quite attractive, but not every system can be solved with it. The way we shall state the condition is this: in every row the main diagonal term (the one with the same row and column number) must *dominate* the other terms in that row. This means that the absolute value of the main diagonal term must be greater than the sum of the absolute values of all the other terms in the row. We wish to write a program that will examine a two-dimensional array of numbers to see whether it satisfies this condition;

```
      DIMENSION X(25), Y(25)
      ODD = 0.0
      EVEN = 0.0
      DO 51 I = 2, 22, 2
      EVEN = EVEN + Y(I)
   51 ODD = ODD + Y(I+1)
      AREA = (X(2)-X(1))/3.0*(Y(1) + 4.0*(EVEN+Y(24)) + 2.0*ODD +Y(25))
```

Figure 16.7. Another version of a program for numerical integration by Simpson's rule.

```
C CHECK MATRIX  FOR SATISFYING CONVERGENCE CRITERION FOR G-S
C
      DIMENSION D(20, 20)
      OK = 1.0
      DO 100 I = 1, 20
      SUM = 0.0
      IF (I .GT. 1) GO TO 92
      DO 93 J = 2, 20
   93 SUM = SUM + ABS(D(I, J))
      GO TO 100
   92 IF (I .EQ. 20) GO TO 94
      IM1 = I - 1
      DO 95 J = 1, IM1
   95 SUM = SUM + ABS(D(I, J))
      IP1 = I + 1
      DO 96 J = IP1, 20
   96 SUM = SUM + ABS(D(I, J))
      GO TO 100
   94 DO 97 J = 1, 19
   97 SUM = SUM + ABS(D(I, J))
  100 IF (SUM .GT. ABS(D(I, I))) OK = 0.0
      IF (OK .EQ. 1.0) WRITE (6, 110)
      IF (OK .EQ. 0.0) WRITE (6, 111)
  110 FORMAT (1H , 36HMATRIX MEETS CONVERGENCE REQUIREMENT)
  111 FORMAT (1H , 44HMATRIX DOES NOT MEET CONVERGENCE REQUIREMENT)
      STOP
      END
```

Figure 16.8. A program to check whether a given matrix satisfies sufficient conditions for convergence of the Gauss-Seidel iteration method.

we shall assume that the array is 20 by 20 rather than the 3 by 3 of the sample.

A program is shown in Figure 16.8. We first set a variable named OK to 1; if this value is never changed to zero, it will mean that we never found a row that failed to satisfy the criterion. Then comes a DO statement to give I the values 1 through 20. As this program is written, I identifies a (horizontal) row; a (vertical) column is identified by the subscript J. Such a convention, of course, is arbitrary, but it is in keeping with common mathematical notation in which a_{ij} is the element in the ith row and jth column. FORTRAN doesn't care, of course; the issue is readability for us.

After the first DO statement, we are in a section that is carried out for each of the 20 rows. In each row we must form the sum of the absolute values of all elements not on the main diagonal and then compare this sum with the absolute value of the main diagonal element. Accordingly, we set SUM equal to zero before going into the part of the program that accumulates the sum.

The basic idea now is to form the sum of the terms to the left of the main diagonal, then the sum of the terms to the right of the main diagonal.* The strategy is as follows.

One segment will form the sum of the absolute values of elements to the left of the main diagonal; this segment is skipped over for the first row, in which there are no elements to the left of the diagonal. A second segment will form the sum of the absolute values of all terms to the right of the main diagonal; this segment is skipped on the last row.

In the program we see the two segments immediately after the SUM = 0.0 statement. The IF statement skips to the second segment on the first row. Unless this is the first row, we set up a variable named IM1 to be the number of the last element in the row before the diagonal term. A DO statement calls for the summation of the absolute values of all terms from the first up to the one just before the diagonal. After checking to see if this is the last row and, if so, skipping down, a second DO calls for the summation of the absolute values of all terms from the one to the right of the diagonal up to the twentieth.

Now we have an IF statement to check whether the criterion has been met for this row: if SUM is

---

* It might seem attractive to form the sum of *all* terms in the row and at the end subtract off the main diagonal

term. This procedure, which would produce the same value in infinite precision, can cause a serious loss of accuracy in a digital computer in which we can keep only a finite number of digits and in which a main diagonal term is very much larger than the others. Review Case Study 12 for a reminder of this problem.

greater than the absolute value of the diagonal term, we set OK to zero to denote the failure. The main diagonal term, as we noted, is the one with the same row and column number; because this is row I we can call for the element in row I and column I by writing D(I, I).

This IF statement is the one named as the end of the range of the first IF statement, the one that picked a row. This statement will thus be executed for each row. If it is found that the first row fails the test, this does mean that we will still go on to test all the others, which is a waste of time. We could readily enough rewrite the program to avoid this waste, but, because it is presumably rather unlikely, the possible small computer time saving is hardly worth the programming effort. (A decision on this tradeoff must frequently be made and the outcome is by no means always of this sort. In other problems, as we shall see, it is often worth considerable programming time and trouble to speed up the execution of the program.)

Finally, we turn to the actual solution of a system of equations, using the method of Gauss-Seidel iteration. Let us state the method in terms of a system of three equations in three unknowns.

$$a_{11}x_1 + a_{12}x_2 + a_{13}x_3 = b_1$$

$$a_{21}x_1 + a_{22}x_2 + a_{23}x_3 = b_2$$

$$a_{31}x_1 + a_{32}x_2 + a_{33}x_3 = b_3$$

Suppose we make guesses at the values of x_2 and x_3—it doesn't matter whether they are good guesses; zeros will work. We solve the first equation for x_1, writing a prime to indicate that it is a new approximation:

$$x_1' = \frac{b_1 - a_{12}x_2 - a_{13}x_3}{a_{11}}$$

Now, using the new value for x_1 and the initial guess at x_3, we solve the second equation for x_2:

$$x_2' = \frac{b_2 - a_{21}x_1' - a_{23}x_3}{a_{22}}$$

Finally, using the new approximations for x_1 and x_2, we solve the third equation for x_3:

$$x_3' = \frac{b_3 - a_{31}x_1' - a_{32}x_2'}{a_{33}}$$

This process of computing a new value for each of the variables constitutes one *iteration*. Now we perform another iteration, always using the most recently computed value of each variable. If the system of equations satisfies the criterion that the diagonal terms dominate their rows, the iteration scheme is guaranteed to converge to a solution, regardless of the initial guesses used. (The condition we have stated is *sufficient* to guarantee convergence; a discussion of the *necessary* condition, that the matrix coefficients be positive definite, is considerably beyond the scope of this book. In other words, there are systems of equations in which the main diagonal terms do not dominate their rows, for which the Gauss-Seidel method nevertheless converges to a solution.)

We wish to write a program that will apply the Gauss-Seidel method to any system up to 50 equations. The program must include the reading of the coefficients, of which there would be 2500 in the largest case, and the constant terms. It must include a test for convergence to decide when the approximations are close enough and an iteration counter to stop the process if for any reason it fails to converge. We shall employ the counter method to determine whether the system of equations satisfies the criterion of convergence rather than any explicit test of the sort illustrated in the preceding program.

Figure 16.9 is a flowchart of the method, drawn to correspond quite closely to the FORTRAN techniques we shall be using. We begin by clearing the arrays for the coefficients, the constant terms, and the unknowns. The idea here is that with many hundreds of data values to enter it might be a considerable convenience not to enter zero values. This is especially true, for some systems of equations in practice have a high proportion of zero coefficients. Next we read a card containing the number of equations, the convergence test value, and the maximum number of iterations to be permitted. We then read the coefficients and constant terms, using a card layout and sentinel scheme that we shall describe in connection with the program. Finally, in the preliminary phase, we set the iteration counter ITER to one.

A connector (the circled 1) takes us to the part of the flowchart that begins an iteration. The convergence test is as follows: if the maximum absolute difference between the last two approximations to any variable is less than ϵ, the iteration scheme will be stopped. This requires keeping a variable that represents the largest difference found so far; this is BIG which we set to zero initially. The line $i = 1$ represents part of a DO statement action. The SUM = 0.0 box is the start of the actions for

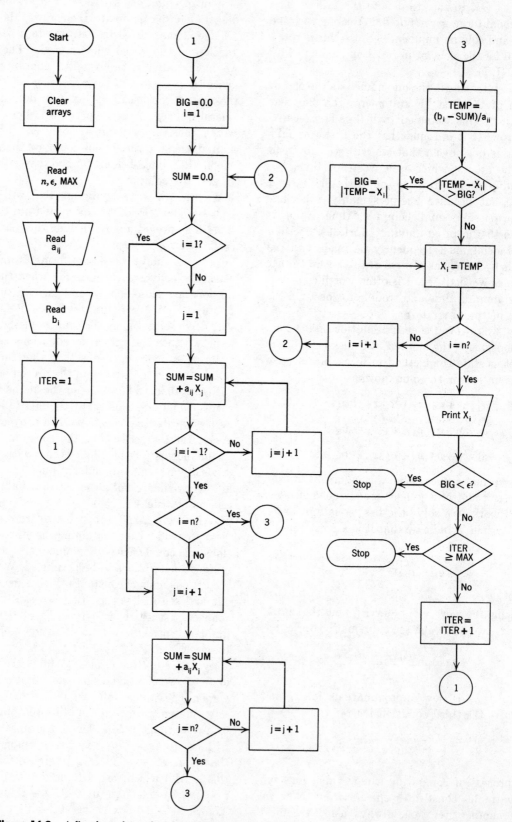

Figure 16.9. A flowchart of a method for solving a system of simultaneous equations by the Gauss-Seidel method.

```
C CASE STUDY 16
C GAUSS-SEIDEL METHOD FOR SOLVING SIMULTANEOUS EQUATIONS
C SOLVE N SIMULTANEOUS EQUATIONS IN N UNKNOWNS, WHERE N IS NOT
C GREATER THAN 50. ONLY THE NON-ZERO ELEMENTS NEED BE ENTERED.
C
      DIMENSION A(50, 50), B(50), X(50)
C PRINT A HEADING LINE
      WRITE (6, 999)
  999 FORMAT (1H1, 2H I, 8X, 4HX(I))
C CLEAR ARRAYS
      DO 50 I = 1, 50
      B(I) = 0.0
      X(I) = 0.0
      DO 50 J = 1, 50
   50 A(I,J) = 0.0
C READ N, EPSILON, AND THE MAXIMUM NUMBER OF ITERATIONS
      READ (5, 20) N, EPS, MAXIT
   20 FORMAT (I2, F10.0, I2)
C READ THE COEFFICIENTS
   31 READ (5, 21) I, J, A(I,J), NEXT
   21 FORMAT (2I2, F10.0, I2)
      IF (NEXT .EQ. 0) GO TO 31
C READ THE CONSTANT TERMS
   32 READ (5, 22) I, B(I), NEXT
   22 FORMAT (I2, F10.0, I1)
      IF (NEXT .EQ. 0) GO TO 32
C BEGIN THE ITERATION SCHEME
      ITER = 1
C STATEMENT 99 IS EXECUTED ONCE PER SWEEP
   99 BIG = 0.0
C INDEX I SELECTS A ROW
      DO 100 I = 1, N
C STATEMENT 102 IS EXECUTED ONCE PER ROW
  102 SUM = 0.0
C SEGMENT FROM HERE THROUGH STATEMENT 107 GETS THE SUM OF THE TERMS
CC IN A ROW, EXCLUDING THE MAIN DIAGONAL TERM
      IF (I .EQ. 1) GO TO 105
      LAST = I - 1
      DO 106 J = 1, LAST
  106 SUM = SUM + A(I,J)*X(J)
      IF (I .EQ. N) GO TO 103
  105 INITL = I + 1
      DO 107 J = INITL, N
  107 SUM = SUM + A(I,J)*X(J)
C COMPUTE NEW VALUE OF A VARIABLE
  103 TEMP = (B(I) - SUM) / A(I,I)
      RESID = ABS(TEMP - X(I))
C AT END OF SWEEP, THIS STATEMENT HAS PUT LARGEST RESIDUAL IN BIG
      IF (RESID .GT. BIG) BIG = RESID
  100 X(I) = TEMP
C ONE SWEEP HAS NOW BEEN COMPLETED - PRINT VARIABLES
      WRITE (6, 991)
  991 FORMAT (1H0)
      DO 42 I = 1, N
   42 WRITE (6, 200) I, X(I)
  200 FORMAT (1H , I2, 1PE16.6)
C IF LARGEST RESIDUAL IS LESS THAN EPSILON, PROCESS HAS CONVERGED
      IF (BIG .LT. EPS) STOP
C IF ITERATION COUNTER EXCEEDS MAXIMUM ALLOWABLE, GIVE UP
      IF (ITER .GT. MAXIT) STOP
      ITER = ITER + 1
      GO TO 99
      END
```

Figure 16.10.   A program for solving a system of simultaneous equations by the Gauss-Seidel method.

one line, that is, the computation of a new approximation to one variable. This involves accumulating a sum of products in SUM, which we accordingly set to zero.

The rest of the middle column of boxes, beginning with the decision $i = 1$?, forms the sum of the off-diagonal terms in one row. The actions here very closely parallel those in the program in Figure 16.8, with the exception that here we are forming a sum of products rather than a sum of absolute values.

At connector 3 we carry out certain operations that apply at the end of the accumulation of the sum of products. We set TEMP equal to the new approximation, then ask whether the absolute value of the difference between the new and old values is greater than BIG, and, if so, set BIG equal to the difference. BIG thus always contains the largest difference found so far in this iteration. TEMP then becomes the new approximation, and we go back to connector 2 if more equations remain.

If $i = n$ at this point, we have completed an iteration; that is, we have computed a complete set of approximations to the unknowns. These approximations are printed so that we may see the convergence. If BIG is now less than ϵ, the convergence test constant, we are finished. Otherwise, we ask whether the iteration counter exceeds the limit value MAX and either stop or increment the iteration counter and go back for another iteration.

The program in Figure 16.10 follows the flowchart so closely in most regards that we need only point out a few features of the program.

The first two DO statements form a nest with a common terminal statement, but the inner DO has a smaller range than the outer DO.

In the READ statement at 31 we read the coefficient cards, which have the following format:

Columns	Contents
1–2	Row number of coefficient
3–4	Column number of coefficient
5–14	Coefficient
15	Sentinel: zero on all cards except last

The READ statement reads the row and column numbers and then uses them immediately—in the same READ list—as subscripts.

In the sentinel scheme we punch some nonzero digit in column 15 of the last card. Thus we can write IF (NEXT .EQ.0) GO TO 31 to loop back on all cards except the last.

A similar scheme reads the constant terms with a slight difference in format, for we must specify only the row number. The rest of the program is an almost literal transcription of the flowchart.

The program was run with the system of equations shown in Figure 16.11 and using $\epsilon = 0.00001$. Figure 16.12 is the output. We see that after seven iterations the method has produced a set of approximations quite close to the exact solution 1, -2, 3, -4, 5, -6. This rapid convergence is a direct result of the fact that in most cases the diagonal terms dominate quite strongly. That rapid convergence in such a case should be expected can be seen by noting the limiting case: all off-diagonal terms are zero. The value of each unknown is then simply the constant term divided by the diagonal term.

Many physical problems lead to systems of equations that can be guaranteed by theoretical considerations to satisfy the conditions for convergence of this method; this method is, in fact, commonly used, as we shall see illustrated in later case studies,

$$17.031\,x_1 - 0.615\,x_2 - 2.991\,x_3 + 1.007\,x_4 - 1.006\,x_5 \qquad\qquad = \qquad 0.230$$

$$-1.000\,x_1 + 34.211\,x_2 - 1.000\,x_3 - 2.100\,x_4 + 6.300\,x_5 - 1.700\,x_6 = -22.322$$

$$0.500\,x_2 + 13.000\,x_3 - 0.500\,x_4 + 1.000\,x_5 - 1.500\,x_6 = 54.000$$

$$4.501\,x_1 + 3.110\,x_2 - 3.907\,x_3 - 61.705\,x_4 + 12.170\,x_5 + 8.999\,x_6 = 240.236$$

$$0.101\,x_1 - 0.812\,x_2 - 0.017\,x_3 - 0.910\,x_4 + 4.918\,x_5 + 0.100\,x_6 = 29.304$$

$$1.000\,x_1 + 2.000\,x_2 + 3.000\,x_3 + 4.500\,x_4 + 5.000\,x_5 + 21.803\,x_6 = -117.818$$

Figure 16.11. The system of equations used to test the program in Figure 16.10.

especially No. 26. For systems to which the Gauss-Seidel iteration method does not apply we can use the Gauss elimination method, which is discussed in Case Study 24.

EXERCISES

*1. Two one-dimensional arrays named A and B each contain 30 elements. Write a program segment using a DO statement to compute

$$D = \left(\sum_{i=1}^{30} (A_i - B_i)^2 \right)^{1/2}$$

2. Given a one-dimensional array named X that contains 50 elements, write a program segment using a DO statement to compute the 49 elements of another array, named DX, from

$$DX(I) = X(I + 1) - X(I) \qquad I = 1, 2, \ldots, 49$$

3. A two dimensional array named AMATR contains 10 rows and 10 columns. A one-dimensional array named DIAG contains 10 elements. Write a program segment to compute the elements of DIAG from

$$DIAG(I) = AMATR(I, I) \qquad I = 1, 2, \ldots, 10$$

*4. A one-dimensional array named M contains 20 integers. Write a program segment using a DO statement to replace each element by itself, multiplied by its element number. In other words, replace m_i by $i \cdot m_i$, $i = 1, 2, \ldots, 20$.

*5. Two one-dimensional arrays named R and S have a *maximum* of 40 elements each. The *actual* number of elements is given by the value of a previously computed integer variable M. Compute the first M elements of an array named T, which also has a maximum of 40 elements, according to

$$T(i) = R(i) + S(i) \qquad i = 1, 2, \ldots, M$$

6. Two one-dimensional arrays, A and B, have a maximum of 18 elements each. N is an integer, the value of which does not exceed 18. Compute

$$C = \sum_{k=1}^{N} A_k B_k$$

*7. A one-dimensional array named F contains at most 50 elements. Each of the first M elements, except the first and Mth, is to be replaced by

$$F_i = \frac{F_{i-1} + F_i + F_{i+1}}{3}$$

This is an example of techniques for *smoothing* experimental data to reduce the effect of random errors.

*8. A one-dimensional array named B contains 50 elements. Place the largest of these elements in BIGB and the element number of BIGB in NBIGB.

9. Two one-dimensional arrays named X and Y contain 50 elements each. A variable named XS is known to be equal to one of the elements in X. If $XS = X_i$, place Y_i in YS.

I	X(I)
1	1.350479E-02
2	-6.520854E-01
3	4.178926E 00
4	-4.189778E 00
5	5.089769E 00
6	-6.222032E 00
1	1.272242E 00
2	-1.996795E 00
3	2.960053E 00
4	-3.992127E 00
5	5.000772E 00
6	-6.009086E 00
1	9.926802E-01
2	-2.001492E 00
3	2.999252E 00
4	-4.001735E 00
5	4.599765E 00
6	-5.999013E 00
1	9.999035E-01
2	-2.000039E 00
3	3.000067E 00
4	-3.999916E 00
5	4.999991E 00
6	-6.000017E 00
1	1.000005E 00
2	-1.999992E 00
3	3.000002E 00
4	-4.000003E 00
5	5.000001E 00
6	-6.000001E 00
1	1.000001E 00
2	-2.000000E 00
3	3.000000E 00
4	-4.000000E 00
5	5.000000E 00
6	-6.000000E 00
1	9.999999E-01
2	-2.000000E 00
3	3.000000E 00
4	-4.000000E 00
5	5.000000E 00
6	-6.000000E 00

Figure 16.12. The output produced when the system of equations in Figure 16.11 was run with the program in Figure 16.10.

This kind of *table search* has a wide variety of applications, such as finding a value in a table of electric utility rates from a rate code or finding the numerical code corresponding to an alphabetic name.

*10. A two-dimensional array A contains 15 rows and 15 columns. A one-dimensional array X contains 15 elements. Compute the 15 elements of a one-dimensional array B according to

$$B_i = \sum_{j=1}^{15} A_{ij} X_j \qquad i = 1, 2, \ldots, 15$$

This is multiplication of a matrix and a vector.

11. Three two-dimensional arrays A, B, and C have 15 rows and 15 columns each. Given the arrays A and B, compute the elements of C from

$$C_{ij} = \sum_{k=1}^{15} A_{ik} B_{kj} \qquad i, j = 1, 2, \ldots, 15$$

This is matrix multiplication.

*12. A two-dimensional array RST has 20 rows and 20 columns. Compute the product of the main diagonal elements of RST and store it in DPROD. A main diagonal element is one that has the same row and column number, so that

$$DPROD = \prod_{i=1}^{20} RST(I, I)$$

13. A solution to the following specialized system of equations is to be found:

$$a_{11}x_1 \qquad\qquad\qquad = b_1$$
$$a_{21}x_1 + a_{22}x_2 \qquad\qquad = b_2$$
$$a_{31}x_1 + a_{32}x_2 + a_{33}x_3 \qquad = b_3$$
$$\cdot \quad \cdot \quad \cdot \quad \cdot \quad \cdot \quad \cdot \quad \cdot \quad \cdot \quad \cdot$$
$$a_{n1}x_1 + a_{n2}x_2 + a_{n3}x_3 + \cdots + a_{nn}x_n = b_n$$

First write a program to solve this system on the assumption that the a's are contained in a two-dimensional array that will have about half zeros for the missing elements. This is a moderately simple program: first solve for x_1, substitute this result into equation 2, and so on.

The difficulty, from the standpoint of computer solution, is that there is a great deal of wasted space in the array, which uselessly restricts the size of the system that can be solved. Devise a method of storing the coefficients in a *one*-dimensional array and write a program to find the unknowns. Assume that it must be possible to handle a maximum of 100 equations in 100 unknowns. The actual number of equations is given by the value of N.

14. Same as Exercise 13, except that the system of equations is

$$a_{1,1}x_1 + \cdots + a_{1,n-2}x_{n-2} + a_{1,n-1}x_{n-1} + a_{1,n}x_n = a_{1,n+1}$$
$$\cdot \quad \cdot \quad \cdot \quad \cdot \quad \cdot \quad \cdot \quad \cdot \quad \cdot \quad \cdot \quad \cdot \quad \cdot \quad \cdot \quad \cdot$$
$$a_{n-2,n-2}x_{n-2} + a_{n-2,n-1}x_{n-1} + a_{n-2,n}x_n = a_{n-2,n+1}$$
$$a_{n-1,n-1}x_{n-1} + a_{n-1,n}x_n = a_{n-1,n+1}$$
$$a_{n,n}x_n = a_{n,n+1}$$

Note that it is *not* possible to write a statement like

$$DO\ 12\ I = N, 1, -1$$

which would be handy here. An equivalent "IF loop" must be written to work backward through the x's. Alternatively, we may write

$$DO\ 12\ II = 1, N$$

$$I = N - II + 1$$

I is then used as the subscript; it will vary from N downward to 1.

15. Given a two-dimensional array named C, with 10 rows and 11 columns, compare C(1, 1) with all other elements in the first column, looking for the element with the largest absolute value; make the value of L equal to the row number of the element in column 1 with the largest value. If at the end of these operations L = 1, do nothing more; otherwise exchange the elements in row 1 with the elements in row L, whatever it is.

CASE STUDY 17 MATRIX MULTIPLICATION; OBJECT-TIME FORMATS

A FORTRAN two-dimensional array can be thought of as the computer form of a matrix in mathematics; a great many problems in applied mathematics are most easily formulated in ways that require operations with matrices. We accordingly present several case studies in which various matrix processes are the heart of the program. In this case study we consider the multiplication of two matrices.

Within this framework we shall discuss ways in which format information can be determined at the time of execution of the object program rather than when the source program is written. For instance, in a program that will be executed many times we might want to be able to decide on the format of input data only when the data is punched. We shall learn how to do this and how to "generate" a FORMAT statement, in our case for output.

Let us begin our investigation of the matters of this case study with matrix multiplication. We have two matrices A and B, and wish the product C. If A has m rows and n columns, then B must have n rows and may have any number of columns r. The product C will have m rows and r columns. We see that the product of two matrices is defined only if the number of columns of the first is equal to the number of rows of the second. Thus, even if AB is defined, BA will not necessarily be defined, and even if it is the two products will be identical only in very special cases.

The product matrix is defined in terms of its elements. If a representative element of C is called c_{ik}, it is defined by

$$C_{ik} = \sum_{j=1}^{n} a_{ij}b_{jk} \qquad \begin{array}{l} i = 1, 2, \cdots, m \\ k = 1, 2, \cdots, r \end{array}$$

We can also describe the process verbally: element c_{ik} is the scalar product of row i in the A matrix, with column k in the B matrix.

This is a simple matter to program, with the help of three DO statements, as we see in Figure 17.1. The first DO runs through the rows of A, from 1 to M; the second runs through the columns of B, from 1 to R. The second DO is executed M times, of course, once for each row of A. The two variables I and K, in effect, pick an element of C. We zero this element in preparation for accumulating a sum there, then encounter the third DO, which forms the scalar product. For two 20 x 20 matrices statement 3 would be executed 8000 times. We see here the power of nested DO statements in specifying extensive procedures in a simple, easy-to-follow manner.

Now we return to the beginning of the program to consider the question of input and output.

The basic idea is that instead of writing a FORMAT statement containing information on data arrangement and spacing we shall put that same information into an array; then in the READ or WRITE statement, in which we usually write the statement number of a FORMAT statement, we shall write the name of the array that contains the format-

```
C CASE STUDY 17
C MATRIX MULTIPLICATION AND OBJECT-TIME FORMATS
C
      INTEGER R
      DIMENSION A(20, 20), B(20, 20), C(20, 20),
     1    FORMA(12), FORMB(12), FORMC(22)
C READ FORMATS
      READ (5, 1) (FORMA(I), I = 1, 12)
      READ (5, 1) (FORMB(I), I = 1, 12)
      READ (5, 1) FORMC(1), FORMC(22), FORMZ, FORMNZ, BLANK
    1 FORMAT (12A6)
C READ DIMENSIONS
      READ (5, 2) M, N, R
    2 FORMAT (3I2)
C READ DATA
      READ (5, FORMA) ((A(I, J), J = 1, N), I = 1, M)
      READ (5, FORMB) ((B(I, J), J = 1, R), I = 1, N)
C MATRIX MULTIPLICATION
      DO 3 I = 1, M
      DO 3 K = 1, R
      C(I, K) = 0.0
      DO 3 J = 1, N
    3 C(I, K) = C(I, K) + A(I, J) * B(J, K)
C OUTPUT SECTION
      DO 4 J = 2, 21
    4 FORMC(J) = BLANK
      DO 7 I = 1, M
      DO 6 J = 1, R
      IF (C(I, J) .EQ. 0.0) GO TO 5
C NON-ZERO IF HERE
      FORMC(J+1) = FORMNZ
      GO TO 6
C ZERO IF HERE
    5 FORMC(J+1) = FORMZ
      C(I, J) = BLANK
    6 CONTINUE
    7 WRITE (6, FORMC) (C(I, J), J = 1, R)
      STOP
      END
```

Figure 17.1. A program to multiply two matrices, using object-time formats.

ting information. We see that in the DIMENSION statement three variables named FORMA, FORMB, and FORMC have been named as one-dimensional variables with 12, 12, and 22 elements, respectively. They will hold the formats for matrices A, B, and C, following the rather obvious naming. The number 12 was chosen from force of habit to provide enough space to hold the information in 72 ($= 12 \times 6$) card columns, although in this case there is very little chance we would need so much space.

As always, the DIMENSION statement provides space but does not insert anything in it. The first three executable statements take care of that. The first READ fills that array FORMA with whatever is on the first card in the data section of the input deck. This card should contain exactly what we would otherwise put in a FORMAT statement, with the exception of the word FORMAT itself. For ex-

ample, if we had chosen to use the statement

FORMAT (4FI0.0)

the card read by this first READ should contain

(4F10.0)

We note here a new feature of the list of an input or output statement: it is permissible to use *indexing* in a manner that closely parallels a DO loop. The meaning of the statement is probably obvious at a glance: the 12 elements of the array FORMA are to be filled in succession from the card. The form of such a list must be as indicated: there must be a comma between the array designation and the indexing information and the entire list must be enclosed in parentheses. FORMB is handled in a similar way.

Next comes a READ to obtain the information needed in the output operations that we shall discuss shortly and a READ to get the sizes of the arrays.

The fifth READ actually reads data under control of the format information that has now been entered into FORMA. The array name FORMA has simply replaced the FORMAT statement number that normally would be written in the READ. (Incidentally, even if all of the format information could be contained in one word, the reference here must nevertheless be to an *array*, although the array would have only one element.)

The list indexing here is a little more complex, although the parallel with nested DOs is quite direct. In fact, the idea of an "inner" and an "outer" DO is more graphic here, with the inner DO actually being the one in the inside parentheses. To be explicit, this list indexing calls for the two-dimensional array A to be read in *row order*, that is, taking all the elements in row 1 in order, then all elements in row 2, etc. We note that list indexing can involve variables as well as constants, and, although this example does not exhibit it, we may also use indexing by some increment other than 1 by writing a third parameter as in a DO statement.

The output section is somewhat more complex. We not only wish to leave the decision on format open until the time of execution but also to perform certain operations in the interest of a more attractive ouput to give some idea of the power of object-time formats. Accordingly we shall do two things:

1. Make it possible to specify the format in which ordinary output values are to be printed, much as we have done with input.

2. Arrange that zero values result in blank space on the output rather than the standard form, which consists of a zero followed by a decimal point and which some consider unattractive. This is not a criticism of the designers of the language, who probably made the best decision in this instance; the point is to show how we can modify the output to suit our needs and preferences.

The format information for each line is contained in the array FORMC which we established as having 22 elements. This time we shall not read in the whole array from a card because we wish to carry out special operations for zero or near-zero elements. Instead we read in five items:

1. A left parenthesis into FORMC(1), together with five blanks. Blanks are ignored here as in most instances in FORTRAN. The slight waste of storage space is much less important than the strenuous efforts that would be required to pack the format information exactly as we would ordinarily write it in a program.

2. Two slashes and a right parenthesis into FORMC(22); the two slashes will produce double spacing. The official description of the FORTRAN language in the American Standards Association document states that stored format information may not contain Hollerith fields. Some compilers *do* permit them, thus making possible stored carriage control characters, but we shall defer to the ASA document and not use this feature.

3. The field specification that should be used for zero results.

4. The field specification that should be used for nonzero results.

5. A word of all blanks for two uses that we shall discuss.

In planning the output, our first problem is that we do not know the size of the arrays that may have to be printed; they could have anything from 2 to 20 rows and columns. We shall decide to print all the elements in one row in one line and leave it to the writer of the object-time format to make a choice that is consistent with the number of elements. We have accordingly set up for a maximum of 20 numbers to be printed and have allowed correspondingly for 20 field specifications in the array FORMC, plus the opening and closing parentheses, which total 22.

There is nothing, actually, that dictates having one field specification in one word. The program we shall write is simpler that way, but at the expense of limiting the field specifications to five characters, for we must provide the comma that goes between field specifications. Some specifications, naturally, are longer than five characters: 1PE15.5, for instance, has seven. After study of the sample provided by this problem, the reader will realize that if necessary the program could be rewritten to accommodate such conditions.

We allow for as many as 20 elements, but we do not know that there will be that many, and we must have control over the contents of any positions of the FORMC array that are not otherwise used. Therefore the first action in the output section is to clear to blanks the 20 elements of the array FORMC that will potentially contain field specifications.

Then we have a DO statement that carries out the rest of the program for each line of the output matrix. The next DO inspects each element of row I to determine whether it is zero. If it is nonzero, we place in the appropriate position of FORMC the field specification that was entered into FORMNZ. If it is zero, we must not only place the field specification contained in FORMZ into the format array but somehow blank out the contents of that element; this is done by placing alphabetic blanks there.

The final WRITE, 7, involves list indexing in a final—slightly different—form. The output variable has two subscripts; values are supplied to I by the earlier DO statement and to J by list indexing.

This program was tested with the matrix multiplication

$$A = \begin{pmatrix} 1.237 & 2.500 & -3.000 \\ 4.625 & -0.500 & 10.000 \\ -2.7369 & 0.9123 & -1.732 \end{pmatrix}$$

$$B = \begin{pmatrix} 1.771 & 0.000 & 2.000 & 2.000 \\ 0.173 & 3.000 & -1.500 & 6.000 \\ 1.013 & 2.500 & 1.000 & 0.000 \end{pmatrix}$$

The format information read into FORMA was

$$(12F5.3)$$

which expects all elements of A to be on one card. The format information read into FORMB was

$$(4F10.0)$$

which not only involves a different format for the data values but requires that there be only four values on each of the three cards.

For FORMNZ we entered

$$F10.4,$$

and for FORMZ we entered

$$A10,$$

The A10 tells the output routines to expect alphabetic data in a field 10 positions wide. The 10 was chosen to be identical to the width of the F10.4 fields.

The output is shown in Figure 17.2. We see that the program has performed largely as expected, with the zero elements of the product matrix properly blanked.

One surprise awaits the careful reader, however. How about the last element in the third row? Why did it print as zeros, instead of being blanked? Looking at A and B more closely, we see that the elements in the expression are

$$(-2.7369)(2.0) + (0.9123)(6.0) + (-1.732)(0.0)$$

which is equivalent to $-5.4738 + 5.4738$. Is this not in fact zero?

The answer is to be found in the fact that this program was run on a binary computer, and, as we have noted before, simple decimal fractions often have no exact binary equivalents. The actual result was very small, compared with the numbers that went into it—but *not* zero.

EXERCISES

*1. Modify the program of Figure 16.10 so that it reads an object-time format, giving the desired format of each line of output, and then uses that information in place of statement 200.

2. Modify the program of Figure 14.5 so that statements 57 and 70 are replaced by formats read at object time. It is possible to put two object-time formats on one card or to read to separate cards. The latter is probably easier and certainly less likely to lead to error.

*3. DATA is a one-dimensional array of at most 10 elements. A card is punched with a value of N in columns 1 and 2 and with 1 to 10 elements of DATA in succeeding columns. The number of elements is given by the value of N. Each number is punched with a decimal point but no exponent in seven columns. Write statements to read such a card.

4. Same as Exercise 3, except that the numbers are the *odd-numbered* elements of DATA; there are therefore at most five of them. N is the *element* number of the last one, not the total number of elements.

*5. Describe in words what card format and deck make-up would be required for each of the following groups of statements to be meaningful:

```
a.    DIMENSION X(10)
      READ (5, 69) (X(I), I = 1, 7)
   69 FORMAT (10F4.0)
b.    DIMENSION X(10)
      READ (5, 70) N, (X(I), I = 1, N)
   70 FORMAT (I2, 10F4.0)
c.    DIMENSION X(10)
      READ (5, 71) N, (X(I), I = 1, N)
   71 FORMAT (I2/10F4.0)
```

-0.4158		-4.2760	17.4740
18.2344	23.5000	20.0000	6.2500
-6.4437	-1.5931	-8.5742	-0.0000

Figure 17.2. The output of the program in Figure 17.1.

d.   DIMENSION X(10)
       READ (5, 72) N, (X(I), I = 1, N)
    72 FORMAT (I2/(F4.0))

6. Same as Exercise 5.

a.   DIMENSION Y(10, 10)
       READ (5, 79) K, (Y(K, I), I = 1, 10)
    79 FORMAT (I2, 10F5.0)

b.   DIMENSION Y(10, 10)
       READ (5, 80) K, M, (Y(K, I), I = 1, M)
    80 FORMAT (2I2, 10F5.0)

c.   DIMENSION Y(10, 10)
       READ (5, 81) M, N, ((Y(I, J), J = 1, N),
       I = 1, M)
    81 FORMAT (2I2/(10F8.0))

d.   DIMENSION Y(10, 10)
    83 READ (5, 82) I, J, Y(I, J), L
    82 FORMAT (2I2, F10.0, I1)
       IF (L .EQ. 0) GO TO 83

7. A one-dimensional array named CVG contains a maximum of 40 elements. The input deck to be read has one element per card; each card contains the element number in columns 1 and 2 and the element itself in columns 3 to 12, punched with a decimal point but without an exponent. The cards cannot be assumed to be in sequence on the element numbers. It is not known how many cards there are, but the last card of the deck is blank, which will look like an element number of zero. Write a program segment to read the deck and place each value in the correct location in the array.

*8. A two-dimensional array named STL is named in the statement

DIMENSION STL(10, 13)

The actual number of rows and columns is given by the values of the variables M and N, respectively. Write a program to punch on cards as many elements as there actually are in the array, in row order. Each element should be punched on a separate card along with its row and column numbers. Use I2 for the integers and 1PE20.7 for the real numbers.

*9. Given an integer variable I, with $1 \leq I \leq 12$, set up a program that will print in three printing positions one of the abbreviations JAN, FEB, MAR, APR, etc., depending on the value of I. Assume a computer in which one alphameric variable can hold six characters.

10. Given an integer variable J, with $1 \leq J \leq 7$. Set up a program to print one of the words MONDAY, TUESDAY, etc., depending on the value of J. Assume a computer in which one alphameric variable can hold six characters, which, of course, is not enough to hold a word like TUESDAY.

*11. Given the following program segment.
       DIMENSION FMT(12), X(10)
    101 READ (5, 102) (FMT(I), I = 1, 12)
    102 FORMAT (12A6)
       .
       .
       .
       WRITE (6, FMT) (X(I), I = 1, 10)

Show exactly what should be punched on the card read by the foregoing READ statement in order to print the 10 numbers in each of the following ways:

a. All 10 on one line, using F10.2 for each.
b. Five on each of two lines, using 1PE20.6 for each number.
c. X(1) on first line, using F10.2;
   second line blank;
   X(2), X(3), and X(4) on third line, 1PE20.6 for each;
   X(5), X(6), X(7), on fourth line, 1PE20.6 for each;
   X(8), X(9), and X(10) on fifth line, 1PE20.6 for each.

12. Given the following program segment.
       DIMENSION FMT(12), A(80, 80)
    201 READ (5, 202) (FMT(L), L = 1, 12)
    202 FORMAT (12A6)
    203 READ (5, FMT) I, J, T
       IF (I .EQ. 0) GO TO 204
       A(I, J) = T
       GO TO 203
    204 whatever follows.

State what should be on the card read by the READ at 201 to permit cards of the following types to be read by the READ at 203.

a. Columns 1–2: I
   Columns 3–4: J
   Columns 5–14: T, in form suitable for use with F10.0
b. Columns 1–2: I
   Columns 3–4: J
   Columns 5–14: T, in form suitable for use with F10.4
c. Columns 1–2: I
   Columns 3–4: J
   Columns 5–18: T, in form suitable for use with E14.7
d. Columns 1–2: I
   Columns 3–4: J
   Columns 5–16: T, for use with F12.0

There are five such groups per card; the second starts in column 17, and so on.

e. Columns 1–2: I
   Columns 3–4: J
   Columns 5–10: T, for use with F6.3

There are eight such groups per card; the second starts in column 11, and so on.

f. Columns 21–22: I
   Columns 31–32: J
   Columns 47–61: T, for use with E15.6

Ignore the contents of all other columns on the card.

13. Modify the program of Figure 17.1 so that a zero value is replaced with the word ZERO instead of blank.

CASE STUDY 18   GRAPHING OF A DAMPED OSCILLATION AND A DIFFERENTIAL EQUATION SOLUTION; THE **DATA** STATEMENT

It not infrequently happens that what a person wants from a computer, in the early stages of a study, is a *general* idea how a proposed system would operate. Great accuracy may be of little importance. In such cases a graphical presentation may be the best solution.

At other times the nature of the problem may be such that a pictorial presentation gives the best intuitive understanding of the system, even though considerable accuracy may be required in the calculation.

In this case study we shall explore methods of preparing graphs by using the normal printer associated with the computer. Case Study 26 also uses these techniques for a different kind of graph, and in Case Study 28 we shall see some of the techniques for preparing output to be plotted using an off-line plotting table when the rather gross kinds of graphs that we can get with the printer are not adequate.

For the first example let us produce a plot of a few cycles of a damped oscillation. The formulation here is in terms of an electrical circuit, but the same formula describes many other physical systems.

The current flowing in a series circuit containing resistance, inductance, and an initially charged capacitor, but no voltage source, is given by

$$i = i_m e^{-Rt/2L} \sin 2\pi f_1 t$$

where

$$i_m = \frac{2\pi f_0{}^2 Q}{f_1}$$

$$f_0 = \frac{1}{2\pi} \sqrt{\frac{1}{LC}}$$

$$f_1 = \frac{1}{2\pi} \sqrt{\frac{1}{LC} - \frac{R^2}{4L}}$$

and i = current flowing at time t, amperes
i_m = maximum current, amperes
R = resistance, ohms
t = time since closing switch, seconds
L = inductance, henrys
f_0 = frequency of undamped circuit ($R = 0$), cycles per second
f_1 = frequency of damped circuit, cycles per second
C = capacitance, farads
Q = initial charge on capacitor, coulombs

We wish to compute a number of points on the curve of instantaneous current versus time in order to draw a graph. One of the inputs to the program, along with the physical parameters, is an integer that will give the number of cycles of the curve desired (CYCLES) and another number that will give the number of points per cycle to be computed (NPERCY).

The program is shown in Figure 18.1. The type statements here are a considerable con-

```
C CASE STUDY 18
C DAMPED OSCILLATION, PLOTTING
C
      REAL I, IM, Q, R, C, L, F0, F1, C1, C2, T, DELT, TEMP, LINE
      INTEGER CYCLES, NPERCY, LIMIT, J, J1
      DIMENSION LINE(81)
      DATA BLANK, DOT, STAR/1H , 1H., 1H*/
C READ PARAMETERS
  611 READ (5, 612) Q, R, C, L, CYCLES, NPERCY
  612 FORMAT (4F10.0, 2I2)
C SKIP TO TOP OF NEW PAGE, PRINT PARAMETERS, SPACE 3 LINES
      WRITE (6, 613) Q, R, C, L
  613 FORMAT (1H1, 4HQ = , F10.6, 6X, 4HR = , F9.3, 6X, 4HC = ,
     1    F10.6, 6X, 4HL = , F7.3/////)
C PRINT A LINE OF DOTS, WHICH WILL BE VERTICAL AXIS WHEN PAPER IS TURNED
      DO 101 J = 1, 81
  101 LINE(J) = DOT
C WRITE THE LINE--NOTE THAT INDEXING IS NOT NEEDED, SINCE ENTIRE ARRAY
C IS PRINTED
      WRITE (6, 102) LINE
  102 FORMAT (1H , 81A1)
C BLANK THE LINE
      DO 103 J = 1, 81
  103 LINE(J) = BLANK
C PUT A DOT IN LINE(41), TO PRODUCE THE HORIZONTAL AXIS
      LINE(41) = DOT
C COMPUTE INTERMEDIATE VARIABLES
      F0 = 0.1591549 / SQRT(L*C)
      F1 = 0.1591549 * SQRT(1.0/(L*C) - R**2/(4.0*L**2))
      TEMP = NPERCY
      DELT = 1.0 /(TEMP * F0)
      IM = 6.2831853 * F0**2 * Q / F1
      C1 = R / (2.0 * L)
      C2 = 6.2831853 * F1
C START T AT ZERO AND SET UP LOOP CONTROL
      T = 0.0
      LIMIT = CYCLES * NPERCY
      DO 11 J1 = 1, LIMIT
C COMPUTE CURRENT
      I = IM * EXP(-C1*T) * SIN(C2*T)
C COMPUTE DESIRED LOCATION OF PLOTTING SYMBOL
      J = 40.0 * (I/IM + 1.0) + 1.5
C PUT ASTERISK IN SELECTED POSITION
      LINE(J) = STAR
C WRITE THE LINE
      WRITE (6, 102) LINE
C PUT A BLANK IN SELECTED POSITION, WHICH MIGHT HAVE BEEN ON AXIS
      LINE(J) = BLANK
C PUT A DOT BACK IN AXIS LOCATION, IN CASE IT WAS BLANKED OUT
      LINE(41) = DOT
C INCREMENT T AND RETURN (VIA DO) TO COMPUTE AND PLOT ANOTHER POINT
   11 T = T + DELT
C GO BACK FOR ANOTHER DATA CARD AND PRODUCE NEW PLOT
      GO TO 611
      END
```

Figure 18.1. A program to produce a plot of a damped oscillation.

venience: electrical engineers always write i for current and, using the REAL statement, we can do the same. In the DIMENSION statement we establish LINE as the name of an array of 81 elements that will hold the line of information to be plotted, as we shall investigate a little later.

First, however, let us digress long enough to study the DATA statement that appears next and which we shall find useful in many subsequent case studies. The problem here is that we need to set up three alphanumeric characters for use in plotting: a blank, a period, and an asterisk. There is no such thing as an alphabetic constant; we cannot write a statement like

$$STAR = 1H*$$

to give the variable STAR the "value" of the asterisk symbol (at least not in all FORTRANS). We can read a card that contains the symbol and the others, but it is an annoyance to have to do so because that same card must be inserted at the front of the data deck every time the program is executed, even though the values on it never change.

The DATA statement permits us to enter data into the program at the time of compilation. The data so defined is then loaded into the object computer with the object program and is available for any use we wish. It must be emphasized, however, that the data values are established at the time the object program is loaded. *The DATA statement is not executable.* It is permissible, if we should wish, to assign new values to the variables that were loaded with the DATA statement, but having done so it is NOT possible to "re-execute" the DATA statement to put the variables back to their original values.

The general form of the DATA statement is

$$DATA \; list/d_1, d_2, \ldots, d_n/, list/d_1, d_2,$$
$$k*d_3, \ldots, d_m/ \ldots$$

In this symbolic description a "list" contains the names of variables to receive values; the d's are the values and k, if it is used, is an integer constant. For example, we might write

$$DATA \; A, B, C/14.7, 62.1, 1.5E - 20/$$

This statement would assign the value 14.7 to A, 62.1 to B, and $1.5 \cdot 10^{-20}$ to C. The following statement would have exactly the same effect:

$$DATA \; A/14.7/, B/62.1/, C/1.5E - 20/$$

The choice is a matter of personal preference.

The following statement assigns the value zero to A and 21.7 to the other five:

$$DATA \; A, B, C, D, E, F/0.0, 5*21.7/$$

The list may employ implied DO-indexing.

$$DATA \; ZERO, (A(I), I = 1,5), A(6)/0.0,$$
$$5*1.0, 100.5/$$

This would give ZERO the value zero, put 1.0 in the first five elements of A, and 100.5 in A(6). We recall that parentheses are required to enclose a list element with implied DO-indexing. The list of

a DATA statement is formed like the list of an input or output statement, with the one exception that in implied DO-indexing the parameters must always be integer *constants*, never variables.

Returning to the program of Figure 18.1, we see that the constants in a DATA statement may be of the Hollerith type, which is what we need here.

The scheme of graphing we shall use in this example is as follows. We first produce a row of dots across the top of the page; this will be the y-axis when the page is turned a quarter-turn for viewing. Then we print a succession of lines, each containing a dot in the middle that will be part of the x-axis and an asterisk representing a point on the curve. The reader may wish to glance at Figure 18.2 to get a clearer picture of what we are doing. Figure 18.2 is a page of output, slightly reduced, oriented as it was produced by the printer. Figure 18.3 shows three other graphs, turned on the page for normal viewing, and reduced somewhat more.

We have chosen to provide 81 printing positions in each line—a line, remember, being parallel to the y-axis as we shall interpret the graph. The choice of 81 was largely arbitrary. There can be as many lines as there are points; in other words, the graph is of indefinite length in the x-direction (down the page).

Inspecting the program further, we see some fairly standard preliminaries. The parameters are read and printed at the top of the page. This printed line will be sideways when the graph is viewed, but we cannot be unduly concerned about that; it is more important to have some identification than to be overnice about dubious esthetics.

Next we put all dots in the array LINE and print it, which produces the y-axis. After blanking out the 81 dots, we put one of them back in position 41 to get the x-axis. The details of the calculation need not detain us, for we are concerned only with the graphing aspect. After computing the current I, we are ready to plot a point. Here we take advantage of physical knowledge of the formulation of the problem: the maximum value that I can attain is IM, as noted in the problem statement. Therefore the quotient I/IM ranges between -1 and $+1$; (I/IM + 1.0) then lies between 0. and 2.0; 40.0(I/IM + 1.0) lies between 0.0 and 80.0; 40.0*(I/IM + 1.0) + 1.5 lies between 1.5 and 81.5; when this value is truncated to an integer, we get an integer value between 1 and 81, inclusive. The extra 0.5 in 1.5 was for rounding. The numerical value I has thus been transformed into a value that

Q = 0.000010   R = 1.000   C = 0.000010   L = 0.002

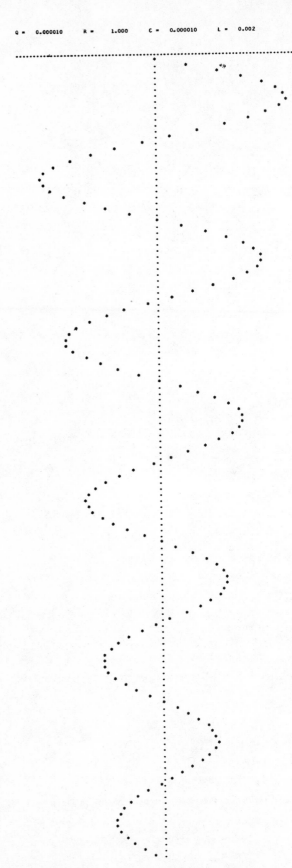

Figure 18.2. The output of the program in Figure 18.2.

we can use as a subscript to choose the position of an asterisk in the array LINE.

We have taken essential advantage of the fact that I can never be greater than IM in absolute value, to guarantee that the subscript we produce is not less than one nor greater than the maximum value stated in the DIMENSION statement, as required. In other cases, as we shall see in the next example, we cannot be so sure about the sizes of these subscripts, and we test them accordingly.

After printing the line with its asterisk, x-axis dot, and 79 blanks, we blank out the asterisk, then put a dot into the axis position against the possibility that the asterisk was itself on the axis and had therefore replaced the axis dot. The rest of the program is involved with the production of the full range of I values and reading more parameter cards. One line is printed for each value of I.

This program takes no explicit account of the question of scales on the axes. It is left to the user to choose the number of points per cycle of the curve in such a manner that the graphs "look right." In many cases this kind of nonchalance about scale, of course, is not acceptable, for it requires a closer examination of how the spacing of lines and characters on the printer can be taken into account to produce something more precise. We shall not delve into the matter, which gets more complicated than is worth our study. In Case Study 26, however, we shall see a sample of what can be done at least to make the scales in the two directions consistent.

In our second example of graphing we shall produce a plot that appears on the page in normal orientation. The point is not so much to save the user the labor of turning the page but to make printer plotting available in situations in which the preceding method breaks down. For instance, it was fundamental that the points were produced in a strictly increasing sequences of x-values; without this assurance, we might have found ourselves having to move back up the page—which is quite impossible. Furthermore, it may in some cases simply be an advantage to be able to use the familiar rectangular shape of the usual computer forms, which is the case in our example.

In Figure 18.4 we start with two arrays of 600 elements each; Z contains values along the horizontal axis that range from zero to about 12π; V contains values along the vertical axis that range between approximately −25 and +25. We are not sure of the values until we compute them. These arrays were produced in the course of integrating

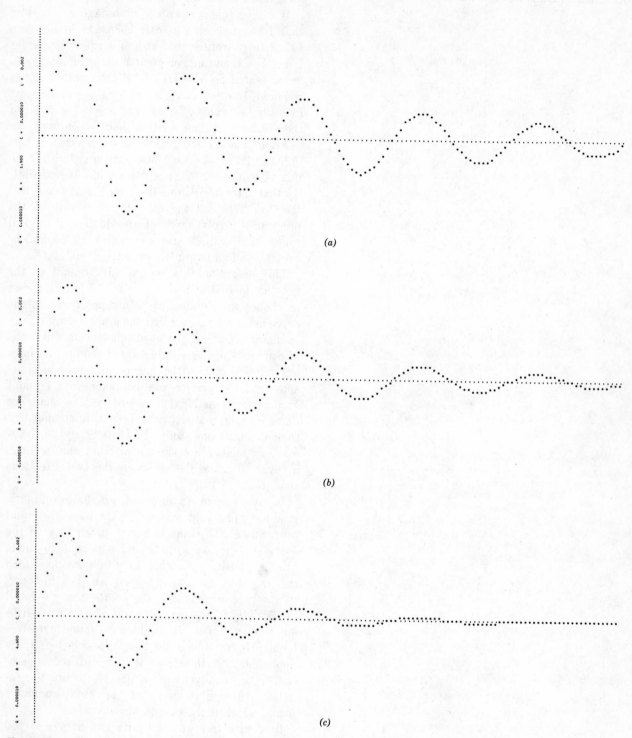

Figure 18.3. The output of the program in Figure 18.1 for three other values of resistance. The plots have been turned on the page for normal viewing.

```
C CASE STUDY 18
C ION TRAJECTORIES IN A MONOPOLE MASS SPECTROMETER
C PRINTER PLOT VERSION
C A DESCRIPTION OF THE PROBLEM AND THE SOLUTION APPEAR IN CASE STUDY 28
C
      DIMENSION Z(1200), V(1200), DV(1200), GRAPH(50, 130)
      DATA DOT/1H./, BLANK/1H /
C THE FOLLOWING SECTION SOLVES THE DIFFERENTIAL EQUATION AND MAY BE
C IGNORED FOR THIS CASE STUDY
      DERIV(Z, V) = - (A + 2.0*Q*COS(2.0*Z))*V
   12 READ (5, 13) A, Q
   13 FORMAT (2F10.0)
      Z(1) = 0.0
      V(1) = 0.0
      DV(1) = 1.0
      H = 3.14159265 / 50.0
      H2 = H / 2.0
      DO 20 I = 1, 600
      A1 = H * DERIV(Z(I), V(I))
      A2 = H * DERIV(Z(I) + H2, V(I) + H2*DV(I) + H/8.0*A1)
      A3 = H * DERIV(Z(I) + H, V(I) + H*DV(I) + H2*A2)
      V(I+1) = V(I) + H*(DV(I) + (A1 + 2.0*A2)/6.0)
      DV(I+1) = DV(I) + (A1 + 4.0*A2 + A3)/6.0
   20 Z(I+1) = Z(I) + H*FLOAT(1)
C THIS ENDS THE SECTION OF THE PROGRAM THAT SOLVES THE EQUATION
C CLEAR THE ARRAY THAT HOLDS THE GRAPH
      DO 63 I = 1, 50
      DO 63 J = 1, 130
   63 GRAPH(I, J) = BLANK
C SKIP TO THE TOP OF THE PAGE
      WRITE (6, 64)
   64 FORMAT (1H1)
C SET UP DO LOOP TO DEAL WITH ALL 600 POINTS
      DO 65 L = 1, 600
C CONVERT DISTANCES INTO SUBSCRIPT VALUES
      I = 25.0 + V(L)
      J = 1.0 + 3.0*Z(L)
C CHECK THAT POINT LIES WITHIN GRAPH, AS EXPECTED, AND ENTER A DOT IF SO
   65 IF (I .GE. 1 .AND. I .LE. 50 .AND. J .GE. 1 .AND. J .LE. 130)
     1 GRAPH(I, J) = DOT
C PRINT THE GRAPH
      DO 70 I = 1, 50
      I1 = 51 - I
   70 WRITE (6, 71) (GRAPH(I1, J), J = 1, 130)
   71 FORMAT (1H , 130A1)
      GO TO 12
      END
```

Figure 18.4.   A program to produce printer plots of the trajectory of an ion in a monopole mass spectrometer.

a differential equation that we shall explore more fully in Case Study 28, where we shall produce a much more attractive and accurate plot with plotting equipment that is separate from the computer. The plot here was produced in the course of checking out the program of Case Study 28.

The scheme this time is to set up a two-dimensional array with 50 rows and 130 columns, these values being not quite the largest that could be accommodated on the printer and paper to be used. Given a pair of values from the arrays Z and V, we translate each value into a subscript to pro-

duce a graph that approximately fills the page. The subscript I picks a row and is therefore the vertical dimension. The translation formula

$$I = 25.0 + V(L)$$

takes a value of $V(L)$ that we said could be as large as 25 and as small as -25 and translates it into a subscript ranging from about 1 to about 50, which is in the range of permissible subscripts. The values of $Z(L)$ range from zero to 12π; the formula shown produces a value between 1 and 113. Because we are not actually sure of the values that might be

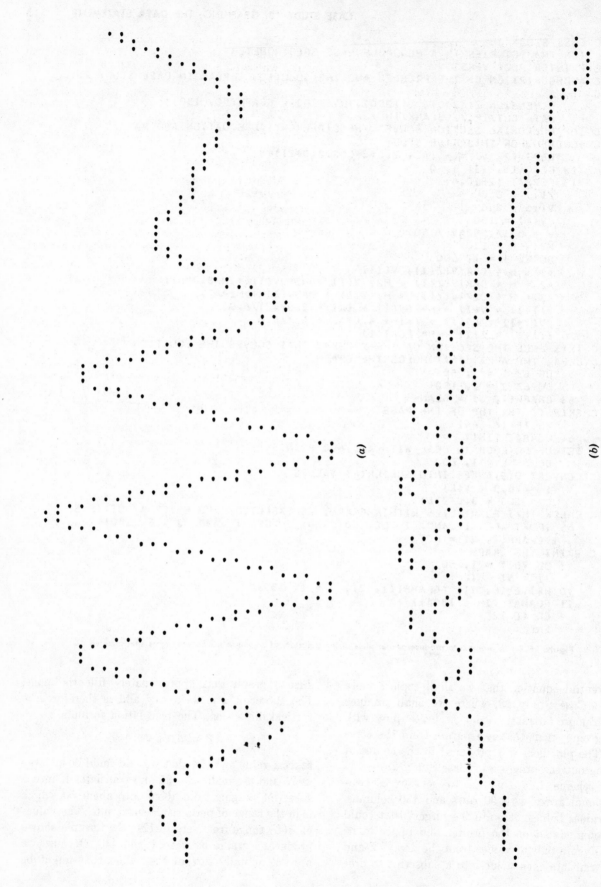

(a)

(b)

Figure 18.5. The output of the program in Figure 18.4. Compare with Figure 28.3.

produced by the program, we insert an IF statement to plot the point only if it does lie within the limits of the graph.

The plotting action is carried out for all 600 pairs of values, after which the entire graph is printed, from the top down. (This requires a bit of manipulation, for negative increments are not permitted in the DO statement.) No axis lines were provided for, since the intent was merely to get a rough idea whether the program was working correctly, which would have been quite difficult to do by looking at the 600 values for each of many dozens of calculations. Figure 18.5 shows two of the plots that were produced, somewhat reduced; each plot here was on a separate page of output.

EXERCISES

1. Modify the program in Figure 18.1 so that the vertical scale is variable. ("Vertical" means as the final graph is normally viewed; that is, after turning the page from the way it was printed.) Make a suitable change in the DIMENSION statement and in the conversion formula so that there may be between 20 and 100 points vertically.

*2. Write a routine to plot a unit circle on a graph. Use 48 lines vertically and 80 horizontally; assuming a printer with 6 lines per inch vertically and 10 printing positions per inch horizontally, such a graph will be square. Print rows of dots for the axes. Given the x-y coordinates of a point, it will be necessary to use two different conversion formulas to find the element in which to store the character for plotting. Let 1 in. = 1 unit on both axes.

Generate and plot 40 points equally spaced around a circle of radius 1 with center at the origin.

3. Modify the program of Exercise 2 so that it reads two numbers from a data card, one of which gives the number of vertical lines that are to represent one unit and the other the number of horizontal printing positions. These numbers become parameters in the formulas that locate the element in which to store the plotting character.

4. Modify the program of Exercise 3 so that points in the first quadrant are plotted with the character "1," points in the second quadrant are plotted with the character "2," and so on.

5. Produce a plot of the unit circle with each point plotted as its quadrant number. Consider the graph to be the complex plane. For each point $x + iy$ plot also the point $CEXP(x + iy)$. If $x + iy$ is in quadrant j, plot both the point itself and $CEXP(x + iy)$ with the character j, even though in this case all of the function points will be in the first quadrant.

CASE STUDY 19 SOLUTION OF LARGE SYSTEMS OF SIMULTANEOUS EQUATIONS; MAGNETIC TAPE OPERATIONS

In this case study we shall notify the program of Case Study 16 to make it possible to solve systems of as many as 300 equations in 300 unknowns. There are 90,000 coefficients in such a system, which are too many to contain in the high speed storage of all but a very few computers. We shall accordingly write the coefficients onto magnetic tape as they are read and during the iteration process read them back into the computer, one row at a time. It is one of the advantages of the Gauss-Seidel iteration method that it requires only one row of coefficients to be in main storage at one time.

To prepare for the programming techniques we shall need we pause to study the use of magnetic tapes.

In everything that we have done up to this point we have assumed the use of tape 5 for input and tape 6 for output and off-line operations. To recapitulate, this means that punched-card information is transferred to magnetic tape before it is needed, in a manner that does not tie up the computer. The point of this off-line transfer to magnetic tape is to avoid slowing down the high-speed computer to the much slower pace of the card reader, which is limited by considerations of mechanical motion of the cards. Thus when we write something like

READ (5,112) A, C, G

we are calling for values for A, C, and G to be read from tape 5 on which they have been previously placed in the card-to-tape operation. An analogous description applies on output: the results are written on tape 6 and later printed with a separate small computer or in some other manner that does not tie up the main computer.

Tapes 5 and 6 are usually, but not necessarily universally, assigned for these purposes. Whatever the numbering system at a particular installation, there are in any case other magnetic tapes available for these and other purposes. Typically, there might be a total of eight tapes available to the programmer, plus others that are reserved for other purposes such as holding the compiler and the operating system.

Not only have we restricted ourselves to tapes 5 and 6, but we have assumed that FORMAT statements are always used. This, too, must now be relaxed because there is another mode of tape operation: the use of *binary* tapes. The full implications of the term binary are not intended here; all we mean by the word is that the information is written on the tape in a manner that makes it impossible to print it. The only meaningful use of a binary tape is to write something on it with the program, then read the information back into the computer for further use at a later time in the same or another program.

The opposite, so to speak, of binary is "BCD." The wording, again, refers to details of a particular class of computers, and is of no great interest to us. Instead of "BCD" and "binary," we could better substitute "formatted" and "nonformatted," respectively.

The numbering of magnetic tape functions varies greatly from one installation to the next; the programmer must get the list of functions for his system. The conventions used on the computer employed to run the programs in this book are more or less representative, as given in Table 19.1.

TABLE 19.1

Fortran Tape Unit	Mode	Function
1	Binary	Input or output
2	Binary	Input or output
3	Binary	Input or output
4	Binary	Input or output
5	BCD	Input
6	BCD	Output
7	Binary	Output
8	BCD	Input or output

Using binary tapes means that we do not reference a FORMAT statement in the READ or WRITE. The form of such a statement is simply

$$\text{READ (n) list}$$

or

$$\text{WRITE (n) list}$$

where n is one of the numbers 1, 2, 3 or 4. Tape 7 is reserved for writing the object programs, which are not formatted.

These tape designations are symbolic; this means that a given FORTRAN tape unit may correspond to any of the physical tape units of the computer, depending on the operating system specifications, which may be changed by control cards. The symbolic designations we have listed may also be changed; those we have given are to a certain extent conventional but not fixed.

The designation of the input or output unit desired may be specified at the time of execution of the object program by writing an integer variable in place of the unit number and assigning a suitable value to the variable. We might write, for instance,

$$\text{READ (N, 106), A, B, C}$$

Earlier in the program we might assign to N the value 5 or 8 to select the corresponding BCD tape unit.

There are three FORTRAN statements written specifically for use with magnetic tape. In each case the verb (END FILE, REWIND, or BACK-SPACE) is followed by an integer constant or variable designating the FORTRAN unit number.

The END FILE statement places an end-of-file mark on the designated tape. Its primary use, from our standpoint, is to serve as a signal to the off-line printer that no more valid information will follow. Without an end-of-file mark on a tape it is possible to get an output listing with a great deal of "garbage" following the valid output because the operator could not tell when the information ended.

The REWIND statement returns a tape to its beginning. This step is necessary, for instance, when a binary tape has been written with intermediate results and we want to read the data back into the computer. Tape cannot be read backward in FORTRAN, so we rewind the tape and use a binary READ to get the information.

The BACKSPACE statement backs up the specified tape by one "record." A record, for our purposes, may be defined as the information corresponding to one card or one printer line in the case of a BCD tape and the information written by one WRITE statement in the case of a binary tape.

The object program is set up automatically to test for inadvertent attempts to destroy the standard input and output tapes. For instance, a BACKSPACE statement that names FORTRAN tape 6 will be ignored because it could destroy some of the information already on the tape. Similarly, we are prevented from writing on tape 5 for the same reason.

We are now ready to return to a discussion of the program for solving large systems of simultaneous equations. The logic of the solution is quite close to that of Case Study 16, and the modifications do not greatly complicate the logic. Therefore we shall dispense with a flowchart this time. The program shown in Figure 19.1 can be examined directly.

The array for the coefficients is now *one*-dimensional; we shall never have the entire array in storage at once. We see two small arrays set up to hold the input and output formats. With thousands of coefficients to be read and hundreds of results to be printed, it is almost essential that some flexibility be permitted in card and printing formats.

After clearing the X array to zeros, reading the formats, and reading the input and output formats and the values of N, EPS, and MAX, we come to a REWIND statement, which may seem pointless: we have not yet written anything on the tape. The problem is that we cannot know for sure that the

```
C CASE STUDY 19
C LARGE SIMULTANEOUS EQUATIONS WITH TAPE
C
      DIMENSION A(300), B(300), X(300), INFMT(12), OUTFMT(12)
C CLEAR THE X ARRAY
      DO 50 L = 1, 300
   50 X(L) = 0.0
C READ THE OBJECT-TIME FORMATS
      READ (5, 87) INFMT
      READ (5, 87) OUTFMT
   87 FORMAT (6X, 12A6)
C READ N, EPSILON, AND THE MAXIMUM NUMBER OF ITERATIONS
      READ (5, 101) N, EPS, MAX
  101 FORMAT (I3, F10.0, I3)
C INITIAL REWIND TO ASSURE PROPER TAPE POSITION
      REWIND 1
C READ THE COEFFICIENTS, A ROW AT A TIME, AND WRITE ON TAPE
      DO 60 I = 1, N
      READ (5, INFMT) (A(L), L = 1, N)
   60 WRITE (1) (A(M), M = 1, N)
C WRITE END OF FILE TO MARK THE END OF VALID INFORMATION
      END FILE 1
C REWIND TO POSITION TAPE FOR READING
      REWIND 1
C READ THE CONSTANT TERMS
      READ (5, INFMT) (B(L), L = 1, N)
      ITER = 1
   99 BIG = 0.0
      DO 100 I = 1, N
  102 SUM = 0.0
C READ A ROW OF COEFFICIENTS FROM TAPE
      READ (1) (A(L), L = 1, N)
      IF (I .EQ. 1) GO TO 105
      LAST = I - 1
      DO 106 J = 1, LAST
  106 SUM = SUM + A(J)*X(J)
      IF (I .EQ. N) GO TO 103
  105 INITL = I + 1
      DO 107 J = INITL, N
  107 SUM = SUM + A(J)*X(J)
  103 TEMP = (B(I) - SUM) / A(I)
      IF (ABS(TEMP - X(I)) .GT. BIG) BIG = ABS(TEMP - X(I))
  100 X(I) = TEMP
      REWIND 1
      IF (BIG .LT. EPS) GO TO 752
      IF (ITER .GE. MAX) GO TO 752
      ITER = ITER + 1
      GO TO 99
  752 WRITE (6, 147) ITER, EPS, BIG
  147 FORMAT(1H0, I5, 2E14.7   )
      WRITE (6, OUTFMT) (X(L), L = 1, N)
      STOP
      END
```

Figure 19.1. A program using magnetic tapes to solve a system of as many as 300 simultaneous equations by the Gauss-Seidel method.

tape is at its initial point, even though it ordinarily will be.

If by chance it is positioned somewhere else, we will have no trouble in writing, but when we later rewind the tape and try to read the information we have written we will encounter tape area before the point at which we began writing. Usually there will be information on it from some previous use of the tape, and we will be in trouble. It is good programming practice to rewind every tape before using it.

The DO loop that follows is executed N times. On each repetition of the loop one row of coefficients is read from card records on tape, under control of the object-time formats read earlier. As each row is read, it is immediately written on binary tape 1.

When this process is completed, we end-file and rewind the tape in order to be ready to read it from the beginning.

After reading the constant terms we go into the iteration scheme, which is similar to the program in Figure 16.10. The major difference, of course, is that for each row we must read the binary tape into the one-dimensional array A. We write statements such as

$$SUM = SUM + A(J)*X(J)$$

with the knowledge that the proper row will be occupying the A-array at every point. We must rely on the fact that the rows are being used in the same order in which they were read. Prudence would normally dictate that the coefficient array be printed for checking, since so much depends on a correct deck make-up.

After each iteration we rewind the coefficient tape so that it is correctly positioned for the next iteration.

This program was tested with the same system of equations used to test the earlier program. The input format card contained in columns 1 to 14

IN   (6F10.0)

Notice that in FORMAT statement 87 we have a 6X field specification, to skip over whatever identification might be punched in columns 1 to 6. The format information was set up to accommodate the coefficients from one row on one card, which was a convenient way to arrange the data in this case. The output format card read

OUT   (I5, 1PE16.6)

which is the same as we used earlier.

The output from the modified program was identical with the earlier results.

EXERCISES

*1. BCD tape 8 contains records produced from a deck of cards. The cards were punched in a format for reading with F10.0 field specifications, eight numbers per card. The first field, columns 1–10, is guaranteed to be nonzero on all cards before a sentinel consisting of a blank card. This tape is to be read, and the information on it written on binary tape 3, in the form of eight-word binary records. When the sentinel is detected, an end-of-file mark should be placed on the binary tape and both tapes rewound.

2. Revise the program of Exercise 1 so that the binary records will have 400 words in them, which is the information from 50 card records. You may not assume that the number of card records will be a multiple of 50; if the last binary record contains fewer than 50 eight-word groups, fill out the rest of the record with zeros. (This is done most easily by zeroing the array in which the binary record is built up before starting to read BCD records.) Each tape has a sentinel consisting of a stock number of zero.

3. Binary tapes 1 and 2 each contain 10,000 10-word records. Each tape is in sequence on the contents of the first word of each record. As an example of what is meant here, each of the records might be an inventory record, with the first word as the stock number. It is then stated that each tape is in ascending sequence on the stock numbers.

Write a program that reads these tapes and *merges* them onto tape 3, which at the conclusion of the program will have 20,000 records, all in ascending sequence on the first words.

4. BCD tape 8 contains an unknown number of card records representing meteorological readings produced by automatic recording equipment that failed randomly in an experiment during which it was untended for some weeks. Let us suppose that there are four numbers in each record, suitable for reading with F5.2 field specifications; they are, in order, time, temperature, wind, and humidity. You are to read this tape and inspect each record for validity by performing the following tests. The whole-number portion of time must be between zero and 23 inclusive, and the fraction part must be less than 0.60. The temperature must be between +40 and +110. The wind must be between zero and 60. The humidity is in the form of a wet-bulb temperature which must be between +30 and +110. Any record that passes these validity tests is to be written on binary tape 4 in the form of a four-word record; those that fail are to be written on standard BCD output tape 6 for later printing.

The last record on tape 8 is a sentinel signaled by a negative time.

The output tape 4 should not have a sentinel but rather be written with a first record consisting of one integer word that will give the number of records it contains. This is obviously not known at the start of writing; you will have to write the information on another tape, counting good records, then write the number of records on tape 4, and finally read from the "scratch" tape and write on 4.

CASE STUDY 20 EVALUATING A BESSEL FUNCTION BY NUMERICAL INTEGRATION; STATEMENT FUNCTIONS

It often happens that a programmer will find a relatively simple computation recurring through his program, making it desirable to be able to set up a function to carry out the computation. It would conceivably be possible to set up a new function of the cosine and square root type, but that requires considerable effort and the result might be of no value to anyone else.

The answer in such a case is a statement function, which can be defined for the purpose of one program and which is very simple to write and to incorporate in the program. It has no effect on any other program.

We shall consider the application of the statement function technique in a numerical integration that gives the value of a Bessel function. The method of programming the numerical integration is of some interest in itself.

Case Studies 21–23 develop techniques that may be viewed as extensions of the ideas introduced with statement functions, so that we are here laying the foundation for the study of an important part of programming.

A statement function is *defined* by writing a single statement of the form $a = b$, where a is the name of the function and b is an expression. The name, which is invented by the programmer, is formed according to the same rules that apply to a variable name: one to six letters or digits, the first of which must be a letter. If the name of the statement function is mentioned in a prior type statement, there is no restriction on the ini-

tial letter; if the name is not mentioned in a type statement, the initial letter distinguishes between real and integer in the usual way. The name, of course, must not be the same as that of any supplied function.

The name of the function is followed by parentheses enclosing the argument(s), which must be separated by commas if there is more than one. The arguments *in the definition* must not be subscripted.

The right-hand side of the definition statement may be any expression not involving subscripted variables. It may use variables not specified as arguments and it may use other functions (except itself). All function definitions must appear before the first executable statement of the program. If the right-hand side of a statement function uses another statement function, the 6ther function definition must have appeared *earlier* in the program.

As an illustration, suppose that in a certain program it is frequently necessary to compute one root of the quadratic equation, $ax^2 + bx + c = 0$, given values of a, b, and c. A function can be defined to carry out this computation, by writing

ROOT(A, B, C) =
    (−B+SQRT(B**2 − 4.*A*C))/(2.*A)

The compiler will produce a sequence of instructions in the object program to compute the value of the function, given three values to use in the computation.

This is *only* the definition of the function; it does not cause computation to take place. The variable names used as arguments are only dummies; they may be the same as variable names appearing elsewhere in the program. The argument names are unimportant, except as they may distinguish between integer and real.

A statement function is *used* by writing its name wherever the function value is desired and substituting appropriate expressions for the arguments. "Appropriate" here means, in particular, that if a variable in the definition is real the expression substituted for that variable must also be real, and similarly for the other types of variables. The values of these expressions will be substituted into the program segment established by the definition and the value of the function computed. The actual arguments may be subscripted if desired.

Suppose, now, that we wish to use this function with 16.9 for *a*, R — S for *b*, and T + 6.9 for *c*; the value of the function (root) is to be added to the cosine of *x* and the sum stored as the new value of ANS. All this can be done with the statement

ANS = ROOT(16.9, R — S, T + 6.9) + COS(X)

Suppose that later in the program it is necessary to compute the function with DATA(I) for *a*, DATA(I + 1) for *b*, and 0.087 for *c*; the function value is to be cubed and stored as the value of TEMP:

TEMP = ROOT(DATA(I),
　　　　　　　DATA(I + 1), 0.087)**3

It must be emphasized that the variables A, B, and C in the function definition have no relation to any variables of the same names that may appear elsewhere in the program. To illustrate, suppose that the value of the root is needed for the equation

$$22.97x^2 + ax + b = 0$$

where *a* and *b* are variables in the program. The root may be found by writing

VAL = ROOT(22.97, A, B)

The A and B that appear here in the *use* of the function are completely unrelated to the A and B in the *definition* of the function. In summary, the definition variables are dummies that establish how the expression values in the use should be substituted into the object program set up from the definition.

For another example of the usefulness of statement functions, suppose that in a certain program it is frequently necessary to evaluate the function

$$E = \frac{1}{x^5\left(e^{\frac{1.432}{Tx}} - 1\right)}$$

The argument this time is to be just *x*. This is easily set up as a function:

E(X) = 1.0/(X**5*(EXP(1.432/(T*X)) — 1.0))

The X here, as always in a function definition, is only a dummy variable that defines a computational procedure; when an expression is later written in using the function, the same actions are carried out on the actual value of the argument as are shown being done with X in the definition.

There is no prohibition against using X as an actual variable, of course. The function just defined could be used in statements like these:

SUM4 = SUM4 + E(X)

SUM2 = SUM2 + E(X + H)

EFFIC = 64.77*H/3.0*(4.0*SUM4 + 2.0*SUM2
　　　　　+ E(A) + 4.0*E(B — H) + E(B))/T**4

We see in this function an example of something that was mentioned earlier: the use of a variable in the function definition that is not an argument. The only argument here is X; this is a dummy. T, however, since it is not an argument, is *not* a dummy: it is the same T that presumably appears elsewhere in the program. This use of variables that are not arguments is perfectly legal; as we see, it saves the effort of making arguments out of variables for which we shall never want to substitute anything but their own values. (It is *logically* possible to think of a statement function with *no* arguments: all the variables would simply take on their current values in the program, as T did in the last example. This is not permitted with statement functions. We shall see that it *can* be done with a SUBROUTINE subprogram.)

To illustrate the use of a statement function in a complete program we turn to one way of defining the Bessel function $J_1(x)$:

$$J_1(x) = \frac{1}{\pi} \int_0^\pi \cos(t - x \sin t)\, dt$$

This is a slightly awkward way of evaluating a Bessel function, of course. For practical computations we would turn to the series expansion

$$J_1(x) = \tfrac{1}{2}x - \frac{(\tfrac{1}{2}x)^3}{1^2 \cdot 2} + \frac{(\tfrac{1}{2}x)^5}{1^2 \cdot 2^2 \cdot 3} - \frac{(\tfrac{1}{2}x)^7}{1^2 \cdot 2^2 \cdot 3^2 \cdot 5} + \cdots$$

and use techniques along the lines of those in Case Study 12. However, the example is quite suitable for our purposes—to see how to use a statement function in a practical numerical integration technique.

We recall from Case Study 16 that Simpson's rule for performing the integration

$$I = \int_a^b f(t)\, dt$$

is

$$I = \frac{h}{3}[f(a) + 4f(a + h) + 2f(a + 2h) + 4f(a + 3h) \\ + 2f(a + 4h) + \cdots + 4f(b - 3h) \\ + 2f(b - 2h) + 4f(b - h) + f(b)]$$

where h is the width of the strips into which the area under the curve $f(t)$ is divided.

The error in this approximation can be made as small as we please by taking h small enough; that is, by using enough intervals. (Eventually roundoff error will *increase* the error as h is reduced, but we will not be making h small enough to get into that problem this time.) An error formula is available to provide an estimate of the error of the approximation, but unfortunately it requires evaluation of the fourth derivative of the integrand, which in this case is quite involved. Even if the programmer could write it down without error, he would then be faced with the task of estimating its maximum value. If a program for formula manipulation such as FORMAC is available, all of this (including the differentiation) can be done by the

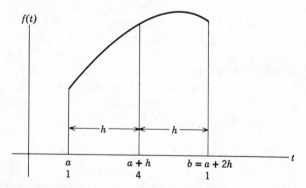

Figure 20.1. A two-interval version of integration by Simpson's rule.

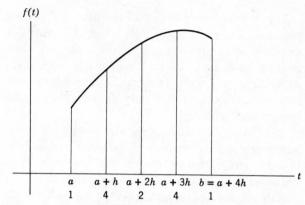

Figure 20.2. A four-interval version of integration by Simpson's rule.

computer—but it is still a chore. A simpler way is available, in which we simply keep improving the approximation until it is good enough.

To see how this is done, consider the sketch in Figure 20.1, in which we have used just two strips. The t-axis has been labeled with the t-values at the ordinates and with the multipliers in Simpson's rule formula for this simple case:

$$I = \frac{h}{3}[f(a) + f(a + h) + f(b)]$$

This becomes the value of a variable named INT (for integral). It is probably not a very good approximation. We continue, by dividing h by 2 and thus providing twice as many strips as sketched in Figure 20.2. The form of Simpson's rule that now applies is

$$I = \frac{h}{3}[f(a) + 4f(a + h) + 2f(a + 2h) + 4f(a + 3h) + f(b)]$$

The h here is, of course, the new value, after dividing by 2. This new value of the integral is called NEWINT; it ought to be a considerably better approximation because the error in the process is significantly reduced by reducing the width of the interval. We compare INT and NEWINT; if they are sufficiently close we accept NEWINT as the value of the integral. Otherwise we divide the interval by 2 again, set INT equal to NEWINT, and repeat the process. The situation is sketched in Figure 20.3. This can be done as many times as necessary to achieve convergence or until an iteration counter is exceeded.

The question is, are we really going to evaluate the complete Simpson's rule formula every time we go back for a new approximation? It is quite unnecessary to do so, in fact. Look at the three

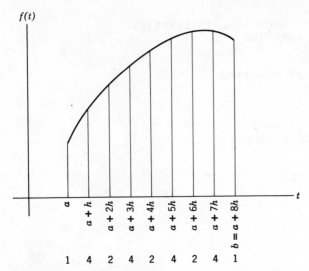

Figure 20.3. An eight-interval version of integration by Simpson's rule.

sketches but ignore the end points (which never change) and observe the pattern. The interior point in Figure 20.1 is in the center of the interval; that same point is needed in Figure 20.2, in which it is multiplied by 2. The three interior points in Figure 20.2 are all needed in Figure 20.3, in which they are all values multiplied by 2.

This same procedure can be carried out systematically. The basic scheme is to accumulate separately the ordinates that must be multiplied by two and those that must be multiplied by four; in the program these are called, not very imaginatively, TWO and FOUR. If it develops that the approximation is not sufficiently accurate, we can add the two together: they all become points that must be multiplied by 2 the next time.

The attraction of the integration method is thus that we compute just as many ordinates as necessary to get the required accuracy and no ordinate is ever recomputed.

Figure 20.4 presents a flowchart of the method. The logic is essentially what we have just described, but some of the details should be amplified.

The starting value for h is just half the interval of integration, $(b - a)/2$. Then we initialize to 1 an iteration counter that will be used to set a limit on the number of intervals. TWO and FOUR are started at their proper first values and INT is computed for the two-interval case.

Now we enter the part of the processing that is repeated. We halve h, double n, combine in TWO the previous contents of TWO and FOUR, set FOUR to zero, set t to h, and enter the loop that

accumulates the new ordinates. Here we have used a graphical indication of the work that will be done by a DO loop in running through all values of i from 1 to n as we get the new ordinates. Finally, we compute NEWINT and check for convergence and for failure to converge in the maximum allowable number of strips.

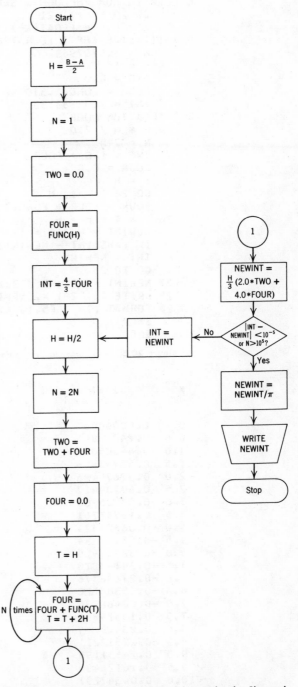

Figure 20.4. Flowchart of a method of integration by Simpson's rule with automatic interval halving until desired accuracy is achieved.

```
C CASE STUDY 20
C EVALUATING A BESSEL FUNCTION BY NUMERICAL INTEGRATION,
C USING AN ARITHMETIC STATEMENT FUNCTION
C
      REAL INT, NEWINT
C THIS IS THE ARITHMETIC STATEMENT FUNCTION
      FUNC(T) = COS(T - X*SIN(T))
C WRITE COLUMN HEADINGS
      WRITE (6, 24)
   24 FORMAT (1H1, 15H   X          J1(X)////)
C LOOP TO RUN THROUGH A SERIES OF X VALUES
      DO 29 K = 1, 21
      X = 0.5 * FLOAT(K-1)
C INITIALIZE THE INTEGRATION PROCESS
      H = 1.57079632
      N = 1
      TWO = 0.0
      FOUR = FUNC(1.57079632)
      INT = 1.33333333 * FOUR
C EVALUATION LOOP
   25 H = H / 2.0
      N = 2*N
      TWO = TWO + FOUR
      FOUR = 0.0
      T = H
      DO 26 I = 1, N
      FOUR = FOUR + FUNC(T)
   26 T = T + H + H
      NEWINT = H/3.0 * (2.0*TWO + 4.0*FOUR)
      IF (ABS(INT - NEWINT) .LT. 1.0E-6 .OR. N .GT. 10000) GO TO 27
      INT = NEWINT
      GO TO 25
   27 NEWINT = NEWINT / 3.14159265
   29 WRITE (6, 28) X, NEWINT
   28 FORMAT (1H , F5.1, F12.8)
      STOP
      END
```

Figure 20.5. A program for integration by Simpson's rule with automatic interval halving.

X	J1(X)
0.	0.00000001
0.5	0.24226848
1.0	0.44005062
1.5	0.55793656
2.0	0.57672486
2.5	0.49709417
3.0	0.33905903
3.5	0.13737761
4.0	-0.06604324
4.5	-0.23106034
5.0	-0.32757905
5.5	-0.34143810
6.0	-0.27668376
6.5	-0.15384120
7.0	-0.00468270
7.5	0.13524855
8.0	0.23463649
8.5	0.27312211
9.0	0.24531195
9.5	0.16126459
10.0	0.04347291

Figure 20.6. The output of the program in Figure 20.5.

The program in Figure 20.5 is so close to a literal transcription of the flowchart that only a few notes should be needed to make it clear, beyond pointing out the use of the statement function. This appears just after the REAL statement. Actually, the only requirement is that the statement function be defined before it is used, but most programmers tend to place them all at the beginning of the program. The definition here is in terms of T, the same variable name that is employed later when the function is called into operation, but we have seen that this is legal. The variable X in the definition is itself, for it is not listed as an argument.

The arithmetic statement function is used twice in the program. The first time the argument is a constant, $\pi/2$, and the second it is a variable. These are, of course, just two examples; much more complex expressions might have been used if we had needed them. The convergence test constant was 10^{-6} and the limit on the number of intervals, 10^5.

When the IF statement stops the process, we divide by π, as required in the problem statement, and print the result.

The program is written to take advantage of the fact that $f(0) + f(\pi) = 0$.

Figure 20.6 shows the output produced when the program was run.

EXERCISES

*1. Define a statement function to compute

$$\text{DENOM}(X) = X^2 + \sqrt{1 + 2X + 3X^2}$$

Then use the function to compute

$$\text{ALPHA} = \frac{6.9 + Y}{Y^2 + \sqrt{1 + 2Y + 3Y^2}}$$

$$\text{BETA} = \frac{2.1Z + Z^4}{Z^2 + \sqrt{1 + 2Z + 3Z^2}}$$

$$\text{GAMMA} = \frac{\sin Y}{Y^4 + \sqrt{1 + 2Y^2 + 3Y^4}}$$

$$\text{DELTA} = \frac{1}{\sin^2 Y + \sqrt{1 + 2\sin Y + 3\sin^2 Y}}$$

2. Define a statement function to compute

$$\text{SLG}(A) = 2.549 \log\left(A + A^2 + \frac{1}{A}\right)$$

Then use the function to compute

$$R = X + \log X + 2.549 \log\left(A + A^2 + \frac{1}{A}\right)$$

$$S = \cos X + 2.549 \log\left(1 + X + (1 + X)^2 + \frac{1}{1 + X}\right)$$

$$T = 2.549 \log\left[(A - B)^3 + (A - B)^6 + \frac{1}{(A - B)^3}\right]$$

$$U = [B(I) + 6]^2 + 2.549 \log\left[\frac{1}{B(I)} + \frac{1}{B(I)^2} + B(I)\right]$$

*3. Define a logical statement function to compute the "exclusive or" of two logical variables. The exclusive or is true when exactly one of the inputs is true; it is false when both inputs are false and when both inputs are true. Then write the statements that would be needed to test whether the exclusive or is associative, that is, whether

$$(A \circ B) \circ C = A \circ (B \circ C)$$

in which a circle has been used to indicate the exclusive or.

4. Given three logical variables, A, B, and C, a full binary adder accepts three inputs and provides two outputs, the sum and the carry. The sum is true if any one of the three inputs is true and the others are false or if all three inputs are true. The carry is true if any two or more of the inputs is true. Write logical statement functions to compute these two functions.

*5. Given two one-dimensional arrays of N logical elements each, named A and B, that are to be viewed as the digits of two binary numbers. The two numbers are to be added in binary to produce the $N + 1$ elements of an array named C. Use the functions developed in Exercise 4 in a loop.

CASE STUDY 21 ROOTS OF A QUADRATIC EQUATION; SUBPROGRAMS; THE **COMMON** AND **EQUIVALENCE** STATEMENTS; ADJUSTABLE DIMENSIONS

Useful as a statement function often is, it does have two rather serious restrictions: the definition is limited to one statement and it can compute only one value. The FUNCTION and SUBROUTINE subprograms remove these restrictions.

This is only half the story, however. The outstanding feature of these two types of functions is that they are *subprograms;* they can be compiled independently of the main program of which they are a part. Their variable names are completely independent of the variable names in the main program and in other subprograms. They may have their own DIMENSION statements (and the other specification statements described below). In short, FUNCTION and SUBROUTINE subprograms can be completely independent of the main program—yet it is quite easy to set up "communication" between the main program and the subprogram(s). This means that a large program can be divided into parts that can be compiled independently, making possible two important kinds of flexibility in writing programs.

The ability to compile a subprogram independently of the main program of which it is a part means that one subprogram can be used with different main programs. For instance, many different programmers in an installation may need a subprogram to solve a system of simultaneous equations. Since arrays are involved and since many statements are required, statement functions are out of the question. However, a SUBROUTINE subprogram can be written to do the job. It can be compiled by itself, and the compiled object program can be combined with *any* main program. All that is necessary is for the main program to have been written with the conventions of the subprogram in mind.

The other flexibility provided by separate compilation is the freedom to compile and run segments of one program independently of each other. This means that parts of a program can be checked out as they are written, which can be an important advantage. For example, if there are many subprograms, all called by one main program, individual subprograms can be checked out and tested before all of the others are finished.

Subprograms thus have three major advantages. One is the primary motivation of any function: a group of statements written in one place in a program can be called into action from anywhere else in the program, thus avoiding wasteful duplication of effort in writing the source program and the waste of storage space that would be caused by the duplication of segments of the object program. A second advantage is that of avoiding duplication of effort by making it possible for a subprogram to be used with many different main programs. The third advantage is that of separate checkout.

Whatever the motivation for their use, subprograms are a powerful feature of the FORTRAN language. We shall study them quite carefully, along with some new statements that are meaningful only in connection with subprograms. The example that we shall use to illustrate the application of these ideas is mathematically almost trivial; we choose a simple illustration because so much new FORTRAN material must be introduced before we can arrive at the full-scale illustration.

As with the arithmetic statement functions, we must distinguish carefully between the definition and the use. The computation desired in a FUNCTION subprogram is *defined* by writing the necessary statements in a segment, writing the word FUNCTION and the name of the function before the segment, and writing the word END after it. The name is formed as for variables and statement functions: one to six letters or digits, the first of which must be a letter. The letter must be chosen according to the naming convention in the absence of a type declaration. If the naming convention is to be overridden, or if the type is other than real or integer, the word FUNCTION is preceded by one of the five types.

The name of the subprogram is followed by parentheses enclosing the argument(s), which are separated by commas if there is more than one.

The name of the function must appear at least once in the subprogram as a variable on the left-hand side of an assignment statement or in the list of an input statement. In other words, the name of a FUNCTION subprogram is associated with a value; a value must therefore be given to it in the subprogram.

As before, the arguments in the subprogram definition are only dummy variables. The arguments in the function definition must be distinct nonsubscripted variables or array names. Within the subprogram itself, however, subscripted variables may be used freely. The subprogram must contain at least one RETURN statement, for reasons that we shall see shortly.

To *use* a FUNCTION subprogram it is necessary only to write the name of the function where its value is desired, with suitable expressions for arguments. The mechanics of the operation of the object program are as follows: the FUNCTION subprogram is compiled as a set of machine instructions in one place in storage, and wherever the name of

```
FUNCTION S(R)
IF (R .LT. 120.0) S = 1.7E4 - 0.485*R**2
IF (R .GE. 120.0) S = 1.8E4/(1.0 ± R*R/1.8E4)
RETURN
END
```

Figure 21.1. A FUNCTION subprogram to compute the safe loading of a column.

the subprogram appears in the source program a transfer to the subprogram is set up in the object program. When the computations of the subprograms have been completed, a transfer is made back to the section of the program that brought the subprogram into action. The RETURN statement(s) in the subprogram results in object program instructions to transfer back to the place in the main program from which the subprogram was called. (This is actually quite similar to the way a statement function is set up, except that in that case there can be only one statement in the definition and there is no question when the function's operations are complete.)

As a simple example of what can be done with a FUNCTION subprogram, suppose we rewrite the program of Case Study 6. It may be recalled that the task was to find the safe loading S in a certain column as a function of the slimness of ratio R, according to the formula

$$S = \begin{cases} 17,000 - 0.485R^2 & \text{for } R < 120 \\ \dfrac{18,000}{1 + \dfrac{R^2}{18,000}} & \text{for } R \geq 120 \end{cases}$$

Setting up a FUNCTION to do this computation is a simple matter, as shown in Figure 21.1.

This defines the function. Now to use it we might have something like that shown in Figure 21.2.

With this function available, we can use it to compute a value in many other types of situations other than just preparing a table. For example, it might be necessary to check that a value named PLANND is less than or equal to the safe load for a slimness ratio named RATIO. We assume that this check is part of a much larger program that contains at

```
      REAL R, S
   23 READ (5, 1) R
    1 FORMAT (8F10.0)
      SAFE = S(R)
      WRITE (6, 2) R, SAFE
    2 FORMAT (1H , 1P10E13.5)
      GO TO 23
      END
```

Figure 21.2. A program that uses the FUNCTION of Figure 21.1.

```
      FUNCTION DIAGPR(A, N)
      DIMENSION A(10, 10)
      DIAGPR = A(1, 1)
      DO 69 I = 2, N
   69 DIAGPR = DIAGPR * A(I, I)
      RETURN
      END
```

Figure 21.3. A FUNCTION subprogram to find the product of the main diagonal elements of an array.

an appropriate point the statement

IF (PLANND .GT. S(RATIO)) STOP

A FUNCTION subprogram can be set up to have many arguments, including arrays. For example, suppose that it is necessary to find the product of the main diagonal elements (those having the same row and column number) of square arrays. The arrays from which this product is computed must have been mentioned in a DIMENSION statement in the "calling" program, as always, and all the arrays must have the same dimensions. (We shall see shortly how to remove the same-dimension restriction.) The array names in the FUNCTION argument list and subprogram will be dummies, but the dummy array names must still be mentioned in a DIMENSION statement in the subprogram. Suppose that the arrays in question are all 10 x 10 but that they are not necessarily full; the value of an integer variable gives the number of rows and columns. The subprogram could be as shown in Figure 21.3.

Now, if we want the product of the main diagonal elements of a 10 x 10 array named X, in which the actual size is 8 x 8, with the extra elements containing nothing, we can write

DET = DIAGPR(X, 8)

To find the square of the product of the main diagonal elements of an array named SAM, in which the number of rows and columns containing meaningful data is given by the value of an integer variable named JACK, we could write

EIG = DIAGPR (SAM, JACK)**2

A FUNCTION subprogram is seen to be quite similar to a statement function, except that it can use many statements instead of just one and it can use any of the FORTRAN statements instead of just an assignment statement. A subprogram can call on other subprograms as long as it does not call itself and as long as two subprograms do not call each other.*

A FUNCTION subprogram has been described as computing just one value, the one associated with the name of the FUNCTION. Actually, there can be any number of output values: any of the arguments may refer to output. For an example of how this can be useful, consider the following extension of the requirements of the program in Figure 21.3. Suppose that if the product of the main diagonal elements is less than or equal to 100 we wish to go to statement 12; if it is between 100 and 1000, we wish to go to statement 123; if it is greater than or equal to 1000, we wish to go to statement 1234. All of these statement numbers refer to statements in the *main* program. This could, of course, be done with IF statements in the main program, but if it has to be done frequently we prefer a simpler way.

To accomplish the simplification let us first add a type specification to make the value associated with the name of the integer type. This is done by writing the word INTEGER in front of the word FUNCTION in the subprogram definition. We then write the modified subprogram so that the value of

*A subprogram that calls itself is said to be *recursive*. This *is* permitted in the ALGOL language and others, in which it finds greatest utility in non-numerical applications, such as compiler programs, processing natural languages (e.g., English), and in operations on the *symbols* of mathematics as distinguished from their values. Recursiveness can be accomplished in some cases in FORTRAN by "stacking" arguments in arrays.

```
      INTEGER FUNCTION DIAGPR(A, N, PROD)
      DIMENSION A(10, 10)
      PROD = A(1, 1)
      DO 69 I = 2, N
   69 PROD = PROD * A(I, I)
      IF (PROD .LE. 100.0) DIAGPR = -1
      IF (PROD .GT. 100.0 .AND. PROD .LT. 1000.0) DIAGPR = 0
      IF (PROD .GE. 1000.0) DIAGPR = 1
      RETURN
      END
```

Figure 21.4. The FUNCTION subprogram in Figure 21.3 modified to store the product as an argument and to give an integer value for the function.

DIAGPR is negative, zero, or positive, depending on whether the product is less or equal to 100, between 100 and 1000, or greater or equal to 1000. We also add PROD as an argument and within the subprogram give it the value of the product of the main diagonal elements. The modified program is shown in Figure 21.4.

Now suppose that we want the product of the main diagonal elements of a 10 x 10 array named BETA in which seven rows and seven columns are used; the product is to be called GAMMA. We are to transfer to one of the three statements numbered 12, 123, or 1234, as already described.

IF (DIAGPR(BETA, 7, GAMMA)) 12, 123, 1234

The appearance of the name of the function, written with appropriate arguments, causes the function to be called into operation, in the course of which a value is given to its name and also to the other output parameter, which is GAMMA in this case. Control returns from the subprogram to the arithmetic IF statement in the main program, where the proper transfer is made. At any of these locations the newly computed value of GAMMA may be used.

The basics of a SUBROUTINE subprogram, although quite similar to those of a FUNCTION subprogram, show three differences.

1. A SUBROUTINE has no value associated with its name. All outputs are defined in terms of arguments; there may be any number of outputs.

2. A SUBROUTINE is not called into action simply by writing its name, since no value is associated with the name. Instead, we write a CALL statement to bring it into operation; this specifies the arguments and results in storing all the output values.

3. Since the output of a SUBROUTINE may be any combination of the various types of values, there is no type associated with the names and likewise no convention attached to the first letter of the name. The naming of a SUBROUTINE is otherwise the same as the naming of a FUNCTION.

In all other respects the two subprograms are entirely analogous.

The essential features of the SUBROUTINE subprogram are illustrated in the following example. Suppose that in a certain program it is frequently necessary to find the largest element (in absolute value) in a specified row of a 50 x 50 array. The input to the SUBROUTINE is therefore the array name and the row number. The output will be the absolute value of the largest element in that row and its column number. The SUBROUTINE could be as shown in Figure 21.5.

Now suppose that the largest element in the third row of a 50 x 50 array named ZETA is needed. The absolute value of the element is to be called DIVIS and column number is to be called NCOL. We write the statement

CALL LARGE (ZETA, 3, DIVIS, NCOL)

This brings the subprogram into operation, stores the values of DIVIS and NCOL found by the subprogram, and returns control to the statement following the CALL. If, later, it is necessary to find the largest element in row $M+2$ of an array named DETAIL, storing its absolute value in SIZE and the column number in KW, we can write

CALL LARGE (DETAIL, M + 2, SIZE, KW)

To emphasize the independence of the variable names between the main program and any subprograms, we note that it would be possible and legal to write the statement

CALL LARGE (ARRAY, I, BIG, J)

If this is done, all of the input variables to the subprogram must be defined and given values in the calling program and all output variables from the subprogram must be defined in the calling program.

```
      SUBROUTINE LARGE(ARRAY, I, BIG, J)
      DIMENSION ARRAY(50, 50)
      BIG = ABS(ARRAY(I, 1))
      J = 1
      DO 69 K = 2, 50
      IF (ABS(ARRAY(I, K)) .LT. BIG) GO TO 69
      BIG = ABS(ARRAY(I, K))
      J = K
   69 CONTINUE
      RETURN
      END
```

Figure 21.5. A SUBROUTINE to find the largest element in a specified row of an array.

The name I in the calling program and the name I in the subprogram are unrelated. And this must logically be so: the name I in the subprogram tells *what to do with* a value from the calling program, whereas the name I in the calling program must *specify a value*, one that has already been computed by the calling program. In the case of output variables, J, for instance, the variable J in the subprogram identifies a value that the subprogram computes, whereas J in the calling program identifies a result transmitted from the subprogram.

Adjustable Dimensions

It is possible for a subprogram to be defined in terms of arrays that are of adjustable size. We do this by writing in the subprogram a DIMENSION statement in which we write integer variable names instead of integer constants. The integer variables must appear in the argument list and be given values by the calling program.

This is much easier to understand in an example. Consider the program in Figure 21.5 which we rewrite with adjustable dimensions in Figure 21.6. It will be noted that another variable, N, has been added to the argument list, which gives the number of rows and columns in the array to be searched. The DIMENSION statement says that the array is N x N, and the DO statement says to inspect rows 2 through N.

With this revision, the subroutine can be used to find the largest element in a specified row of *any* square array. We might write, for instance,

CALL LARGE (ALPHA, 49, 6, BIGGST, M)

This will find the largest element in row 6 of a 49 x 49 array named ALPHA, placing the largest elements in BIGGST and its row number in M. Or we might write

CALL LARGE (GAMMA, L, M, DELTA, K98)

The array this time is L x L, where L would have to have been given a value before the call.

The adjustable dimension facility has a number of advantages. One, of course, is that a very general sort of subprogram can be adapted to the needs of a particular main program, without the awkwardness and potentially wasted storage of specifying a maximum size and then using only some of the elements. A given subprogram can be called many times from one main program, each time if necessary with arrays of different sizes. Looking at another aspect of the usefulness of subprograms, a prewritten subprogram can serve the rather different requirements of many programmers.

We have noted that it is permissible for one subprogram to call another. When this is done, adjustable dimension information may be passed "through" subprograms. For instance, a main program might call a subprogram with array dimension information given by the value of a variable in the argument list. The subprogram called might in turn call another subprogram, "passing" the dimension information to the subprogram *it* called.

The restrictions on the usage of adjustable dimensions are reasonable. The subprogram must not use a subscript value greater than specified in the call; the subprogram must not redefine the value of an adjustable dimension; the calling program must not leave the value of a dimensioned variable undefined. We may also note at this point that an array mentioned in a COMMON statement must not have adjustable dimensions.

Summary of the Four Types of Functions

FORTRAN provides for four types of functions: those supplied with the system, statement functions, FUNCTION subprograms, and SUBROU-

```
      SUBROUTINE LARGE(ARRAY, N, I, BIG, J)
      DIMENSION ARRAY(N, N)
      BIG = ABS(ARRAY(I, 1))
      J = 1
      DO 69 K = 2, N
      IF (ABS(ARRAY(I, K)) .LT. BIG) GO TO 69
      BIG = ABS(ARRAY(I, K))
      J = K
   69 CONTINUE
      RETURN
      END
```

Figure 21.6.   The SUBROUTINE in Figure 21.5 is modified to accept arrays of adjustable dimension.

TABLE 21.1

	Supplied	Statement	FUNCTION	SUBROUTINE
Naming	1–6 characters, first of which is a letter	1–6 characters, first of which is a letter	1–6 characters, first of which is a letter	1–6 characters, first of which is a letter
Type	Implied by first letter; can be overridden in some FORTRAN's by REAL, INTEGER, COMPLEX, LOGICAL, or DOUBLE PRECISION	Implied by first letter unless overridden by REAL, INTEGER, COMPLEX, LOGICAL, or DOUBLE PRECISION	Implied by first letter unless overridden by REAL, INTEGER, COMPLEX, LOGICAL, or DOUBLE PRECISION	None—no value associated with name
Definition	Provided with the compiler	One arithmetic statement before first usage of function	Any number of statements after word FUNCTION	Any number of statements after word SUBROUTINE
How called	Writing name where function value is desired	Writing name where function value is desired	Writing name where function value is desired	CALL statement
Number of arguments	One or more, as defined	One or more, as defined	One or more, as defined	Any number, including *none*, as defined
Number of outputs	One	One	One is associated with function name; others may be specified as arguments	Any number

TINE subprograms. The salient features of the four types are summarized in Table 21.1.

The EQUIVALENCE and COMMON Statements

These two nonexecutable statements make possible certain conveniences in the naming of variables and the assignment of storage locations to them.

The EQUIVALENCE statement causes two or more variables to be assigned to the same storage location, which is useful in two rather different ways.

In one usage the EQUIVALENCE statement allows the programmer to define two or more variable names as meaning the same thing. It might be that after writing a long program the programmer will realize that he has inadvertently changed variable names and that X, X1, and RST7 all mean the same thing. Rather than going back and changing the variable names in the program, a time-consuming and error-prone process, he can write

EQUIVALENCE (X, X1, RST7)

and the mistake is corrected.

The other application is in making use of the same storage location to contain two or more variables that are different but are never needed at the same time. Suppose that in a certain program the variable I27 appears in the initial READ statement and in a few subsequent statements but is never used after that. Later in the program a value is given to the variable NPL, which is then used as a DO parameter. Later the variable JJM2 is applied to a similar purpose. At the end of the variable NEXT1 is given a value and then used in the final WRITE statement. As things now stand, four storage locations will be allocated to these variables, which is pointless, since their usage never overlaps. If the programmer is short of storage space, he can assign all four variables to one location by writing

EQUIVALENCE (I27, NPL, JJM2, NEXT1)

The same thing could, of course, be accomplished by changing the variable names, but using an EQUIVALENCE is obviously simpler.

These two applications of EQUIVALENCE differ only in viewpoint; the statement and its treatment by the compiler are the same in either case.

One EQUIVALENCE statement can establish equivalence between any number of sets of variables. For instance, if A and B are to be made equivalent, as are X, Y, and Z, we can write

EQUIVALENCE (A, B), (X, Y, Z)

Seldom is storage so "tight" that the EQUIVALENCE statement is *really* needed for nonsubscripted variables. The value comes in establishing equivalences between arrays.

In many versions of FORTRAN array names must be mentioned with a single constant subscript, regardless of the actual dimensionality of the array. We might, for instance, have statements like

DIMENSION A(50), B(5, 10), C(2, 5, 5)
EQUIVALENCE (A(1), B(1), C(1))

This EQUIVALENCE statement causes storage to be assigned so that the elements of the three arrays occupy the same 50 storage locations. It is not required, however, that the arrays made equivalent have the same number of elements. If they do not, and if the arrays in the EQUIVALENCE statements all have element number 1, as above, then the extra elements at the end of the longer array will simply not be shared locations.

It is also permissible to specify that equivalence be established between element locations other than the first. If, for instance, X is a one-dimensional array with 10 elements, and D, E, and F are all nonsubscripted variables, we might write

EQUIVALENCE (X(1), D),

(X(2), E), (X(10), F)

The single variable D would be assigned to the same location as the first element of X, E would be assigned to the same location as the second element of X, and F would be assigned to the same location as the tenth element of X.

Arrays can be overlapped by the same techniques. We might write

EQUIVALENCE (A(1), B(20))

The general idea is that the storage assignments are made in such a manner that the specified equivalence is established, and both arrays are stored in consecutive locations just as arrays always are. The question we need to be able to answer, often, is, "In a two- or three-dimensional array, which element *is* the twentieth or the Nth?"

The answer is given by the *element successor* rule, which tells where a given element is stored in the linear sequence of storage locations. Table 21.2 gives the needed information. A, B, and C in this table are dimensions, as given by a DIMENSION, COM-MON, or type statement; a, b, and c are the values of subscript expressions. By "subscript declarator" we mean the dimensioning information given in the DIMENSION, or COMMON, or type statement. For instance, in

DIMENSION X(3, 12)

the subscript declarator is (3, 12).

Consider an example. In the statement just given, in terms of the notation of Table 21.2, $A = 3$, $B = 12$. Now where, for instance, is the element in row 2, column 9, that is, X(2, 9)? Table 21.2 says that the "value" of this subscript is $2 + 3 \cdot (9 - 1) = 26$. In other words, we are to think of the 36 elements of the array arranged in a linear sequence in storage. Then, if element (1, 1) is in position 1 in this string, element (2, 9) is in position 26.

Study will show that Table 21.2 is a formal expression of the rule given earlier: arrays are stored in such a way that the first subscript varies most rapidly and the last varies least rapidly. This was the convention described in connection with using array names in input or output statements without subscripts.

All of the preceding has assumed that the variables in question were real, integer, or logical and furthermore that each takes up one storage location. (There are local exceptions to the latter.) In the case of complex and double precision variables two storage locations usually are required for each element, and when a single element is made equivalent to a double element it is the first part of the double element that is involved.

The reader is cautioned that the various versions of FORTRAN vary considerably on the matter of storage assignments for arrays, in the effect of complicated EQUIVALENCE and COMMON statements, in the presence or absence of special rules governing complex and double precision arrays, etc., etc. Different FORTRANs vary as much in this area as anywhere. The description given here is intended to conform to the proposed American Standards Association FORTRAN, but every programmer must get the details for his system.

TABLE 21.2

Dimensionality	Subscript Declarator	Subscript	Subscript Value
1	(A)	(a)	a
2	(A, B)	(a, b)	$a + A \cdot (b - 1)$
3	(A, B, C)	(a, b, c)	$a + A \cdot (b - 1) + A \cdot B \cdot (c - 1)$

The COMMON statement. It has been stated that each subprogram has its own variable names: the name X in the main program is not necessarily taken to be the same as the name X in a subprogram. However, if the programmer *wishes* them to mean the same thing, he can write

COMMON X

in *both* the main program and the subprogram. The compiler will then assign the two variables (and they still are distinct, in principle) to the same storage location, which as a practical matter makes them the same.

But the statement is not limited to this kind of use. Suppose we write

| main program: | COMMON X, Y, I |
| subprogram: | COMMON A, B, J |

Then X and A are assigned to the same storage location, as are Y and B and I and J.

EQUIVALENCE and COMMON have a somewhat similar function. What is the difference between them?

EQUIVALENCE assigns two variables *within the same main program or within the same subprogram* to the same storage location; COMMON assigns two variables *in different subprograms or in a main program and a subprogram* to the same location.

When an array is named in an EQUIVALENCE statement *and* in a COMMON statement, the equivalence is established in the same general way as described earlier. This may increase the size of the COMMON block of storage and thus change the correspondences between the COMMON block described and some other COMMON block in another program.

For instance, consider a program containing the following three statements:

DIMENSION A(4), B(4)
COMMON A, C
EQUIVALENCE (A(3), B(1))

Without the EQUIVALENCE statement, the COMMON block would contain five storage locations in the sequence

A(1), A(2), A(3), A(4), C

With the EQUIVALENCE statement, the B array is brought into COMMON, so to speak, and requires the following sequence of storage locations

A(1), A(2), A(3), A(4), C
B(1), B(2), B(3), B(4)

COMMON is now six storage locations long.

COMMON may be lengthened in this way, but it may *not* be lengthened by any attempt to push the start of a COMMON block forward. For instance, with the same DIMENSION and COMMON statement just considered, the following EQUIVALENCE would be illegal.

EQUIVALENCE (A(1), B(2))

The storage assignment in COMMON would need to be

A(1), A(2), A(3), A(4), C
B(1), B(2), B(3), B(4)

Since B is not mentioned in the COMMON statement, but is brought into COMMON by the EQUIVALENCE, the first element of B now precedes the start of this block of COMMON. This is not permitted.

As we have mentioned several times in passing, a variable named in a COMMON or type statement may have subscripting information. We might write

COMMON A(23), J(2, 8) LOGIC (3,3,7)
REAL X(10)
DOUBLE PRECISION VARNCE(5, 20)

A variable that is written with subscripting information in a COMMON or type statement must *not* be mentioned in a DIMENSION statement. On the other hand, it is still permissible to name a variable in a DIMENSION statement and also to name it, without subscripting information, in a COMMON or type statement.

Two variables in COMMON must not both be named in an EQUIVALENCE statement. The reason for this is instructive. EQUIVALENCE says that two or more variables in *one program* (main program or subprogram) are to be assigned to the same storage location. COMMON says that variables in *different programs* are to be assigned to the same location. The way COMMON works is that all the variables named in a COMMON statement are assigned to storage in the sequence in which the names appear in the COMMON statement. This is true even when there is only one COMMON statement, in which case COMMON does not cause multiple assignments at all. But then if there are two or more COMMON statements, correspondences are established simply because the COMMON statement is treated the same way wherever it appears.

For example, if in a main program we write

$$COMMON\ A, B, C, D$$

the four variables named are assigned to storage locations in the order named, in a special section of storage called "COMMON storage." Thus A is a specific storage location, followed by B, etc. Now suppose that in some subprogram we have

$$COMMON\ W, X, Y, Z$$

This means that W is assigned to the first storage location in COMMON, X is assigned to the next one, and so on, and the "COMMON block" used by the subprogram is the same as that used in the main program. Ergo, A and W have been assigned to the same location—without the compiler ever knowing about more than one COMMON statement at a time. Indeed, the two programs may very well have been compiled entirely separately.

Now suppose that we had the combination

$$COMMON\ A, B, C, D$$
$$EQUIVALENCE\ (A, B)$$

The net effect is a contradiction: the COMMON says to put the four variables named into a special area of storage, in the order named and *in separate locations*, whereas the EQUIVALENCE says that A and B are to be assigned to the *same* location.

In the description just given we have said that there is only one COMMON block in storage. This is actually too limited: we can establish as many distinct blocks of COMMON storage as we please by *labeling* COMMON. What we have been discussing so far is in fact called *blank COMMON* to signify that it has no label.

Each COMMON block—blank COMMON and as many labeled COMMON blocks as there may be—is set up as described above. That is, variables and arrays are assigned storage locations in the order in which they are listed. Any rearrangements made necessary by EQUIVALENCE statements are made.

Labeled COMMON may be used, if there is some need to do so, simply to guarantee a particular arrangement of storage locations. This is probably rare in normal usage, being limited to what might be called "extra-legal" programming. (For instance, in a particular compiler and computer it may be possible to refer to the data adjacent to an array by using subscripts larger than the maximum given in dimensioning information.)

In "normal" or "legal" usage, the value of labeled COMMON is to have two COMMON blocks with the same name in two programs that are executed together. When this is done, the two blocks must be the same length. Assuming this to be the case, the variables in the two blocks are assigned to the same storage locations, just as with blank COMMON.

Labels are written between slashes in front of the variable names. We might write, for instance

$$COMMON/X/A, B, C$$

If a single COMMON statement includes labeled COMMON and blank COMMON, the blank COMMON portion may either be written first without a name, as we have done heretofore, or the name may be omitted between slashes.

For a final example, suppose we were to write the following two statements in a main program and in a subprogram.

$$COMMON\ A, B, C\ /B1/D, E/B2/F(20), G(2, 5)$$
$$COMMON\ R, S, T\ /B1/U, V/B2/X(10), Y(10, 2)$$

Blank COMMON would contain A, B, and C, in that order, in the program containing the first COMMON, and R, S, and T in the program containing the second. A and R would thus be assigned to the same storage location, as would B and S, and C and T. The COMMON block labeled B1 would establish D and U in the same location and E and V in the same. We assume that all of the foregoing variables were not mentioned elsewhere in a DIMENSION statement. B2 in the first program contains the 20 elements of F and the 10 elements of G. The same 30 locations would also contain the 10 elements of X and the 20 elements of Y. The overlap between the four arrays involved would cause the compiler no difficulty—indeed the compiler would never really consider the situation. Such a pair of statements would put $F(11)$ and $Y(1, 1)$ in the same location, for instance. If that is the intended action, then everything will work nicely. If not, naturally there will be some surprises in store for the programmer. As a matter of fact, the intricacies of things like labeled COMMON and the interrelationships between COMMON and EQUIVALENCE account for a disproportionate percentage of the questions programmers have to ask about FORTRAN.

Solution of quadratic equations using subprograms. To see these concepts in action we now turn to the solution of quadratic equations, using subprograms and the COMMON statement.

A program is to be set up to solve the quadratic equation $Ax^2 + Bx + C = 0$. The program must be able to read from cards the coefficients of many such equations—possibly hundreds—and to produce an easily readable report showing for each equation the coefficients and the roots. A data card contains the coefficients of two equations, six values in all. However, each equation is to be written on a line by itself. The roots can be real or complex, although we shall do all arithmetic using FORTRAN real variables. A heading is to be printed at the top of each page, the pages are to be numbered, and the lines are to be counted as they are printed, so that each page will contain only 20 lines of output, double-spaced.

The program will be written to use two SUBROUTINE subprograms. Each subprogram can be called twice for each data card, avoiding duplication and perhaps making the complete program easier to correct and modify. The main routine will handle reading of data cards, printing of page headings, page numbering, line counting, and detecting the end of the deck, which is signaled by a blank card placed at the end of the deck for the purpose. The detection of this card can be set up as a test for a data card in which A = 0. A can never properly be zero with valid data; if A = 0, the equation is not quadratic.

The first subprogram will get the solutions, taking into account that if the discriminant $B^2 - 4AC$ is negative the roots are complex. The input to this subprogram, named SOLVE, consists of the names of the three coefficients; they are named as arguments of the subprogram. The output consists of the real and imaginary parts of the two roots, which are named X1REAL, X1IMAG, X2REAL, and X2IMAG. These four variables are needed in the main program and in both subprograms. They are named in COMMON statements in all three places, making it unnecessary to write them as arguments of the subprograms.

The second subprogram, named OUTPUT, writes the coefficients and the roots. It is desired to print the results in such a way that the reader can tell at a glance if the roots are pure real or pure imaginary. If the roots are real, the space for the imaginary parts is to be left blank, and if they are pure imaginary the space for the real parts is to be left blank. We recall that complex roots always occur

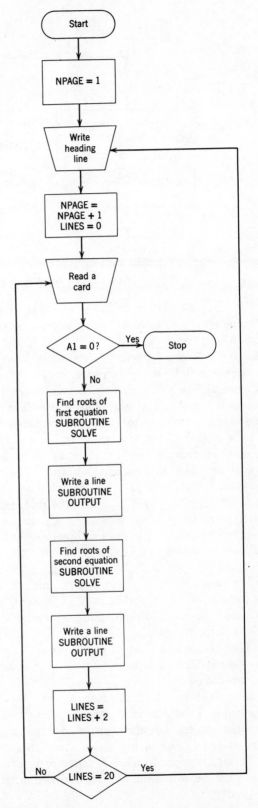

Figure 21.7. Flowchart of the main program for finding the roots of quadratics.

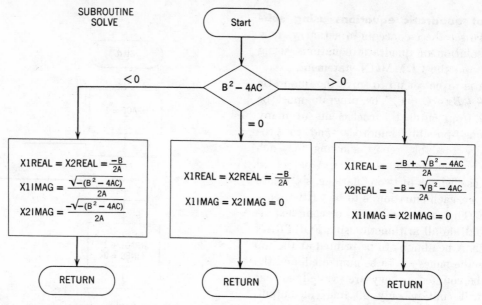

SUBROUTINE
SOLVE

$B^2 - 4AC$

< 0　　　> 0

$= 0$

$$X1REAL = X2REAL = \frac{-B}{2A}$$
$$X1IMAG = \frac{\sqrt{-(B^2 - 4AC)}}{2A}$$
$$X2IMAG = \frac{-\sqrt{-(B^2 - 4AC)}}{2A}$$

$$X1REAL = X2REAL = \frac{-B}{2A}$$
$$X1IMAG = X2IMAG = 0$$

$$X1REAL = \frac{-B + \sqrt{B^2 - 4AC}}{2A}$$
$$X2REAL = \frac{-B - \sqrt{B^2 - 4AC}}{2A}$$
$$X1IMAG = X2IMAG = 0$$

RETURN　　　RETURN　　　RETURN

Figure 21.8. Flowchart of the subprogram for finding the roots of quadratics.

as complex conjugates; it can never happen that only one root is complex or that two complex roots have different real parts.

Figure 21.7 is a flowchart of the main program, which is seen to be straightforward.

Figure 21.8 is a flowchart of the subprogram for finding the roots. The procedure shown steers a middle course between the bare minimum required to distinguish between real and complex roots and the more complicated tests that could be made to take advantage of every special situation. The bare minimum would be to go to the complex section if the discriminant is negative and to the real section if it is zero or positive. Since the arithmetic IF statement automatically gives a three-way branch, it seems reasonable to take special action if the discriminant is zero to avoid computing the square root of zero. We could, however, go further with this testing for special conditions. If C is zero, both roots are real, one being zero and the other — B/A. If B is zero, the formulas simplify slightly. If *B* and C are both zero, then, of course, both roots are zero—but it is hard to see why such a case would ever be entered.

In any such case it is necessary to draw the line somewhere. Time can indeed be saved by taking advantage of special situations, unless testing for them wastes all the saving. Even where there is a net saving, though, a thorough series of tests may simply not be worth the trouble and the program complexity.

The flowchart of the output subprogram, Figure

21.9, is also fairly simple. Note that it is not necessary to test both imaginary components for zero values, since both will always be zero or nonzero. This is not true of the real parts.

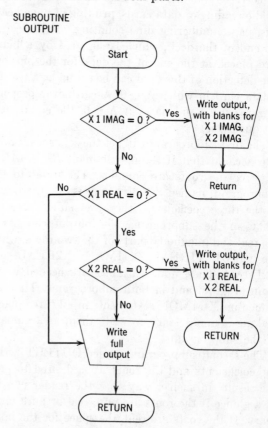

SUBROUTINE
OUTPUT

Start

X 1 IMAG = 0 ?　Yes　Write output, with blanks for X 1 IMAG, X 2 IMAG

No

X 1 REAL = 0 ?　No

Yes

X 2 REAL = 0 ?　Yes　Write output, with blanks for X 1 REAL, X 2 REAL

No

Return

RETURN

Write full output

RETURN

Figure 21.9. Flowchart of the subprogram for producing the output.

The main program is shown in Figure 21.10. We begin with a COMMON statement naming the real and imaginary parts of the two roots. The same statement appears in both subprograms, so that these four variables are assigned to the same storage locations by all three programs, even though the three will be compiled quite separately. This would not be true, of course, without the COMMON statements.

Next we set the page number to 1 and write the heading line, the bulk of which consists of column identifications; in fact, the only variable in the line is the page number itself. The carriage control character of 1 in the first Hollerith field specification causes spacing to the top of a new page. The two slashes at the end of the FORMAT force an extra line space between the heading line and the first line of regular output.

After incrementing the page counter and initializing the line counter, we read a data card. The READ statement must have different names for the three coefficients of the two equations. A test of the first value for A detects the sentinel card and stops if this is the sentinel.

Now we are ready to find the roots for the first set of coefficients, and we call into operation the subprogram that solves the equation. The arguments are the first set of coefficients. When control returns from SOLVE, values will have been given to X1REAL, X1IMAG, X2REAL, and X2IMAG. Now we call the subprogram for writing the output, naming only the three coefficients; the four parts of the roots are communicated via COMMON. Another call of each of the subprograms computes and writes the roots for the second set of coefficients from the data card. After incrementing the line counter by two, an IF statement determines whether the page is full and goes back to write the heading line if it is.

The coding of the two subprograms, Figures 21.11 and 21.12, should not be hard to follow. Two new variables, DISC and S, are set up in the SOLVE subprogram to avoid computing certain expressions twice. Advantage is taken of the fact that complex roots occur only as complex conjugates, once again to avoid computing an expression twice.

The OUTPUT subprogram is not complicated, but the FORMAT statements should be studied carefully. The blank spaces for the two special cases of pure real and pure imaginary roots are intro-

```
C CASE STUDY 21
C QUADRATIC EQUATION SOLUTION WITH SUBROUTINES
C THIS IS THE MAIN PROGRAM
C
      COMMON X1REAL, X1IMAG, X2REAL, X2IMAG
C INITIALIZE PAGE NUMBER
      NPAGE = 1
C PRINT HEADING, INCLUDING PAGE NUMBER
    5 WRITE (6, 8) NPAGE
    8 FORMAT (1H1, 9X, 1HA, 14X, 1HB, 14X, 1HC, 11X, 7HX1 REAL, 8X,
   1    7HX1 IMAG, 8X, 7HX2 REAL, 8X, 17HX2 IMAG      PAGE, I4//)
C INCREMENT PAGE NUMBER
      NPAGE = NPAGE + 1
C INITIALIZE LINE COUNTER
      LINES = 0
C READ A DATA CARD
   14 READ (5, 15) A1, B1, C1, A2, B2, C2
   15 FORMAT (6F10.0)
C CHECK FOR SENTINEL
      IF (A1 .EQ. 0.0) STOP
C SOLVE BOTH EQUATIONS AND PRINT RESULTS
      CALL SOLVE (A1, B1, C1)
      CALL OUTPUT (A1, B1, C1)
      CALL SOLVE (A2, B2, C2)
      CALL OUTPUT (A2, B2, C2)
C INCREMENT LINE COUNTER
      LINES = LINES + 2
C CHECK IF PAGE IS FULL
      IF (LINES .EQ. 20) GO TO 5
      GO TO 14
      END
```

Figure 21.10. The main program for finding the roots of quadratics.

```
C CASE STUDY 21
C QUADRATIC EQUATION SOLUTION WITH SUBROUTINES
C THIS IS THE SUBROUTINE FOR FINDING THE ROOTS
C
      SUBROUTINE SOLVE (A, B, C)
      COMMON X1REAL, X1IMAG, X2REAL, X2IMAG
C MAKE INTERMEDIATE VARIABLE OF THE DISCRIMINANT
      DISC = B**2 - 4.0 * A * C
C TEST DISCRIMINANT
      IF (DISC) 50, 60, 70
C HERE IF THE DISCRIMINANT IS NEGATIVE -- ROOTS ARE COMPLEX
   50 X1REAL = -B / (2.0 * A)
      X2REAL = X1REAL
      X1IMAG = SQRT(-DISC) / (2.0 * A)
      X2IMAG = - X1IMAG
      RETURN
C HERE IF THE DISCRIMINANT IS ZERO -- ROOTS ARE REAL AND EQUAL
   60 X1REAL = - B / (2.0 * A)
      X2REAL = X1REAL
      X1IMAG = 0.0
      X2IMAG = 0.0
      RETURN
C HERE IF THE DISCRIMINANT IS POSITIVE -- ROOTS ARE REAL AND UNEQUAL
   70 S = SQRT(DISC)
      X1REAL = (-B + S) / (2.0 * A)
      X2REAL = (-B - S) / (2.0 * A)
      X1IMAG = 0.0
      X2IMAG = 0.0
      RETURN
      END
```

Figure 21.11. The subprogram for finding roots of quadratics.

duced by 15X field specifications. Note the carriage control characters to get double spacing.

The main program of Figure 21.10 and the subprograms of Figures 21.11 and 21.12 were submitted to the computer center as three separate programs at different times. When the three object decks were returned, they were combined into one deck, data cards were added, and the object program was executed. The compiler had arranged the object program decks so that when they were subsequently loaded the computer knew where to start executing the main program, and the interrelationships be-

```
C CASE STUDY 21
C QUADRATIC EQUATION SOLUTION WITH SUBROUTINES
C THIS IS THE SUBROUTINE FOR PRINTING THE RESULTS
C
      SUBROUTINE OUTPUT (A, B, C)
      COMMON X1REAL, X1IMAG, X2REAL, X2IMAG
C CHECK IF ROOTS ARE PURE REAL
      IF (X1IMAG .NE. 0.0) GO TO 90
C HERE IF ROOTS ARE PURE REAL
      WRITE (6, 95) A, B, C, X1REAL, X2REAL
   95 FORMAT (1H0, 1P4E15.4, 15X, 1PE15.4)
      RETURN
C CHECK IF ROOTS ARE PURE IMAGINARY
   90 IF (X1REAL .NE. 0.0 .OR. X2REAL .NE. 0.0) GO TO 100
C HERE IF ROOTS ARE PURE IMAGINARY
      WRITE (6, 103) A, B, C, X1IMAG, X2IMAG
  103 FORMAT (1H0, 1P3E15.4, 15X, 1PE15.4, 15X, 1PE15.4)
      RETURN
C HERE IF ROOTS ARE NEITHER PURE REAL NOR PURE IMAGINARY
  100 WRITE (6, 110) A, B, C, X1REAL, X1IMAG, X2REAL, X2IMAG
  110 FORMAT (1H0, 1P7E15.4)
      RETURN
      END
```

Figure 21.12. The subprogram for producing the output.

A	B	C	X1 REAL	X1 IMAG	X2 REAL	X2 IMAG
1.0000E 00	-2.0000E 00	1.0000E 00	1.0000E 00		1.0000E 00	
1.0000E 00	-1.0000E 01	2.5000E 01	5.0000E 00		5.0000E 00	
1.0000E 00	-3.0000E 00	2.0000E 00	2.0000E 00		1.0000E 00	
2.0000E 00	-6.0000E 00	4.0000E 00	2.0000E 00		1.0000E 00	
1.0000E 00	1.0000E 00	-2.5500E 03	5.0000E 01		-5.1000E 01	
1.0000E 01	-2.0000E 01	1.0000E 01	1.0000E 00		1.0000E 00	
1.0000E 03	-2.0000E 03	1.0000E 03	1.0000E 00		1.0000E 00	
2.0000E-02	-4.0000E-02	2.0000E-02	1.0000E 00		1.0000E 00	
1.4320E 00	9.8760E 00	-5.6780E 00	5.3364E-01		-7.4303E 00	
8.8130E 00	-1.3108E 01	0.	1.4873E 00		0.	
2.3009E 00	1.9917E 00	0.	-0.		-8.6562E-01	
1.0000E 00	0.	-1.0000E 00	1.0000E 00		-1.0000E 00	
1.0000E 00	0.	1.0000E 00		1.0000E 00		-1.0000E 00
9.0000E 00	0.	3.6000E 01		2.0000E 00		-2.0000E 00
1.0000E 00	2.0000E 00	5.0000E 00	-1.0000E 00	2.0000E 00	-1.0000E 00	-2.0000E 00
6.3190E 00	4.3380E 00	2.3294E 01	-3.4325E-01	1.8891E 00	-3.4325E-01	-1.8891E 00
-9.0000E 00	2.3000E 01	3.7000E 01	-1.1189E 00		3.6744E 00	
6.1000E 01	0.	8.7000E 01		1.1942E 00		-1.1942E 00
6.1000E 01	2.0000E 00	8.7000E 01	-1.6393E-02	1.1941E 00	-1.6393E-02	-1.1941E 00
6.1000E 01	1.5900E 02	8.7000E 01	-7.8145E-01		-1.8251E 00	

Figure 21.13. A page of output produced by the programs of Figures 21.10, 21.11, and 21.12.

tween main program and the subprograms were handled automatically, so far as we are concerned.

Figure 21.13 is a page of output produced by the combined program.

EXERCISES

*1. Write a FUNCTION subprogram to compute

$$Y(X) = \begin{cases} 1 + \sqrt{1 + X^2} & \text{if } X < 0 \\ 0 & \text{if } X = 0 \\ 1 - \sqrt{1 + X^2} & \text{if } X > 0 \end{cases}$$

Then write statements to evaluate the following expressions.

$$F = 2 + Y(A + Z)$$

$$G = \frac{Y[X(K)] + Y[X(K + 1)]}{2}$$

$$H = Y[\cos (2\pi X)] + \sqrt{1 + Y(2\pi X)}$$

2. Write a FUNCTION subprogram to compute

$$\text{RHO}(A, B, N) = \frac{A}{2\pi} \sum_{i=1}^{N} B_i$$

in which B is a one-dimensional array of 50 elements ($N \leq 50$).

Then use it to compute $1/2\pi$ times the sum of the first 18 elements of an array named A; call this SOME.

3. Rewrite the program for Exercise 2 so that N represents the number of elements in the array, which may then have adjustable dimensions.

4. A is any 20 x 20 array. Write a FUNCTION subprogram to compute

$$PD(A, I, J) =$$

$$\frac{A(I - 1, J) + A(I + 1, J) + A(I, J - 1) + A(I, J + 1)}{4}$$

Then use it to compute

$$B_{ij} = (1 - \alpha)B_{ij} + \alpha \frac{B_{i-1,j} + B_{i+1,j} + B_{i,j-1} + B_{i,j+1}}{4}$$

(Could a statement function be used here? Why not?)

*5. Write a FUNCTION subprogram for which the argument list contains A, M, and N, where A is an array name, and M and N are the numbers of rows and columns. The function value is to be the sum of the absolute values of all the elements. The dimensions are to be adjustable.

6. Devise a FUNCTION subprogram which could be called in either of the forms

AVER(ROW, L, ARRAY, M, N)
AVER(COLUMN, L, ARRAY, M, N)

The intent is to be able to ask for the algebraic sum of the elements in *row or column* L of an M x N adjustable array.

*7. A is a one-dimensional array with 50 elements. Write a SUBROUTINE subprogram to compute the average of the first N elements and a count of the number of these elements that are zero. Call the subprogram AVERNZ(A, N, AVER, NZ).

Then use the subprogram to get the average of the first 20 elements of an array named ZETA and place the average in ZMEAN and the count of zero elements in NZCNT.

*8. Write a SUBROUTINE subprogram that uses the FUNCTION of Exercise 6 to find the row of an M x N array that has the largest sum; the outputs are the row number and the sum.

Then use the subprogram to operate on an array named OMEGA which has 15 rows and 29 columns; place the largest sum in OMEGAL and the row number in NROW.

This combination of subprograms involves passing adjustable dimensions through a subprogram.

9. Given single variables A, B, X, and L, write a SUBROUTINE subprogram to compute R, S, and T from

$$R = \sqrt{A + BX + X^L}$$

$$S = \cos (2\pi X + A) \cdot e^{BX}$$

$$T = \left(\frac{A + BX}{2}\right)^{L+1} - \left(\frac{A - BX}{2}\right)^{L-1}$$

10. Identify the errors in the following:

a. COMMON A, B(2, 19), R(40),
b. COMMON //R, S, T/LABEL,/U, V, W
c. COMMON G, H, P, Q, Y, Z
   EQUIVALENCE (A, P, R), (B, H, S, Z)
d. COMMON A(12), B, C(14)
   DIMENSION D(9)
   EQUIVALENCE (A(2), D(8), G)
e. EQUIVALENCE (A(3), B(2), C(4)), (A(4), B(6), D(9))

11. Sketch the storage layout that would result from the following:

*a. DIMENSION A(3), B(4)
    EQUIVALENCE (A(2), B(1))
*b. DIMENSION C(2, 3), D(3, 2)
    EQUIVALENCE (C(1), D(3))
c. DIMENSION E(2, 2, 2), F(4)
   EQUIVALENCE (E(5), F(2))

12. Assume a computer in which an integer number and a real number take the same amount of storage space and in which a complex number and a double precision number each take twice as much storage space as a real or integer number. Further, when a complex or double precision number or array element is referenced in an EQUIVALENCE statement, it is always the first of the two storage locations that is meant. Sketch the storage layout that would result from the following:

*a. DIMENSION I(2), R(4), C(3)
   INTEGER I
   REAL R
   COMPLEX C
   EQUIVALENCE (I(1), C(1)), (R(1), C(2))
*b. DIMENSION D(4), I(5)
   DOUBLE PRECISION D
   REAL R1, R2
   INTEGER I
   EQUIVALENCE (D(1), R1), (D(2), R2),
   (D(3), I(2))
 c. DIMENSION C(5), D(3), S(1), T(1)
   COMPLEX C
   DOUBLE PRECISION D
   EQUIVALENCE (C(1), S(1)), (C(2), T(2)),
   (C(2), D(1))

 d. DIMENSION C(3, 3), D(2, 2, 2), R(20)
   COMPLEX C
   DOUBLE PRECISION D
   REAL R
   EQUIVALENCE (C(4), D(7), R(12))

13. Consider the following plan. We wish to make the diagonal elements of a 4 x 4 array A the same as the four elements of a one-dimensional array DIAG. It is proposed to do this with the statement

EQUIVALENCE (A(1), DIAG(1)), (A(6), DIAG(2)),
   (A(11), DIAG(3)), (A(16), DIAG (4))

This is quite impossible. Formulate a rule of which this would be a violation and explain why the rule is necessary.

CASE STUDY 22   NUMERICAL INTEGRATION OF AN ARBITRARY FUNCTION; THE **EXTERNAL** STATEMENT

In this case study we shall explore just one new programming feature, the EXTERNAL statement. The primary emphasis is on a moderately simple program that puts to practical use many of the ideas we saw in the preceding case studies.

The idea here is that we would like to be able to prepare a general-purpose function subprogram to do numerical integration—without knowing when we write the FUNC-TION what is being integrated; that is, one of the arguments of the FUNCTION will be the name of the function that we want integrated, along with the limits of integration. In a thoroughly practical routine we would make the convergence criterion an argument also, which is readily enough done. In an attempt to keep the program simple, we shall not include the feature here so that we can concentrate on the matters of current interest.

The method of numerical integration is very similar to that used in Case Study 20: Simpson's rule with successive halving of the interval until the difference between two approximations is less than some tolerance. The method is so similar to the earlier one, in fact, that we can dispense with a flowchart and proceed immediately to an investigation of the programming techniques used in the program of Figure 22.1.

We see that there is a main program and three subprograms. The main program calls the numerical integration routine three

times, to perform the following three integrations:

$$\text{SININT} = \int_0^\pi \sin t \, dt$$

$$\text{LOGF} = \int_1^{10} \frac{dx}{x} = \log_e 10$$

$$J_1(x) = \frac{1}{\pi} \int_0^\pi (\cos z - x \sin z) \, dz$$

Let us look first at the program for numerical integration, INTEG. We see that there are three arguments: the lower and upper limits, A and B, and the integrand function, named FUNC. This means that when we call INTEG we shall have to write, as the third parameter, the name of a function. In one case we shall use SIN, a supplied function, and in the other two the names of functions that we have written. The integration scheme is remarkably similar to the program of Figure 20.5, with the difference that the limits and the function are now function arguments and with the change that we now do not know that the functions of the limits will always add to zero or that the lower limit is zero.

The interval is set equal to half the difference between the upper and lower limits; we set ENDS equal to the sum of the functions of A and B, and FOUR equal to the value of the function at A + H; we set OLDINT equal

```
C CASE STUDY 22
C NUMERICAL INTEGRATION OF AN ARBITRARY FUNCTION - THE EXTERNAL STATEMENT
C
C THE MAIN PROGRAM
C
      EXTERNAL SIN, BESS, RECIPX
      REAL INTEG, LOGF
      COMMON X
C FIND THE INTEGRAL OF SIN(X) FROM ZERO TO PI
      SININT = INTEG(0.0, 3.14159265, SIN)
      WRITE (6, 29) SININT
   29 FORMAT (1H0, F15.6)
C FIND THE INTEGRAL OF 1/X FROM 1 TO 10, WHICH IS THE LOG OF 10
C RECIPX IS A FUNCTION TO FIND THE RECIPROCAL OF X
      LOGF = INTEG(1.0, 10.0, RECIPX)
      WRITE (6, 29) LOGF
C FIND THE INTEGRAL OF COS(Z - X*SIN(Z)) FROM ZERO TO PI, WHICH IS J1(X).
C WE PLACE X IN COMMON IN THE MAIN PROGRAM AND IN THE FUNCTION BESS
C TO ASSURE THAT X WILL BE THE SAME VARIABLE BOTH PLACES
      X = 2.5
      BESSEL = INTEG(0.0, 3.14159265, BESS) / 3.14159265
      WRITE (6, 29) BESSEL
      STOP
C THIS IS THE END OF THE MAIN PROGRAM
      END




C THIS IS THE FUNCTION SUBPROGRAM TO FIND A RECIPROCAL
      FUNCTION RECIPX(Q)
      RECIPX = 1.0/Q
      RETURN
      END




C THIS IS A FUNCTION SUBPROGRAM TO EVALUATE THE BESSEL INTEGRAND
      FUNCTION BESS(Z)
      COMMON X
      BESS = COS(Z - X*SIN(Z))
      RETURN
      END




C THIS IS THE FUNCTION SUBPROGRAM THAT INTEGRATES THE ARBITRARY FUNCTION
      REAL FUNCTION INTEG(A, B, FUNC)
C INITIALIZE THE INTEGRATION PROCESS
      H = (B-A)/2.0
      N = 1
      ENDS = FUNC(A) + FUNC(B)
      TWO = 0.0
      FOUR = FUNC(A+H)
      OLDINT = H / 3.0 * (ENDS + 4.0 * FOUR)
C EVALUATION LOOP
   25 H = H / 2.0
      N = 2*N
      TWO = TWO + FOUR
      FOUR = 0.0
      T = A + H
      DO 26 I = 1, N
      FOUR = FOUR + FUNC(T)
   26 T = T + H + H
      INTEG = H / 3.0 * (ENDS + 2.0 * TWO + 4.0 * FOUR)
C CHECK FOR CONVERGENCE OR EXCESSIVE NUMBER OF ITERATIONS
      IF (ABS(OLDINT-INTEG) .LT. 1.0E-5 .OR. N .GT. 10000) RETURN
      OLDINT = INTEG
      GO TO 25
      END
```

Figure 22.1. A program for numerical integration of an arbitrary function, including the main program, a FUNCTION for getting the reciprocal of a number, a FUNCTION for evaluating a Bessel integrand, and the FUNCTION for integration.

to the Simpson's rule approximation to the integral for the two-interval case. Now we proceed to the integration loop, which is very much as before. We halve the interval, double the number of intervals, set TWO equal to the old values of TWO and FOUR, and set FOUR equal to zero. A simple DO loop gets the sum of the values of the integrand at the new points. We compute the new approximation and set it equal to INTEG, the name of the function. If the process has converged, we exit immediately via a RETURN statement in the logical IF. Otherwise we set OLDINT equal to INTEG and go back for another iteration. Note that although the rule says that the name of a FUNCTION must appear at least once either on the lefthand side of an assignment statement or in the list of a READ statement, it is perfectly permissible also to use the name of the FUNCTION on the right-hand side of a statement.

The integration process will eventually either converge or the number of intervals will exceed 10,000; either way we return. Naturally in practice we would like to have a warning of the difference between the two cases; we shall in Case Study 23 see how the nonstandard RETURN makes this possible in a simple way.

Now let us consider the main program and how it calls this integration routine. The first statement in the program is the EXTERNAL statement. The problem is that we are going to be naming three routines that are named nowhere else in this program. Without a warning that something unusual is afoot, the compiler will flag the names of these routines as undefined. The purpose of the EXTERNAL statement is to say, in effect, "Hold off; I'll be telling you about these names later on." In the case of RECIPX and BESS the compiler and loading routine will figure out that the FUNCTION subprograms were what we were talking about and by inspecting the library of supplied functions figure out what we meant by SIN.

The integration of the sine function is a simple matter of naming the limits and specifying that the function to be integrated is named SIN. We print the result.

To integrate $1/x$ we need a FUNCTION that computes the reciprocal of an argument. The required function is named RECIPX; the body of the subprogram is surely just about as simple as can be imagined. Observe the sequence of events. We warned FORTRAN that RECIPX was the name either of a basic supplied function or the name of

something would be coming along elsewhere in our program. We called INTEG with RECIPX as an argument in the call, so that in the body of INTEG RECIPX is called many times as the integration process proceeds. When the integration is complete, control returns to the main program in which the result is printed.

The integration to get the Bessel function is a little different in an instructive sort of way. First, we need to divide the result of the call of INTEG by π; the statement illustrates that we are by no means limited to the use of INTEG is a simple assignment statement of the $a = b$ sort. Second, we have in this case a variable x that is involved in the function BESS without being an argument of any of the functions.

Let us try to make the situation clear. The main program calls INTEG and specifies that the function to be integrated is named BESS. INTEG will thus be calling BESS repeatedly, probably some hundreds of times. Each time the function BESS will need a value for x. INTEG has no interest in the fact that there is another variable it does not "know" about; all it cares about is that once a value has been given to T the call FUNC(T) should return a value. So, the main program must supply a value of x, which must be "known" to BESS, without there being any way to make x an argument—at least not within the framework of the structure we have established.

The reader who understood the preceding case study thoroughly will have already realized that the COMMON statement was designed for just this purpose, among others. We insert the statement

COMMON X

in the main program and in BESS, thus establishing the fact that X represents the same variable in both places. Now it is a simple matter to give X a value in the main program, for we know that that value will be the one used by BESS when it sees the name X.

Figure 22.2 shows the lines printed. The integral of the sine from zero to π is indeed 2, the natural

```
2.000000

2.302584

0.497094
```

Figure 22.2. The output of the program of Figure 22.1. The values are the integral of the sine from zero to π, the natural logarithm of 10, and $J_1 (2.5)$.

logarithm of 10 is in fact 2.302585, and we can turn back to Figure 20.6 for a confirmation that $J_1(2.5)$ is in fact 0.497094.

We should try to understand the significance of the technique demonstrated here.

Let us suppose that the numerical integration routine has been fleshed out to make it a complete, accurate, and rapid scheme, in wide demand at an installation. Now, what is required to make use of it? Answer: write a tiny main program to call it and provide a FUNCTION subprogram to compute the value of the integrand. If the integrand involves variables other than the variable of integration, such variables must be named in COMMON statements in the main program and in the FUNCTION. The main program must name the FUNCTION in an external statement. With these simple requirements readily met, any programmer can make immediate use of the integration routine.

CASE STUDY 23 FINDING THE ROOT OF A FUNCTION; USE OF PREPROGRAMMED SUBROUTINES; THE **ENTRY** STATEMENT; NONSTANDARD RETURNS

In Case Study **22** we saw a subprogram that could have wide applicability, if suitably extended and broadened, and might then be attractive to a large number of programmers in an installation. This is a typical circumstance. We have, of course, been using such programs since we started writing SQRT to call into operation a square root routine written by someone else. It is generally the case that effort and error can be saved by judicious use of programs written by others.

In fact, in a well-run computer installation there will be a concerted effort to make such borrowing effective. There will be, usually, a library of perhaps a hundred or more quite standard programs to do such things as matrix inversion and solution of certain standard differential equations. Beyond this, there is available in the organization of users of the particular computer a very much larger choice of routines written for specialized problems. These routines are submitted by the author along with a brief write-up that describes the problem and program, as well as the method of solution, accuracy, and such essentials as how to call the program and in what form it is available.

In this case study we shall utilize a preprogrammed routine for computing Bessel functions of all types and orders in double precision. Writing such a routine is an undertaking that requires at least some weeks, if not months, and should obviously not have to be done by every programmer who needs the capability represented by the routine.

There are, in fact, quite a number of Bessel function routines. The one we shall investigate is entirely suitable for our purposes, especially in that it illustrates two new programming techniques, but no claim is made that it is necessarily the best of all those available; no such evaluation was attempted. The program was written by Stephen Richter of the Electrical Engineering Department of Columbia University, based on earlier work by Professor Paul Diament and with contributions by P. Korn, J. Rosenbaum, and B. Appelman. The work was supported by Air Force and National Science Foundation grants.

The program computes J_n, I_n, Y_n, or K_n, depending on how it is called. With appropriate calls it is also possible to request a combination of functions to be computed. All functions are computed with an accuracy of about one part in 10^{12} and somewhat better in many cases. If the programmer inadvertently attempts to compute a nonexistent function, such as $Y_n(x)$ for n negative, the routine returns to an error location in the calling program by a technique that we shall look into shortly.

146

We begin consideration of the use of this program by inquiring how it is called. The program is a subroutine and a CALL statement is required to bring it into operation. The general form of the call is as follows:

CALL JBES(KORD, ARGBS, XBES, $N)

where KORD is the order of the Bessel function, ARGBS is the double-precision argument, XBES is the double-precision function value returned by the program, and N is a statement number in the calling program. Calling the program in the form shown results in the computation of the function

$$J_{KORD} (ARGBS)$$

This value is assigned to XBES.

To compute the other Bessel functions we make use of the alternative entry points into the program. For instance, in our case we shall want $Y_1(X)$; to get it we write

CALL YBES(1, XM, FXM, $500)

The name YBES specifies the function Y, the 1 names Y_1, the XM says that we want Y_1 computed for the value XM, the FXM is the name of the variable that receives the result, and 500 is the statement number of a WRITE statement in which we print a comment about the error condition that would develop if by error we somehow gave a negative value to KORD.

The calls for computing I_n and K_n are similar, the names being IBES and KBES.

The listing of the source program for JBES appears in Figure 23.1, in which we may see how the error return and the various entries are handled. Looking near the end of the program we see the statements

ENTRY IBES(KORD, ARGBS, XBES, *)

and

ENTRY YBES(KORD, ARGBS, XBES, *)

These inform the FORTRAN compiler that the routine, even though its "official" name is JBES, may also be called by the names of IBES and YBES; these names are then placed in a dictionary that handles the cross references to object programs as the time of their loading. Thus in our case we shall write only CALL YBES, without ever mentioning the "official" name; the compiler and loader

will know that we mean JBES because the ENTRY YBES statement established the correspondence. The ENTRY statement has a second function: to direct that when the routine is called by the alternative name the point of entry is the statement after the corresponding ENTRY. This routine is set up so that after a few preliminary statements, which, in effect, determine the function, it is possible to go to a standard portion of the routine for the rest of the work.

Looking down a few lines from the beginning of the program we see an entry for JBESM, which we note has another argument, named XBESM. When JBESM is called, it places $J_{KORD}(ARGBS)$ in XBES and $J_{KORD-1}(ARGBS)$ in XBESM. Thus it is permissible to have a different number of arguments in the argument list for an ENTRY than for the "primary" call. Furthermore, another entry may specify a new argument that is an array name. For instance near the beginning of the subroutine there is an ENTRY for YBESM, in which BKY must be an array to hold five outputs: Y_{KORD} (ARGBS), $Y_{KORD-1}(ARGBS)$, $Y_{KORD-2}(ARGBS)$, $J_1(ARGBS)$, and $J_0(ARGBS)$. Elsewhere there is an entry for KBESM which produces a similar list of K and I functions.

This subroutine also illustrates the use of nonstandard returns, in this case to signal an error in the use of the subroutine such as calling for Y_n with n negative.

When we wish to write a SUBROUTINE or FUNCTION to provide for nonstandard returns, we place as many asterisks in the argument list in the definition as we need; here, of course, there is only one, but in general there may be any number. Then, if an *ordinary* RETURN statement is encountered, the return is to the statement following the CALL. (In the case of a FUNCTION control returns to the object-program expression evaluation.) If the statement RETURN 1 is encountered, control returns instead to the *first* statement specified in the argument list; if a RETURN 2 is encountered, control returns to the *second* statement specified in the argument list, and so on.

Observe that in the subprogram *definition* we use an asterisk to designate an argument that *in the call* will be a statement number preceded by a dollar sign.

Figure 23.2 is the main program that calls for this subprogram, it uses it to find the root of $Y_1(x)$ which lies between 5 and 6. We begin with the knowledge, derived from tables, that there is exactly

```
      SUBROUTINE JBES(KORD,ARGBS,XBES,*)
      LOGICAL II
      II=.TRUE.
      GO TO 8881
      ENTRY YBESM(KORD,ARGBS,BKY,*)
      IF(KORD.LT.1)GO TO 8899
      II=.FALSE.
      GO TO 8866
      ENTRY JBESM(KORD,ARGBS,XBES,XBESM,*)
      DOUBLE PRECISION       BKIND,BSOR,BSCT,RRRRR,BSKIX,BKINE,BSJK,
     1 BTEST,BSGAM,BSKOX,  BSORD,BSJKK,BKINZ,BSKNX,BKOPD,ARKBS,BSKU,
     2 BSJIX,BSKMX,BSYOX,BSJNX,PASS,         BSYNX,BSYIX,ARGBS,BSJN,
     3BSJN,BSST,BSJOX,BSIOX,XBES,XBESM ,BKY(5)
      II=.FALSE.
      BSJKK=1.
 1000 FORMAT( 12H BES ERROR,=,F10.6)
 8881 BKIND=-1.0
      IF(ARGBS-37.)8884,8884,8805
 8884 BKINE=0.0
      IF(KORD.LT.0)GO TO 8899
      BSORD=KORD
 8885 ARKBS=ARGBS*0.5
      BSJNX=C.0
      BSJN=1.0
      IF(BSORD)8899,8887,8886
 8886 BSOR=BSORD
 8888 BSJN=BSJN*ARKBS/BSOR
      BSOR=BSOR-1.0
      IF(BSOR)8898,8887,8888
 8887 ARKBS=ARKBS*ARKBS*BKIND
      BSJK=1.0
 8890 BSJNX=BSJNX&BSJN
      BSJN=BSJN*ARKBS/(BSJK*(BSJK&BSORD))
      BSJK=BSJK&1.0
      IF((BSJNX&BSJN)-BSJNX)8890,8891,8890
 8891 IF(BKINE.NE.0.)GO TO 8869
      IF(II)GO TO 1C2
      IF(BSJKK.GT.0.)GO TO 100
      XBESM=-BSJNX*2.*(BSJKK&.5)
      RETURN
  100 XBES=BSJNX
      IF(KORD.GT.0)GO TO 101
      BSORD=1.
      BSJKK=0.
      GO TO 8885
  101 BSORD=KORC-1
      BSJKK=-1.
      GO TO 8885
  102 XBES=BSJNX
      RETURN
 8805 BKINE=0.
      BSORD=0.
      IF(KORC.EQ.1.AND.II)BSORD=1.
 8801 BSJK=1.
      BSJKK=1.
```

Figure 23.1. A subroutine to compute Bessel functions in double precision, illustrating the ENTRY statement and nonstandard RETURNs.

```
          BSKU=1.
          PASS=4.*BSORD*BSORD
          BSST=DSIN(ARGBS-(BSORD&.5)*1.5707963267948966)
          BSCT=DCOS(ARGBS-(BSORD&.5)*1.5707963267948966)
          BSJN=BSCT
8809      BSKU=BSKU*BSJKK*(PASS-(2.*BSJK-1.)*(2.*BSJK-1.))/(8.*ARGBS*BSJK)
          IF(BSJKK)8803,8898,88C2
8802      BTEST=BSJN-BSST*BSKU
          BSJKK=-1.
          GO TO 8808
8803      BTEST=BSJN&BSCT*BSKU
          BSJKK=1.
8808      IF(BTEST-(BTEST-BSKU))8804,88C0,8804
8804      BKINZ=2.*ARGBS-.5-BSJK
          BSJN=BTEST
          BSJK=BSJK&1.
          IF(BKINZ)8800,88C9,88C9
8800      BSKU=BSKU*BSJKK*(PASS-(2.*BSJK-1.)*(2.*BSJK-1.))/(16.*ARGBS*BSJK)
          IF(BSJKK)8810,8898,8811
8810      BSJN=BSJN-BSST*BSKU
          GO TO 8812
8811      BSJN=BSJN&BSCT*BSKU
8812      BSJNX=.79788456C8C286535          *BSJN/DSQRT(ARGBS)
          IF(BKINE.NE.0.)GO TO 8869
          IF(KORD.LE.1.AND.II)GO TO 102
8814      IF(BSORD)8898,8815,11C5
8815      BSJIX=BSJNX
          BSORD=1.
           GO TO 8801
1105      IF(KORD-1)104,103,8816
 104      XBES=BSJIX
          XBESM=-BSJNX
          RETURN
8816      DO 8817 KORDB=2,KORD
          BSORD=(KORDB-1)*2
          BSJOX=BSJIX
          BSJIX=BSJNX
8817      BSJNX=-BSJOX&BSORD*BSJIX/ARGBS
 103      XBES=BSJNX
          XBESM=BSJIX
          RETURN
8898      RRRRR=-2.22227
          GO TO 8880
8899      RRRRR=-2.22226
8880      WRITE(6,1000)RRRRR
          RETURN 1
8869      IF(BKINE)8897,8898,8892
8892      IF(BSORD)8898,8895,8896
8866      BKIND=-1.0
          IF(ARGBS-3C.)8893,87CC,87C0
8893      BKINE=1.0
8819      BSORD=0.0
          IF(BKIND)8820,8898,8885
8820      IF(ARGBS-37.)8885,8885,88C1
8895      BSGAM=0.5772156649C153286
          BSKNX=-(((DLOG(ARGBS*0.5))*BSJNX)&BSGAM)
```

Figure 23.1. (Continued)

149

```
            BSJK=2.0
            BSJN=ARKBS
            BSKU=1.0-BSGAM
 8874       BSKMX=BSKNX
            BSKNX=BSKMX&BSJN*BSKU
            IF(BSKNX-BSKMX)8875,8872,8875
 8875       BSJN=BSJN*ARKBS/(BSJK*BSJK)
            BSKU=BSKU&1.0/BSJK
            BSJK=BSJK&1.0
            GO TO 8874
 8872       IF(KORD)8898, 105,8897
  105       IF(BK IND)106,8898,107
  106       XBES=-BSKNX/1.57C7963267948966
            RETURN
  107       XBES=BSKNX
            RETURN
            ENTRY KBES(KORD,ARGBS,XBES,*)
            IF(KORD.LT.0)GO TO 8899
            II=.TRUE.
            GO TO 8883
            ENTRY KBESM(KORD,ARGBS,BKY,*)
            IF(KORD.LT.1)GO TO 8899
            II=.FALSE.
 8883       BKIND=1.0
            IF(ARGBS-12.)8893,8894,8894
 8894       BSKNX=1.2533141373155002          *DEXP(-(ARGBS))/DSQRT(ARGBS)
            BSJN=1.0
            BSKU=1.0
            BSJK=1.0
 8876       BSKU=-BSKU*(2.0*BSJK-1.0)*(2.0*BSJK-1.0)/(8.0*ARGBS*BSJK)
            BTEST=BSJN&BSKU
            IF(BTEST-BSJN)8879,8878,8879
 8879       BK INE=2.0*ARGBS-0.5-BSJK
            IF(BK INE)8878,8877,8877
 8877       BSJN=BTEST
            BSJK=BSJK&1.C
            GO TO 8876
 8700       BSST=DSIN(ARGBS-.785398163395744830)
            BSCT=DCOS(ARGBS-.785398163395744830)
            BSJKK=1.
            BSJN=BSST
            BSKU=1.
            BSJK=1.
 8701       BSKU=-BSJKK*BSKU*(2.*BSJK-1.)*(2.*BSJK-1.)/(8.*ARGBS*BSJK)
            IF(BSJKK)87C2,8898,87C3
 8703       BTEST=BSJN&BSKU*BSCT
            BSJKK=-1.0
            GO TO 8704
 8702       BTEST=BSJN&BSKU*BSST
            BSJKK=1.
 8704       IF(BTEST-(BTEST-BSKU))8705,87C8,8705
 8705       BK INE=2.*ARGBS-.5-BSJK
            IF(BK INE)87C8,87C6,87C6
 8706       BSJN=BTEST
            BSJK=BSJK&1.0
            GO TO 8701
```

Figure 23.1. (Continued)

```
87.08  BSJK=BSJK&1.
       BSKU=-.5*BSJKK*BSKU*(2.*BSJK-1.)*(2.*BSJK-1.)/(8.*ARGBS*BSJK)
       IF(BSJKK)8709,8898,8710
8709   BTEST=BTEST&BSKU*BSST
       GO TO 8711
8710   BTEST=BTEST&BSKU*BSCT
8711   BSYNX=.79788456C8C286535          *BTEST/DSQRT(ARGBS)
       IF(KORD)8899,1C8,8712
 108   XBES=BSYNX
       RETURN
8878   BKINE =(BKINE-0.1)*0.037148C.5
       BSJK=BSJK&1.0
       BSKU=-BKINE*BSKU*(2.C*BSJK-1.C)*(2.0*BSJK-1.0)/(8.0*ARGBS*BSJK)
       BTEST=BTEST&BSKU
       BSKNX=BSKNX*BTEST
       IF(KORD)8899,1C7,8873
8712   BSKNX=-BSYNX*1.57C7963267948966
8873   BKINE=-1.0
       GO TO 8819
8897   BSORD=1.0
       BKINE=1.0
       BSIOX=BSJNX
       IF(BKIND)8820,8898,8885
8896   BSKIX=BSKNX
       BSKNX=(1.C/ARGBS-BSJNX*BSKIX*BKIND)/BSICX
       IF(KORD-1)8898,8868,8871
8871   DO 8870 KORDB=2,KORD
       BSKOX=BSKIX
       BSKIX=BSKNX
       BKORD=KORDB-1
8870   BSKNX=BSKOX*BKIND&BKORD*BSKIX*2.0/ARGBS
8868   IF(II)GO TO 1C5
       BKY(4)=BSJNX
       BKY(5)=BSIOX
       IF(BKIND)8867,8898,69C2
8867   BKY(1)
      1      =-BSKNX/1.57C7963267948966
       IF(KORD-1)8898,8864,8865
8865   BKY(3)
      1      =-BSKOX/1.5707963267948966
8864   BKY(2)
      1      =-BSKIX/1.5707963267948966
       RETURN
6902   BKY(1)=BSKNX
       BKY(2)=BSKIX
       BKY(3)=BSKOX
       RETURN
       ENTRY IBES(KORD,ARGBS,XBES,*)
       II=.TRUE.
       BKIND=1.C
       GO TO 8884
       ENTRY YBES(KORD,ARGBS,XBES,*)
       IF(KORD.LT.C)GO TO 8899
       II=.TRUE.
       GO TO 8866
       END
```

Figure 23.1.   (Continued)

```
C CASE STUDY 23
C FINDING THE ROOT OF Y1(X) BETWEEN 5 AND 6 BY METHOD OF INTERVAL HALVING
      DOUBLE PRECISION XL, XH, XM, FXL, FXM
C SET ITERATION COUNTER
      N = 1
C SET UPPER AND LOWER LIMITS
      XL = 5.0
      XH = 6.0
C GET FUNCTION OF STARTING LOWER LIMIT ALTHOUGH ALL WE REALLY WANT
C IS THE SIGN
      CALL YBES(1, 5.0, FXL, $500)
C ENTER HALVING LOOP
   11 XM = (XL & XH) / 2.0
C GET FUNCTION (SIGN) OF MIDPOINT
      CALL YBES(1, XM, FXM, $500)
C CHECK SIGN OF FUNCTION OF MIDPOINT TO SEE WHETHER THE ROOT LIES
C ABOVE OR BELOW XM
      IF (FXL * FXM .GE. 0.0D0) GO TO 12
C ROOT LIES BELOW XM
      XH = XM
      GO TO 13
C ROOT LIES ABOVE XM
   12 XL = XM
C CHECK CONVERGENCE
   13 IF (XH - XL .LT. 1.0D-13) GO TO 14
C INCREMENT AND CHECK ITERATION COUNTER
      N = N & 1
      IF (N .GT. 50) GO TO 16
C GO BACK FOR ANOTHER ITERATION
      GO TO 11
   16 WRITE (6, 20)
   20 FORMAT (1H0, 34HFAILS TO CONVERGE IN 50 ITERATIONS)
      STOP
   14 WRITE (6, 21) XM
   21 FORMAT (1H0, 39HTHE ROOT OF Y1(X) BETWEEN 5 AND 6 IS..., 1PD20.12)
      STOP
C THIS IS THE ERROR RETURN LOCATION FOR THE SUBROUTINE
  500 WRITE (6, 22)
   22 FORMAT (1H0, 25HSOMETHING AMISS FROM YBES)
      STOP
      END
```

Figure 23.2. A program to find the root of Y_1 between 5 and 6, using the subroutine in Figure 23.1.

one root, no more and no less, in this interval. That means that the sign of $J_1(5)$ will be opposite from the sign of $J_1(6)$. After initializing an iteration counter we set XL ("X lower") equal to 5 and XH ("X higher") equal to 6. Then a CALL statement gets us the value of $Y_1(5)$; we are not really interested in the *value* but in the sign. Now we enter the loop, in which the first action is to compute the location of the midpoint between XL and XH, calling it XM, and get the sign of the function at this point. Now, if FXL and FXM have different signs, which we can determine most easily by looking at the sign of their product, the root lies between XL and XM;

otherwise it lies between XM and XH. We replace either XH or XL with XM, thus halving the interval within which the root must lie. After N iterations of such a process, the interval will be 2^{-N} times what it was to begin with. In the maximum of 50 iterations that we permit, and starting with an interval of one, we could get the root to within about 10^{-15}, assuming the Bessel function routine can produce such accuracy. The convergence criterion that has been used is a somewhat less stringent 10^{-13}.

Three outcomes are possible. The process may for unforeseen reasons fail to converge in 50 iterations

```
THE ROOT OF Y1(X) BETWEEN 5 AND 6 IS...  5.429681040794D 00
```

Figure 23.3. The output of the programs in Figures 23.1 and 23.2.

or the error return from JBES may be taken because of mispunching the program; we may also get a root. There are three paths of program execution at the end which correspond to these three possibilities.

The output is shown in Figure 23.3. This result agrees with published values as far as eight digits, which is the maximum number of digits to which the root can be found in any publications known to the author.

CASE STUDY 24 SOLUTION OF SYSTEMS OF COMPLEX SIMULTANEOUS EQUATIONS BY GAUSS ELIMINATION

We have already examined one method for solving systems of simultaneous equations, the Gauss-Seidel iteration method, in Case Studies 16 and 19, but as we noted when that method was first presented* it applies only when the diagonal term in each row dominates the other terms. Many systems of interest can be guaranteed from their formulation to meet this criterion, but others do not meet it. For these cases, and for other reasons, we turn now to a method that will find a solution, if one exists, for any square system of equations. The method will become the foundation for a related program to find the inverse of a matrix.

The program to be presented is set up to solve systems of equations with complex coefficients and constant terms and for which the solutions are in general also complex. The method applies equally to real or complex cases, with only trivial changes in the program needed to reflect the differences. As a further slight specialization, the program that we shall see was first written to apply to symmetric systems for which it is possible to simplify the preparation of the data slightly. This is no restriction whatever on the generality of the method but merely the kind of thing that is commonly done to lessen the effort and chance of error in preparing data decks.

* Actually, as noted earlier, slightly less stringent conditions are sufficient.

We can illustrate the method of Gauss elimination on a system of three equations in three unknowns.

$$a_{11}x_1 + a_{12}x_2 + a_{13}x_3 = b_1$$
$$a_{21}x_1 + a_{22}x_2 + a_{23}x_3 = b_2$$
$$a_{31}x_1 + a_{32}x_2 + a_{33}x_3 = b_3$$

At least one of the coefficients a_{11}, a_{21}, and a_{31} is not zero or we have a system of three equations with only two unknowns, contrary to assumption. If a_{11} is zero, we rearrange the equations so that the coefficient of x_1 in the first equation is not zero. Interchanging two rows in the system of equations leaves the system essentially unchanged in that it still has the same solution.

Now define a multiplier

$$m_2 = \frac{a_{21}}{a_{11}}$$

We multiply the first equation by m_2 and subtract from the second equation. ("First" and "second" refer to the equations *as rearranged*, if it was necessary.) The result is

$$(a_{21} - m_2 a_{11})x_1 + (a_{22} - m_2 a_{12})x_2 +$$
$$(a_{23} - m_2 a_{13})x_3 = b_2 - m_2 b_1$$

but

$$a_{21} - m_2 a_{11} = a_{21} - \frac{a_{21}}{a_{11}} a_{11} = 0$$

and x_1 has been eliminated from the second equation. This result is, of course, the reason for the choice of m_2. If we now define

$$a_{22}' = a_{22} - m_2 a_{12}$$

$$a_{23}' = a_{23} - m_2 a_{13}$$

$$b_2' = b_2 - m_2 b_1$$

the second equation becomes

$$a_{22}'x_2 + a_{23}'x_3 = b_2'$$

Similarly, we define a multiplier for the third equation:

$$m_3 = \frac{a_{31}}{a_{11}}$$

We multiply the first equation by this multiplier and subtract from the third. Again the vanishing coefficient of x_1 leaves a modified third equation

$$a_{32}'x_2 + a_{33}'x_3 = b_3'$$

where

$$a_{32}' = a_{32} - m_3 a_{12}$$

$$a_{33}' = a_{33} - m_3 a_{13}$$

$$b_3' = b_3 - m_3 b_1$$

Consolidating what we have done, we now have

$$a_{11}x_1 + a_{12}x_2 + a_{13}x_3 = b_1$$

$$a_{22}'x_2 + a_{23}'x_3 = b_2'$$

$$a_{32}'x_2 + a_{33}'x_3 = b_3'$$

These equations, which are completely equivalent to the original system, have the added advantage that x_1 appears only in the first of them. The last two are two equations in two unknowns; if we can solve them for x_2 and x_3, the results can be substituted into the first equation to get x_1. The problem therefore has been reduced from that of solving three equations in three unknowns to that of solving two equations in two unknowns.

We can now proceed to eliminate x_2 from one of the last two equations. Again, if $a_{22}' = 0$, we interchange the last two equations. (If it should happen that $a_{22}' = 0$ *and* $a_{32}' = 0$, the equations are *singular*, that is, they have either no solutions or an infinite number of solutions.)

We define a new multiplier

$$m_3' = \frac{a_{32}'}{a_{33}'}$$

We multiply the (modified) second equation by m_3' and subtract from the (modified) third equation. The result, once again, is that the coefficient of an unknown is eliminated. The third equation becomes

$$a_{33}'' = b_3''$$

where we have let

$$a_{33}'' = a_{33}' - m_3' a_{23}'$$

$$b_3'' = b_3' - m_3' b_2'$$

The original system of equations has now been reduced to a much simpler form:

$$a_{11}x_1 + a_{12}x_2 + a_{13}x_3 = b_1$$

$$a_{22}'x_2 + a_{23}'x_3 = b_2'$$

$$a_{33}''x_3 = b_3''$$

The object of the Gauss elimination method is to arrive precisely at this form, which is called triangular from its appearance.

It is now a straightforward process to solve the last equation for x_3, substitute that result into the second equation to get x_2, and finally substitute both values into the first equation to get x_1. This process is called back substitution.

The generalization of the Gauss elimination procedure to the case of n equations in n unknown is fairly obvious and not worth our while to express in symbolic notation. The method is readily described in a flowchart, as we show in Figure 24.1. When reading this flowchart, it may help to note the meanings assigned to the subscripts i, j, and k:

k refers to the number of the equation being subtracted from other equations; it is also the number of the unknown being eliminated from the last $n - k$ equations.

i refers to the number of the equation from which an unknown is currently being eliminated.

j refers to the number of a column.

The flowchart, with two exceptions, is quite close to a generalization of the procedure as described above. The box marked *, which contains "arrange rows so that $a_{kk} \neq 0$", refers to a process which we shall describe shortly and which improves the accuracy of the results. It turns out that the roundoff errors in the values of the unknowns can be significantly reduced by a judicious choice of rows to interchange.

A second difference between the description and the flowchart is that we have used one symbol m to stand for all the multipliers, for we shall never need more than one of them at a time.

The flowchart of the back substitution is seen in Figure 24.2 to be relatively straightforward. It turns

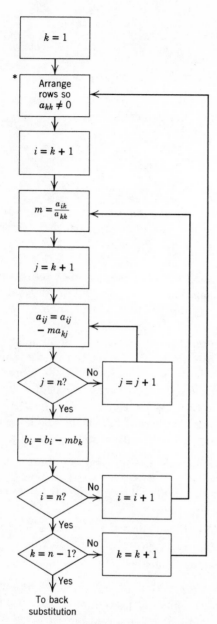

Figure 24.1. Flowchart of the method of Gauss elimination in solving simultaneous equations.

out to be simpler to diagram and program the method if we compute x_n in a separate step at the beginning. It would be possible to draw a more compact flowchart without this separate step, but it would complicate the testing pointlessly in the formation of the sum of the terms after the main diagonal term: the first "sum" would be zero, which would force us to test before accumulating and thereby prevent use of the DO statement. Notice

that although in Figure 24.1 all subscripts *increased* here one of them (i) *decreases*.

Now we must return to the question of the choice of rows to interchange. We recall that it was necessary to ensure that $a_{kk} \neq 0$ to avoid division by zero. Actually, we shall ordinarily wish to interchange rows even when a main diagonal element is already nonzero. The error analysis techniques are beyond the scope of this book in the details, but we can readily enough sketch the method.*

The basic operation in Gauss elimination is the subtraction of a multiple of one row from another.

* The full presentation is available in most works on numerical methods. See, for instance, *Numerical Methods and FORTRAN Programming,* Daniel D. McCracken and William S. Dorn, Wiley, 1964, pp. 238–240.

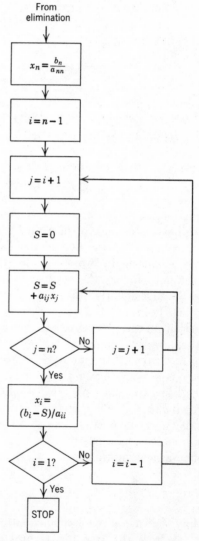

Figure 24.2. Flowchart of the method of back substitution in solving simultaneous equations.

The objective is to reduce the coefficient of one variable to zero; this operation is usually not actually carried out, for nothing would ever need to be done with the zero result. Accordingly we have no concern about accuracy there. For the other coefficients in a row we should like, for reasons that come out of the error analysis, to be subtracting a product as small as possible. This means that the multiplier m should also be small, which in turn means that the denominator a_{kk} should be large. In other words, we should interchange rows so that the largest element on the main diagonal is equal to or larger than all the elements below it.

"Largest" means *in absolute value*, and that statement is true for both real and complex coefficients when by the absolute value of a complex number we mean the positive square root of the sum of the squares of its real and imaginary parts. As we have seen, this is what is done with the FORTRAN function CABS.

A flowchart of the testing and (if necessary) interchanging is shown in Figure 24.3. It should be thought of as replacing the box marked * in Figure 24.1. In studying Figure 24.3, recall from Figure 24.1 that as we enter this phase k has just been given some value. We begin by setting an auxiliary subscript l, equal to k. The first comparison is then between a_{ik}, the element just below the main diagonal term a_{kk}, and a_{lk}, which is a_{kk}. If a_{ik} is found to be larger in absolute value, we set $l = i$. The subscript l therefore always represents the row number of the element in the kth column, the largest one tested so far. The subscript i runs through all values from $k + 1$ through n, inclusive. Thus at the end of the loop l identifies the largest element, the one we want for a_{kk} after interchanging.

Of course, the original a_{kk} might already have been the largest. We therefore immediately test for this possibility and omit the interchanging. The interchanging is done on pairs of values, one from row k and one from row l, whatever l may be. The interchange of each pair requires a three-step process. This operation must be done on all pairs from the main diagonal to the right, which is carried out by a loop using j as a subscript. Finally, the two constant terms are interchanged and the process is complete.

The program shown in Figure 24.4 begins with some fairly standard preliminaries. We read a card containing the line that should be printed as a heading and print it. We read N, an object-time format for the coefficients, the coefficients them-

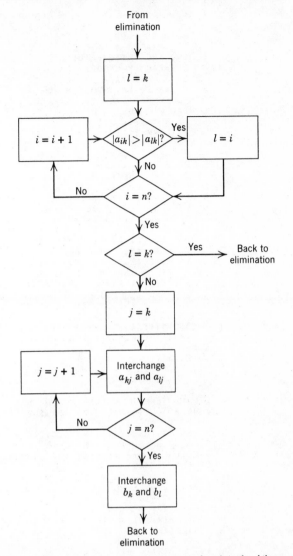

Figure 24.3. Flowchart of the method of choosing the kth row and performing a row interchange, if necessary, in Gauss elimination. All of this flowchart replaces the box marked * in Figure 24.1.

selves, a format for the constants, the constants themselves, and a format for writing the system. We print out the system of equations, and are now ready to solve it.

The program follows the flowchart so closely that we should need to make only a few comments. Observe that in the flowchart of Figure 24.1 the top box $k = 1$, the bottom box $k = n - 1$?, and the incrementing $k = k + 1$ are all implemented by the statement

$$DO\ 610\ K = 1, NM1$$

where NM1 is N − 1. Similarly, the second from the top and second from the bottom boxes are imple-

```
C CASE STUDY 24
C SOLUTION OF SYSTEMS OF COMPLEX SIMULTANEOUS EQUATIONS
C
      REAL FORMA(12), FORMB(12), HEAD(12), FORMX(12)
      COMPLEX A(20, 20), B(20), X(20), TEMP, FACTOR, SUM
C READ AND WRITE HEADING LINE FOR OUTPUT
   20 READ (5, 9902) HEAD
 9902 FORMAT (12A6)
      WRITE (6, 9903) HEAD
 9903 FORMAT (1H1, 12A6/////)
C READ N = NUMBER OF EQUATIONS
      READ (5, 21) N
   21 FORMAT (I2)
C READ OBJECT-TIME FORMAT GIVING ARRANGEMENT OF COEFFICIENTS
      READ (5, 22) FORMA
   22 FORMAT (12A6)
   24 READ (5, FORMA) I, J, A(I , J)
C THIS PROGRAM WAS DESIGNED FOR AN APPLICATION IN WHICH THE MATRICES
C ARE ALWAYS SYMMETRIC, SO ONLY THE TERMS ON AND ABOVE THE MAIN DIAGONAL
C ARE REQUIRED TO BE PUNCHED
      A(J, I) = A(I, J)
      IF (I .NE. N) GO TO 24
C READ OBJECT-TIME FORMAT GIVING ARRANGEMENT OF CONSTANTS
   23 READ (5, 22) FORMB
   29 READ (5, FORMB) I, B(I)
      IF (I .NE. N) GO TO 29
C READ OBJECT-TIME FORMAT GIVING DESIRED ARRANGEMENT OF OUTPUT
      READ (5, 22) FORMX
C WRITE OUT THE SYSTEM OF EQUATIONS
      DO 4055 I = 1, N
 4055 WRITE (6, FORMX) (A(I, L), L = 1, N), B(I)
C START THE PIVOTAL CONDENSATION
C K NAMES THE PIVOTAL ROW
   25 NM1 = N - 1
      DO 610 K = 1, NM1
      KP1 = K + 1
      L = K
C FIND TERM IN COLUMN K, ON OR BELOW MAIN DIAGONAL, THAT IS LARGEST
C IN ABSOLUTE VALUE. AFTER THE SEARCH, L IS THE ROW NUMBER OF THE ELEMENT.
      DO 400 I = KP1, N
  400 IF (CABS(A(I, K)) .GT. CABS(A(L, K))) L = I
C CHECK WHETHER L = K, WHICH MEANS THAT THE LARGEST ELEMENT IN
C COLUMN K WAS ALREADY THE DIAGONAL TERM, MAKING ROW INTERCHANGE
C UNNECESSARY
      IF (L .EQ. K) GO TO 500
C INTERCHANGE ROWS L AND K, FROM DIAGONAL RIGHT
      DO 410 J = K, N
      TEMP = A(K, J)
      A(K, J) = A(L, J)
  410 A(L, J) = TEMP
      TEMP = B(K)
      B(K) = B(L)
      B(L) = TEMP
C ELIMINATE ALL ELEMENTS IN COLUMN K BELOW MAIN DIAGONAL
  500 DO 610 I = KP1, N
      FACTOR = A(I, K) / A(K, K)
      DO 600 J = KP1, N
  600 A(I, J) = A(I, J) -FACTOR * A(K, J)
  610 B(I) = B(I) - FACTOR * B(K)
C BACK SOLUTION
      X(N) = B(N) / A(N, N)
      I = NM1
  710 IP1 = I + 1
      SUM = (0.0, 0.0)
      DO 700 J = IP1, N
  700 SUM = SUM + A(I, J) * X(J)
      X(I) = (B(I) - SUM) / A(I, I)
      I = I - 1
      IF (I .GE. 1) GO TO 710
C FIND ABSOLUTE VALUE AND PHASE OF SOLUTIONS
      DO 901 I = 1, N
      U = CABS(X(I))
      P = 180.0 / 3.14159265 * ATAN2(AIMAG(X(I)), REAL(X(I)))
C WRITE OUT THE SOLUTIONS
  901 WRITE (6, 900) I, X(I), U, P
  900 FORMAT (1H0, I5, 4F10.4)
      GO TO 20
      END
```

Figure 24.4. Program to solve a system of simultaneous equations by Gauss elimination.

mented with the statement at 500,

$$DO\ 610\ I = KP1, N$$

where KP1 = K + 1. Statement 610 in the program is

$$B(I) = B(I) - FACTOR * B(K)$$

which is the end of the range of both DO statements, and gives us an instructive example of how flowchart notation sometimes condenses in writing a program. Some people, in fact, noting the frequent occurrence of such situations, use some special flow-

charting notation to represent the actions carried out by the DO.

The back solution is a relatively straightforward matter. One minor complication is introduced by the requirement that the range of a DO statement must be executed at least once, or, stated in terms of our problem, it is not legal to write

$$DO\ 700\ J = IP1, N$$

where IP1 has a value *greater* than N, and expect the DO range to be skipped entirely. The specifications of the DO statement make it clear that the

REAL TEST CASE

2.0000	-0.	1.0000	-0.	3.0000	-0.
1.0000	-0.	-19.0000	-0.	21.0000	-0.

1	2.0000	0.	2.0000	0.
2	-1.0000	-0.	1.0000	-180.0000

BRIDGE EXAMPLE CASE 2

100.0000	-100.0000	-100.0000	0.	-0.	-0.	50.0000	-0.
-100.0000	0.	150.0000	150.0000	-0.	150.0000	-0.	-0.
-0.	-0.	-0.	150.0000	200.0000	150.0000	50.0000	-0.

1	0.2721	0.2674	0.3815	44.4974
2	0.0394	-0.0047	0.0397	-6.8428
3	0.1435	-0.1372	0.1986	-43.7127

FOUR-EQUATION VERSION OF BRIDGE CURRENTS

100.0000	-100.0000	-100.0000	-0.	-0.	100.0000	-0.	-0.	50.0000	-0.
-100.0000	-0.	150.0000	150.0000	-50.0000	-0.	-0.	150.0000	-0.	-0.
-0.	100.0000	-50.0000	-0.	250.0000	-100.0000	200.0000	-0.	-0.	-0.
-0.	-0.	-0.	150.0000	200.0000	-0.	200.0000	150.0000	50.0000	-0.

1	0.2437	0.2083	0.3206	40.5105
2	0.0111	-0.0638	0.0648	-80.1435
3	-0.0283	-0.0591	0.0655	-115.6208
4	0.1719	-0.0781	0.1888	-24.4440

Figure 24.5. Output of the program in Figure 24.4 for several test cases.

range of the DO is executed first, and then the index is tested. Therefore we must set up the calculation of x_n as a separate step at the beginning. The flowchart was drawn as it was in anticipation of this necessity.

Another minor problem is that the increment of a FORTRAN DO cannot be negative; we should like to be able to write something on the order of

$$\text{DO } 720 \text{ I} = \text{N} - 1, 1, -1$$

which is quite illegal. Accordingly we program the same thing, in order to take the equations from the next-to-the-last down to the first.

This program was developed for use in circuit studies in which the complex variable formulation provides a simple way to represent the varying phase relationships through the circuit. After completing the solution, it is sometimes of interest to get the results in terms of a magnitude and an angle (polar coordinates), instead of the two components (rectangular coordinates). Therefore the program closes with a conversion to polar form, followed finally by writing the solution—which now consists of four numbers for each variable: the real and imaginary part of the rectangular form and the length and angle of the polar form. Figure 24.5 displays three sets of output, including in each case the system that was solved.

The first system is a simple 2 x 2 real test case, to check out the program. The second is a formulation of the current flowing in a certain bridge network. The third is one way of writing four equations for the bridge of the second example; the result is a system of linearly dependent equations which has an infinite number of solutions and ought to lead a division by zero. This did not happen, of course; the reader is reminded of the discussion in Case Study 12.

CASE STUDY 25 A PACKAGE OF MATRIX SUBROUTINES; MATRIX INVERSION; EIGENVALUES AND EIGENVECTORS

A great many problems in applied mathematics lead to formulations in terms of matrix operations, either directly in terms of the physical relations or indirectly from the method of solution of other problems such as differential equations. Any computer installation must accordingly have available a package of subroutines for matrix operations, some of the most common being matrix addition, subtraction, multiplication, and inversion and the finding of eigenvalues and associated eigenvectors, which are usually available to the user either as separate subroutines or as an integrated package. In this case study we shall take the package approach to show how subroutines may profitably call each other. All of the subroutines will employ the technique of adjustable dimensions.

Matrix addition and subtraction are simple operations. The two matrices must be the same size and the operations are defined on an element-by-element basis. Figure 25.1 shows these two subroutines and the one for matrix multiplication. We see in each case that the dimensions of the matrices are included as arguments in the SUBROUTINE statement and that the dimension information in the REAL statement is given in terms of these arguments. This is the heart of the adjustable dimension technique, as we saw in Case Study 21. The actual dimensions will be specified in the CALL statements that bring these subroutines into action, as we shall see.

The program for matrix multiplication is virtually identical with the one in Case Study 17.

Before considering the program for matrix inversion in Figure 25.2, we should pause for a statement of the problem and a sketch of the basis of the method.

The *inverse* of a square matrix A is denoted by A^{-1} and is defined to be such that

$$A*A^{-1} = A^{-1}*A = I$$

I is the *unit matrix*, which has zeros everywhere except on the main diagonal, where there are 1's. The unit matrix is also called the *identity matrix*. The inverse is thus analogous to the reciprocal of a number.

The method we shall use to find the inverse is as follows: by carrying out a series of operations related to the method of Gauss elimination, we shall reduce the matrix A to the unit matrix; every time we do something to A we shall do the same thing to a matrix that was set up initially to be the unit matrix. We claim that when A has been reduced to the unit matrix the other matrix will be its inverse.

That this is so can be seen by considering that every operation we shall carry out on A can be represented as a multiplication by a suitable matrix. Specifically, the operations will consist of row interchange, multiplication of a row by a constant, and subtraction of a multiple of one row from another row. The reader may wish to verify that these

```
C CASE STUDY 25
C A SUBROUTINE FOR MATRIX ADDITION
C
      SUBROUTINE MATADD (A, B, C, M, N)
      REAL A(M, N), B(M, N), C(M, N)
      DO 1 I = 1, M
      DO 1 J = 1, N
    1 C(I, J) = A(I, J) & B(I, J)
      RETURN
      END
```

```
C CASE STUDY 25
C A SUBROUTINE FOR MATRIX SUBTRACTION
C
      SUBROUTINE MATSUB (A, B, C, M, N)
      REAL A(M, N), B(M, N), C(M, N)
      DO 1 I = 1, M
      DO 1 J = 1, N
    1 C(I, J) = A(I, J) - B(I, J)
      RETURN
      END
```

```
C CASE STUDY 25
C A SUBROUTINE FOR MATRIX MULTIPLICATION
C
      SUBROUTINE MATMPY (A, B, C, M, N, R)
      INTEGER R
      REAL A(M, N), B(N, R), C(M, R)
      DO 1 I = 1, M
      DO 1 J = 1, R
      C(I, J) = 0.0
      DO 1 K = 1, N
    1 C(I, J) = C(I, J) & A(I, K) * B(K, J)
      RETURN
      END
```

Figure 25.1. Subroutines for the addition, subtraction, and multiplication of two matrices.

operations can be done by multiplication on the left by a matrix E with these characteristics:

1. Interchange rows i and j of A: multiply A on the left by a matrix E that is the identity matrix with columns i and j interchanged.

2. Multiplication of row i of A by a constant k: multiply A on the left by a matrix E that is the identity matrix except that the diagonal element in row i is k rather than 1.

3. Subtract k times row i of A from row j of A: multiply A on the left by a matrix E that is the identity matrix with $-k$ inserted in row i, column j.

A succession of such operations on any matrix A is equivalent to multiplying A on the left by the product of the successive E matrices. Therefore, the complete set of elementary row operations that we shall perform can be represented as a single matrix which is the product of elementary matrices E, as described in (1), (2), and (3). In fact, the result of applying all the operations to a matrix that starts out as the identity matrix is just the one we have been describing. We shall not perform the row operations by matrix multiplication because it would be too time consuming, but the effect is the same.

We claim that the result *is* the inverse because the operations that produced it, when applied to the matrix A, reduced it to the identity matrix. This can be more clearly stated symbolically. Suppose E_0 is the product of all the elementary matrices E, taken in correct order. Then, because multiplying A on the left by E_0 reduced A to the identity matrix, we have

$$E_0 * A = I$$

But we defined the inverse A^{-1} to be such that

$$A^{-1} * A = I$$

so that E_0 is in fact equal to A^{-1}. Now if we carry out the same elementary row operations represented by E_0 on I, we, of course, get

$$E_0 * I = E_0$$

In short, the elementary operations that reduce A to I, if applied to I produce A^{-1}.

The subroutine parameters are the name of the matrix A, the name of the array in which we want the inverse placed, and N, the number of rows and columns of A. The matrix A is immediately transferred to the left half of a matrix B, which is established in the subroutine, and the identity matrix is placed in the right half of B. All operations are carried out on B, and at the end we move the inverse, which is in the right half of B, to AINV. This approach means that the matrix A is not destroyed by the inversion, and it means that once we have chosen a row operation based on the characteristics of the matrix to be inverted we can carry out that operation on the entire row of $2N$ elements. On the other hand, it is wasteful of storage space to use a working storage location that is twice as large as otherwise necessary merely to simplify programming. In a practical program it would probably be essential to supply the slight extra programming to carry out the inversion operations directly on AINV rather than to set up B to be double size.

Compared with straight Gauss elimination, this program is somewhat more involved, for we must

```
C CASE STUDY 25
C A SUBROUTINE FOR MATRIX INVERSION
C
      SUBROUTINE MATINV (A, AINV, N)
      REAL A(N, N), AINV(N, N), B(50, 100)
C MOVE A TO THE 50 BY 100 MATRIX B, WHICH IS USED FOR WORKING STORAGE.
C THIS METHOD DELIBERATELY WASTES STORAGE IN THE INTEREST OF
C UNDERSTANDABILITY. THE METHOD IS EASILY MODIFIED TO USE LESS SPACE.
      DO 1 I = 1, N
      DO 1 J = 1, N
    1 B(I, J) = A(I, J)
C LOAD RIGHT HALF OF B WITH UNIT MATRIX
      J1 = N + 1
      J2 = 2*N
      DO 2 I = 1, N
      DO 2 J = J1, J2
    2 B(I, J) = 0.0
      DO 3 I = 1, N
      J = I + N
    3 B(I, J) = 1.0
C START THE PIVOTAL CONDENSATION.
C K NAMES THE PIVOTAL ROW
      DO 610 K = 1, N
      KP1 = K + 1
C FIND TERM IN COLUMN K, ON OR BELOW MAIN DIAGONAL, THAT IS LARGEST
C IN ABSOLUTE VALUE. AFTER THE SEARCH, L IS THE ROW NUMBER OF THE ELEMENT.
C THERE IS NO SEARCH FOR THE LARGEST ELEMENT WHEN K = N, BUT THE
C ELIMINATION OF ELEMENTS ABOVE THE DIAGONAL IS STILL NECESSARY
      IF(K .EQ. N) GO TO 500
      L = K
      DO 400 I = KP1, N
  400 IF (ABS(B(I, K)) .GT. ABS(B(L, K))) L = I
C CHECK WHETHER L = K, WHICH MEANS THAT THE LARGEST ELEMENT IN
C COLUMN K WAS ALREADY THE DIAGONAL TERM, MAKING ROW INTERCHANGE
C UNNECESSARY
      IF (L .EQ. K) GO TO 500
C INTERCHANGE ROWS L AND K, FROM DIAGONAL TO RIGHT, INCLUDING IDENTITY
      DO 410 J = K, J2
      TEMP = B(K, J)
      B(K, J) = B(L, J)
  410 B(L, J) = TEMP
C DIVIDE THE ELEMENTS IN ROW K BY A(K, K)
C HOWEVER, DO NOT DIVIDE DIAGONAL TERM BY ITSELF, SINCE SUBSEQUENT
C DIVISIONS WOULD BE INCORRECT, AND DIAGONAL TERM IS NEVER USED
C AFTER THIS ANYWAY
  500 DO 501 J = KP1, J2
  501 B(K, J) = B(K, J) / B(K, K)
C ELIMINATE ALL ELEMENTS IN COLUMN K ABOVE DIAGONAL
C THE ELEMENTS SPECIFICALLY IN COLUMN K ARE NOT, IN FACT, MODIFIED AT ALL,
C SINCE THEIR VALUES ARE NEEDED LATER. WHAT IS DONE IS TO PERFORM THE
C ELIMINATION OPERATIONS ON ALL COLUMNS TO RIGHT OF COLUMN K, WHICH
C ARE BASICALLY ALL WE CARE ABOUT IN OBTAINING THE INVERSE. SEE TEXT.
C MUST SKIP THIS OPERATION FOR FIRST ROW, OF COURSE
      IF (K .EQ. 1) GO TO 600
      KM1 = K - 1
      DO 510 I = 1, KM1
      DO 510 J = KP1, J2
  510 B(I, J) = B(I, J) - B(I, K) * B(K, J)
C ELIMINATE ALL ELEMENTS IN COLUMN K BELOW MAIN DIAGONAL
C SKIP THIS OPERATION WHEN K = N
      IF (K .EQ. N) GO TO 700
  600 DO 610 I = KP1, N
      DO 610 J = KP1, J2
  610 B(I, J) = B(I, J) - B(I, K) * B(K, J)
C MOVE THE INVERSE TO THE OUTPUT MATRIX
  700 DO 701 I = 1, N
      DO 701 J = 1, N
      K = J + N
  701 AINV(I, J) = B(I, K)
      RETURN
      END
```

Figure 25.2. A subroutine for matrix inversion by the Gauss-Jordan method.

eliminate all off-diagonal elements, not just those below the diagonal, and we must reduce the diagonal elements to 1. Thus a DO statement that ran before from 1 to N — 1 now runs from 1 to N, which in turn means that a test has to be set up to handle the special case of the last row. The new procedure is called *Gauss-Jordan elimination*.

The basic scheme is this. Using K to identify the pivotal row, we search for the element of largest absolute value in column K below the main diagonal. Having found it to be in row L, we inter-

change rows K and L unless the largest element was already in row K. The interchange is carried out all the way across to the 2Nth element, that is, including the right half of B where the unit matrix was loaded and where the inverse will be developed. The restriction of the search to elements below the diagonal is necessary to avoid interchanges that would "contaminate" parts of the array already eliminated to zeros.

Having put the largest possible value into the pivotal position to minimize roundoff errors, we

```
C CASE STUDY 25
C A MAIN PROGRAM TO CALL A GROUP OF MATRIX SUBROUTINES
C
      REAL AMATRX(10, 10), BMATRX(10, 10), INVRSE(10, 10),
     1    VECTOR(10), X(10), INFORM(12), OUFORM(12)
      REAL A(4, 4), EIGVCT(4), PROD(4)
C READ THE FORMATS
      READ (5, 2) INFORM
      READ (5, 2) OUFORM
    2 FORMAT (12A6)
C READ THE MATRIX OF WHICH WE WANT THE INVERSE
      READ (5, INFORM) ((AMATRX(I, J), J = 1, 10), I = 1, 10)
      WRITE (6, OUFORM) ((AMATRX(I, J), J = 1, 10), I = 1, 10)
C FIND THE INVERSE
      CALL MATINV (AMATRX, INVRSE, 10)
      WRITE (6, OUFORM) ((INVRSE(I, J), J = 1, 10), I = 1, 10)
C DEMONSTRATE THAT THE PRODUCT OF THE MATRIX AND ITS INVERSE IS THE UNIT MATRIX
      CALL MATMPY (AMATRX, INVRSE, BMATRX,10, 10, 10)
      WRITE (6, OUFORM) ((BMATRX(I, J), J = 1, 10), I = 1, 10)
C DEMONSTRATE THAT THE MULTIPLICATION OF A MATRIX AND ITS INVERSE
C IS COMMUTATIVE, WHICH IS NOT TRUE OF MATRIX MULTIPLICATION IN GENERAL
      CALL MATMPY (INVRSE, AMATRX, BMATRX, 10, 10, 10)
      WRITE (6, OUFORM) ((BMATRX(I, J), J = 1, 10), I = 1, 10)
C READ SOME ARBITRARY VECTORS AND MULTIPLY THEM BY THE INVERSE OF AMATRX.
C FOR EACH SUCH VECTOR, WE THEREBY GET THE SOLUTION OF THE SYSTEM OF
C SIMULTANEOUS EQUATIONS IN WHICH 'AMATRX' IS THE MATRIX OF COEFFIENTS
C AND 'VECTOR' IS THE RIGHT HAND SIDE
      DO 10 I = 1, 3
    9 READ (5, INFORM) VECTOR
      CALL MATMPY (INVRSE, VECTOR, X, 10, 10, 1)
   10 WRITE (6, OUFORM) X
C READ A MATRIX FOR WHICH WE WANT THE LARGEST EIGENVALUE AND ITS EIGENVECTOR
      READ (5, INFORM) ((A(I, J), J = 1, 4), I = 1, 4)
      CALL EIGEN (A, 4, EIGVCT, VALUE)
      WRITE (6, OUFORM) VALUE
      WRITE (6, OUFORM) EIGVCT
C DEMONSTRATE THAT MULTIPLYING THE MATRIX TIMES ITS EIGENVECTOR DOES
C MULTIPLY EACH ELEMENT OF THE EIGENVECTOR BY THE EIGENVALUE
      CALL MATMPY (A, EIGVCT, PROD, 4, 4, 1)
      WRITE (6, OUFORM) PROD
C DEMONSTRATE THAT MULTIPLYING AN EIGENVECTOR BY A SCALAR YIELDS
C ANOTHER EIGENVECTOR
      DO 20 I = 1, 4
   20 EIGVCT(I) = 87.39 * EIGVCT(I)
      WRITE (6, OUFORM) EIGVCT
      CALL MATMPY (A, EIGVCT, PROD, 4, 4, 1)
      WRITE (6, OUFORM) PROD
      STOP
      END
```

Figure 25.3. A program to call a group of matrix subroutines.

10.00000	0.55500	-0.70000	2.00000	2.00000	5.70000	-1.00000	-5.80000	-1.70000	8.20000
2.40000	7.10000	-7.10000	1.60000	0.00800	0.	0.80000	0.	-0.47000	-0.78000
7.55000	2.98000	-1.70000	-10.00000	0.	0.89000	-0.78000	-0.90000	9.00000	-0.01000
1.10000	2.20000	0.	3.30000	4.40000	5.50000	6.60000	7.70000	8.80000	9.90000
4.77700	-5.80000	-1.00000	-100.00000	0.	0.	0.77700	20.80000	-20.80000	20.80000
8.70000	-50.80000	9.80000	9.20000	9.72000	-2.00000	-1.00000	0.	7.85100	-0.70000
1.10000	2.20000	8.80000	3.30000	4.40000	7.70000	5.50000	0.	0.	8.00000
15.00000	16.00000	17.00000	18.00000	19.00000	20.00000	21.00000	0.	-0.01000	-0.02000
10.00000	20.00000	-0.99000	9.25000	9.00700	-4.87000	1.00700	4.50000	-89.00000	0.
0.	0.	1.00100	-0.10000	-0.70000	-0.10000	-0.52000	-0.75000	75.90000	-1.88000

Figure 25.4. The matrix AMATRX used to test the matrix inversion routine.

divide all of row K *to the right of the diagonal* by A(K,K). An early version of the program divided row K, *including* the main diagonal term, by A(K,K). The moment A(K,K) is divided by A(K,K), we are, of course, dividing the rest of the row by 1, which is hardly what we want. We accordingly leave A(K,K) exactly as it is, which is acceptable, for the diagonal element in row K never enters the calculation again. The purist who wants to do *literally* what we described, so that at the end of the program the matrix A has *really* been reduced to the unit matrix, is free to rewrite the program to place A(K,K) in a temporary location and then divide the *entire* row by that value.

A similar situation develops the moment we start to eliminate the elements above the main diagonal. If we *really* eliminate B(I,K), the element in the same column as the diagonal, we no longer have it available to use when deciding by what the diagonal row should be multiplied in subtracting it from the rest of row I. Again, once this "elimination" is carried out, we shall never again need to look at B(I,K), so why not simply leave it there? Purists may rewrite, etc., etc.

At the conclusion of the program the left half of B, the part corresponding to the matrix to be inverted, has not, in fact, been reduced to the unit matrix; but all the operations necessary to have done so have been carried out on the right half, the part that was initially set equal to the unit matrix. This is not to say that the left half of B has not been altered, because it has, but simply that elements that "should" have been eliminated (replaced by zeros) have not been and the main diagonal terms are not 1.

The latter has a side benefit in case we wanted the determinant of the matrix: we can get it by forming the product of the elements on the main diagonal at the end of the process and multiplying by −1 in case an odd number of row interchanges have been carried out. The simplest way to accomplish the last is to set up a variable named, say, SIGN, to be +1 at the beginning of the program and merely reverse its sign every time a row interchanged is carried out. Then multiply the product of the diagonal terms by SIGN, which will be +1 if there have been an even number of interchanges and −1 if odd.

In order to see the matrix inversion program in action, we must now turn to an examination of part of the main program in Figure 25.3. We see that the various matrices (two-dimensional arrays) are listed in REAL statements with numerical dimensions; these are the dimensions that will be used by the adjustable-dimension subroutines when we call them.

The program uses object-time formats in a familiar way. The matrix AMATRX is 10 x 10; it is essentially an arbitrary matrix, having been invented by the author while punching up the input deck. (There is no *guarantee* that an arbitrary matrix *has* an inverse, of course.) This matrix is displayed in Figure 25.4.

Getting the inverse is now a matter of one CALL statement in which we name the matrix of which we want the inverse, the array in which the inverse is to be placed, and N, the number of rows and columns of the matrices. Figure 25.5 shows the inverse as computed. Next we perform the multiplication

$$AMATRX * AMATRX^{-1}$$

which ought to produce the unit matrix; Figure 25.6 is the result, exactly as promised. The minus signs on some of the zeros reflect the fact that the

-0.15243	0.09597	0.37255	0.02358	-0.03834	0.03125	0.22571	-0.06237	0.00538	-0.05717
0.02385	-0.03975	-0.03092	-0.00028	0.00301	-0.01978	-0.03174	0.00873	0.01523	0.02472
-0.09428	-0.06367	0.17735	-0.00568	-0.02004	0.00436	0.14726	-0.03230	0.01431	-0.01204
-0.00223	-0.04109	0.03204	0.03802	-0.01410	-0.00216	-0.01456	-0.00541	0.00205	-0.00975
0.39153	-0.24501	-0.73218	-0.03143	0.08577	-0.04140	-0.57149	0.15744	0.04528	0.17862
0.57537	-0.92110	-0.62113	0.24999	0.05540	-0.14951	-1.14484	0.25480	-0.06551	0.00571
-0.73341	1.14771	0.84059	-0.25433	-0.07697	0.17096	1.36402	-0.26892	-0.00734	-0.12693
0.19863	-0.58638	-0.13584	0.24695	0.00737	-0.07468	-0.59445	0.10109	-0.00730	-0.01051
-0.00107	0.00723	-0.00120	-0.00065	0.00028	0.00094	0.00526	-0.00119	0.00056	0.01405
-0.14589	0.31698	0.17161	-0.06108	-0.01527	0.04637	0.42517	-0.10303	0.02168	0.00186

Figure 25.5.   The inverse AMATRX^{-1} produced by the matrix inversion routine.

off-diagonal products were not in fact zero but rather on the order of 10^{-8} because of roundoff error, as seen in a printout from an earlier version of the program not reproduced here. Some of these small numbers were negative, leading to the minus zeros.

The multiplication

$$AMATRX^{-1} * AMATRX$$

should also produce the unit matrix, according to theory; Figure 25.7 shows that it indeed does.

One of the most common reasons for wanting the inverse of a matrix is that a system of simultaneous equations has to be solved for a large number of right-hand sides. Writing the simultaneous equations problem in matrix notation, we have

$$A * X = B,$$

where A is the N x N matrix of cofficients, X is a N x 1 matrix of unknowns, and B is the N x 1 matrix of right-hand sides. If we have the inverse, we may write

$$A^{-1} * A * X = A^{-1} * B$$

But $A^{-1} * A = I$, and the multiplication $I * X$ leaves X unchanged, so we have simply

$$X = A^{-1} * B$$

meaning that once we have the inverse of the matrix of coefficients we can solve the system of simultaneous equations for *any* given right-hand side by a simple matrix multiplication.

For one simple example the analysis of a mixture of gases in a mass spectrometer requires the solution of a system of simultaneous equations in which the matrix of coefficients depends solely on the calibration of the instrument. The analysis of a given unknown mixture produces a new right-hand side but makes no change in the matrix of coefficients. Thus it is a great time saving to compute the inverse and be able to get the solution of the system by matrix multiplication.

Our main program exhibits this usage next by

1.00000	0.00000	0.00000	-0.00000	-0.00000	0.00000	0.00000	-0.00000	0.00000	0.00000
0.00000	1.00000	-0.00000	0.00000	0.00000	-0.00000	-0.00000	0.00000	0.00000	0.00000
-0.00000	0.00000	1.00000	-0.00000	-0.00000	0.00000	0.00000	-0.00000	0.00000	0.00000
-0.00000	0.00000	0.00000	1.00000	-0.00000	0.00000	0.00000	-0.00000	0.00000	-0.00000
-0.00000	0.00000	0.00000	0.00000	1.00000	0.00000	0.00000	-0.00000	0.00000	-0.00000
-0.00000	0.00000	0.00000	-0.00000	-0.00000	1.00000	0.00000	-0.00000	0.00000	0.00000
-0.00000	0.00000	0.00000	-0.00000	-0.00000	0.00000	1.00000	-0.00000	-0.00000	0.00000
-0.00000	0.00000	0.00000	-0.00000	-0.00000	0.00000	0.00000	1.00000	0.00000	0.00000
-0.00000	0.00000	-0.00000	-0.00000	0.00000	0.00000	0.00000	-0.00000	1.00000	0.00000
0.00000	0.00000	-0.00000	-0.00000	0.00000	-0.00000	-0.00000	0.00000	0.00000	1.00000

Figure 25.6.   The product AMATRX * AMATRX^{-1}, showing that the product of a matrix and its inverse is the unit martix.

1.00000	-0.00000	-0.00000	-0.00000	-0.00000	-0.00000	0.00000	-0.00000	0.00000	0.00000
0.00000	1.00000	0.00000	0.00000	0.00000	-0.00000	-0.00000	-0.00000	-0.00000	-0.00000
0.00000	-0.00000	1.00000	-0.00000	0.00000	0.00000	0.00000	-0.00000	0.00000	0.00000
0.00000	-0.00000	-0.00000	1.00000	-0.00000	-0.00000	-0.00000	-0.00000	-0.00000	-0.00000
-0.00000	0.00000	0.00000	0.00000	1.00000	-0.00000	-0.00000	0.00000	-0.00000	-0.00000
-0.00000	-0.00000	-0.00000	0.00000	-0.00000	1.00000	-0.00000	-0.00000	-0.00000	-0.00000
0.00000	0.00000	0.00000	-0.00000	0.00000	0.00000	1.00000	0.00000	0.00000	0.00000
-0.00000	-0.00000	-0.00000	0.00000	0.00000	-0.00000	-0.00000	1.00000	-0.00000	-0.00000
0.00000	0.00000	0.00000	-0.00000	0.00000	0.00000	0.00000	0.00000	1.00000	0.00000
0.00000	0.00000	0.00000	-0.00000	0.00000	0.00000	0.00000	0.00000	0.00000	1.00000

Figure 25.7. The product AMATRX^{-1} * AMATRX, which is also the unit matrix.

reading in three different 10 x 1 vectors and multiplying them by the inverse. Figure 25.8 displays the three solutions.

Next we come to the section of the main program that calls the eigenvalue subroutine, and again we pause for a statement of the theory and a sketch of the method.*

We are given a square matrix A. An eigenvalue λ and its associated eigenvector X are defined to be such that, if they exist,

$$A * X = \lambda X$$

that is, the effect of multiplying the eigenvector by the matrix is to multiply each element of the eigenvector by the eigenvalue. Zero, being trivial, is excluded as an eigenvalue. Not every matrix has real eigenvalues, but a symmetric matrix has, for reasons that are beyond the scope of this treatment.

It is also beyond our scope to justify the factors that make it a rather simple matter to find the largest eigenvalue and its eigenvector. If the eigenvalues are distinct, *any* vector of the same dimension can be written in the form of a *linear combination* of the eigenvectors $X_1, X_2, \ldots, X_n$, as follows:

$$u = a_1x_1 + a_2x_2 + \cdots + a_nx_n$$

where the a's are real numbers.

* This presentation of the standard subject matter is taken from *Numerical Methods: 1 Iteration, Programming and Algebraic Equations*, Ben Noble, New York, Interscience, 1964, pp. 123–135.

We now look for a way to find the largest eigenvalue in absolute value. To do so we begin with an arbitrary vector u_0, which has a representation as a linear combination of the eigenvalues, as shown above. Now multiply by A to get

$$u_1 = Au_0 = a_1\lambda_1x_1 + a_2\lambda_2x_2 + \cdots + a_n\lambda_nx_n$$

If we continue in this way, we will get

$$u_p = A^pu_0 = a_1\lambda_1{}^px_1 + a_2\lambda_2{}^px_2 + \cdots + a_n\lambda_n{}^px_n$$

$$= \lambda_1{}^p\left[a_1x_1 + \sum_{i=2}^{n} a_i\left(\frac{\lambda_i}{\lambda_1}\right)^p x_i\right]$$

Now suppose that

$$|\lambda_1| > |\lambda_2| \geq |\lambda_3| \geq \cdots \geq |\lambda_n|$$

Then $(\lambda_i/\lambda_1)^p$ tends to zero as p tends to infinity $i > 1$. Thus, as we continually multiply by A, the expression on the right will approach $\lambda_1{}^pa_1x_1$. We can sort out λ_1 from the other eigenvalues.

It must be clearly understood that we do not have to know what the eigenvector x_1 is in order to carry out this procedure. We begin with a nonzero but otherwise arbitrary vector, u_0, and multiply repeatedly by A. Each time we look at the product (which is a vector), find the largest element in it, and divide each element by this largest element. If there *is* a largest eigenvalue, we shall eventually find that that largest element is the same after each multiplication. Such a number is the eigenvalue, and the product A^pu_0 divided by the eigenvalue is the associated eigenvector.

0.04441	-0.00469	0.01153	-0.00172	-0.07629	-0.17608	0.20554	-0.08551	0.00242	0.06584
0.02559	-0.00044	0.00173	-0.00164	-0.02884	-0.09834	0.10900	-0.04753	0.00275	0.03721
0.09219	-0.00923	0.04760	0.00283	-0.16774	-0.25460	0.30168	-0.09600	0.00188	0.07874

Figure 25.8.   The solutions to three systems of equations, each based on the system matrix in Figure 25.4 but with different right-hand sides.

```
C CASE STUDY 25
C SUBROUTINE TO FIND THE LARGEST EIGENVALUE AND ASSOCIATED
C EIGENVECTOR OF A REAL SYMMETRIC MATRIX
C
      SUBROUTINE EIGEN (A, N, VECT, VAL)
      DIMENSION A(N, N), VECT(N), A1(20, 20), A2(20, 20), VWORK(20)
      DO 1 I = 1, N
    1 VECT(I) = 1.0
      CALL MATMPY (A, A, A1, N, N, N)
      CALL MATMPY (A1, A1, A2, N, N, N)
      CALL MATMPY (A2, A2, A1, N, N, N)
      CALL MATMPY (A1, VECT, VWORK, N, N, 1)
      CALL SCALE (VWORK, N, VAL)
      OLDVAL = VAL**0.125
      DO 2 I = 1, 10
      DO 3 J = 1, N
    3 VECT(J) = VWORK(J)
      WRITE (6, 872) VAL, (VECT(K), K = 1, 4)
  872 FORMAT (1H , 1P5E15.5)
      CALL MATMPY (A, VECT, VWORK, N, N, 1)
      CALL SCALE (VWORK, N, VAL)
      IF (ABS(VAL) .LT. 0.000001) RETURN
      IF (ABS((VAL - OLDVAL)/VAL) .LT. 0.0001) RETURN
    2 OLDVAL = VAL
      RETURN
      END
```

Figure 25.9. A subroutine to find the largest eigenvalue and associated eigenvector of a real symmetric matrix.

(It would be better, of course, to say *an* associated eigenvector; once we know an eigenvector, that vector with each element multiplied by any nonzero constant is also an eigenvector for the same eigenvalue.)

It would be well to list the assumptions here and state explicitly exactly what we propose to do. We assume that a single largest real eigenvalue exists. This places serious restraints on the matrix, restraints that are difficult to state in terms of relations among the elements. Usually we know that these conditions are met because of the way the matrix is generated; if we are describing the modes of vibration of a certain physical system, for instance, we know that there are real eigenvalues and we may know them to be distinct.

Our program will find a largest eigenvalue and an associated eigenvector in which the largest element is 1.

This is obviously far from a complete eigenvalue subroutine, which would have to be able to find the other eigenvalues by techniques that are not really beyond us but are not sufficiently instructive for our purposes to be worth investigating.

The subroutine is shown in Figure 25.9. Its arguments are the name of the matrix, the dimension, the location in which the eigenvector should be placed, and the location of the eigenvalue. The dimension of the array is adjustable, but the size of the working storage would limit the maximum to 20. Our first task is to set up the arbitrary starting vector in VECT. Next we would like to raise the

```
C CASE STUDY 25
C SUBROUTINE TO FIND THE ELEMENT IN A ONE-DIMENSIONAL ARRAY
C WITH LARGEST ABSOLUTE VALUE, AND DIVIDE THE ARRAY BY THIS VALUE
C
      SUBROUTINE SCALE (A, N, SC)
      DIMENSION A(N)
      NBIG = 1
      DO 1 I = 2, N
    1 IF (ABS(A(I)) .GT. ABS(A(NBIG))) NBIG = I
      SC = A(NBIG)
      DO 2 I = 1, N
    2 A(I) = A(I) / SC
      RETURN
      END
```

Figure 25.10. A subroutine for scaling a vector, needed in the eigenvalue computation.

```
6.78258E 03      2.26669E-02      1.67782E-01      1.15381E-01      1.00000E 00
2.92893E 00      2.29236E-02      1.68058E-01      1.14563E-01      1.00000E 00
```

```
2.92888

0.02292    0.16806    0.11456    1.00000

0.06756    0.49269    0.33419    2.92888

2.00329   14.68660   10.01167   87.39000

5.90413   43.05595   29.20515  255.95505
```

Figure 25.11. The output of the eigenvalue computation.

given matrix to a moderately high power in hopes of getting the eigenvalue fairly well isolated before starting the iterative process. Working storage arrays A1 and A2 have been set up for the purpose. We multiply A by A and specify that the product is to go into A1; we multiply A1 by A1 and put the product in A2; we multiply A2 by A2 and put the product back in A1; A1 now holds the original matrix raised to the eighth power. Next we multiply the starting arbitrary vector by A^8 and place the result in a working storage array VWORK.

Now we should like to inspect this vector to determine its largest element and divide the vector by that largest element. This is done with a separate subroutine named SCALE, shown in Figure 25.10, which ought to be fairly easy to follow. SCALE modifies the vector supplied to it and delivers the largest value that it found. Note that the dimension N has now been "passed through" EIGEN to SCALE. This is legal.

We are ready at this point to enter the iterative process. We set OLDVAL equal to the eighth root of VAL, on the theory that raising the given matrix to the eighth power and multiplying by the starting vector should have produced an approximation to the eighth power of the eigenvalue. The accuracy of the approximation is of no great importance.

Now we enter a loop that has as its only function to set a limit of 10 on the number of iterations; if we cannot get the eigenvalue by then, there probably is no eigenvalue to find. Next VECT is set equal to the vector produced by the multiplication of A^8 and SCALE. Here we pause to write out the result so far. This, of course, would not be done in an actual program; we display the result for the benefit of readers who would like to watch the iterative process converge. Multiplying VECT by A and calling SCALE again should produce a better

approximation to the eigenvalue. We check to see if the value being improved has turned out to be near zero, which unfortunately is possible, and also if the current approximation and the new one are within 10^{-4} in relative magnitude. A RETURN statement terminates the iteration either way. If the process has not converged, we set OLDVAL equal to VAL and go back for another try.

This procedure has been tried on the following matrix, suggested by Noble in the book cited above.

$$\begin{pmatrix} -0.030 & -0.242 & -0.603 & 0.178 \\ -0.242 & 0.860 & -0.343 & 0.393 \\ -0.603 & -0.343 & 1.350 & 0.251 \\ 0.178 & 0.393 & 0.251 & 2.630 \end{pmatrix}$$

Returning to the main program of Figure 25.3, we see that the subroutine EIGEN was called after reading the matrix just displayed. The output in Figure 25.11 can be interpreted as follows. In the first line we have the result of multiplying the starting approximation to the eigenvector by A^8; 6782.58 is an approximation to the eighth power of the eigenvalue and the eighth root of this number is, in fact, about 3.02, not too far from the eventual final value. The next line says that after one iteration with A as the multiplier, the eigenvalue approximation was 2.92893. The next line was produced by the main program; evidently the subroutine had confirmed convergence, and indeed the latest two values are within 0.0001 in relative magnitude.

In the main program we now demonstrate that multiplying an eigenvector by the matrix multiplies each element of the eigenvector by the eigenvalue. We then multiply the elements of VECT by a random constant to show that any multiple of an eigenvector is also an eigenvector.

CASE STUDY **26** SOLUTION OF LAPLACE'S EQUATION FOR THE TEMPERATURE DISTRIBUTION IN A PIPE

In this case study we shall solve Laplace's equation to find the steady-state temperature distribution in the pipe sketched in Figure 26.1. We know that the fluid inside the pipe is at 400° and that the outside of the pipe is at 0°. These are the boundary conditions. Laplace's equation states that at each "interior" point, that is, within the material of the pipe, the temperature satisfies

$$\frac{\partial^2 u}{\partial x^2} + \frac{\partial^2 u}{\partial y^2} = 0$$

We wish to solve this equation for the given boundary conditions, for each of several lengths of the fin (the part on the right) to get a qualitative picture of the effect of fin length on the temperatures in parts of the pipe some distance from the fin. The results must be presented graphically, for the numerical results will be too voluminous to be meaningful. In this plotting we shall have to interpolate between solution points to produce a square plot because the printer has a different number of printing positions per inch horizontally (10) than the number of lines per inch vertically, (6). Finally, it will be worthwhile to investigate methods of accelerating the solution in any reasonable way we can, since very large amounts of computation are involved.

We shall solve Laplace's equation by replacing the differential equation with a difference equation. We lay out on the cross section of the pipe a square mesh, as suggested in Figure 26.2; in the sketch there are 16 vertical strips, whereas in the program there will be 100, corresponding to a spacing of 0.01 inch between mesh lines. This spacing will be the same in both directions, which simplifies the difference equations somewhat. With this kind of grid, Laplace's differential equation becomes a set of difference equations:

$$u_{ij} = \frac{u_{i,j+1} + u_{i,j-1} + u_{i+1,j} + u_{i-1,j}}{4}$$

where i and j range over all values in the interior of the figure.* The exact number depends on the length of the fin, but in round numbers we may say that there are about 10,000 simultaneous equations in the same number of unknowns; the unknowns are, of course, the temperatures at the interior mesh points. We may solve the system of equations any way we know how and that can be fitted into the computer we have available. We could not use Gauss elimination, unless we used tape or other auxiliary storage, because there would have to be space in high-speed storage for some 100,000,000 coefficients. In any case we would not choose Gauss elimination because the system of equations has a very special characteristic: almost all of the coefficients are zeros. We

---

* For a derivation of this formula, see, for instance, McCracken and Dorn, *Numerical Methods and Fortran Programming*, Chapter 11.

should somehow take advantage of this fact, which Gauss elimination does not.

What we shall do is, in fact, a very close relative of Gauss-Seidel iteration. Let us try to sketch the sequence of operations.

We shall be working throughout with an array of 101 rows and up to 201 columns, each element representing a temperature at a point on the mesh. Some points are on the boundary; these are set to the boundary values at the beginning and never change. Some are in the inside of the pipe, or in the space outside the fin, where in either case there is no metal; we are not solving the equation in these regions, and the program is written to ignore them. Others, upward of 10,000 of them, are solution points. These we initially load with any guesses we may have about the final solution or with arbitrary values. We then iterate on these points, seeking values that satisfy the difference equation at every point. It can be shown that the solution of the difference equation will also be the solution of the differential equation, to a sufficiently close approximation.

(A note on terminology. When we say "interior" points, we are referring to points that are entirely inside the metal of which the pipe is constructed. The cross section has two boundaries: one between the pipe and the outside world and the other between the pipe and the fluid flowing through it. The area

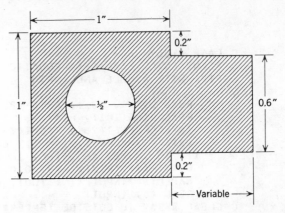

Figure 26.1. Cross-sectional view of a square pipe carrying a hot fluid through a circular opening and having a fin of variable length on one side.

within the fluid is *not* an interior point; it is not even a solution point, for we are searching for the temperatures within the pipe material only. In short, "interior" means within the pipe metal.)

The fact that the hole in the center is circular creates some problems. The square mesh we have laid over the pipe does not, in general, intersect this curved boundary at mesh points. Several approaches to this problem are available. The one chosen here is about the simplest possible and would not always be satisfactory: we simply pick as interior mesh points those that are closest to the circle and let them *define* the boundary. Thus we are not really

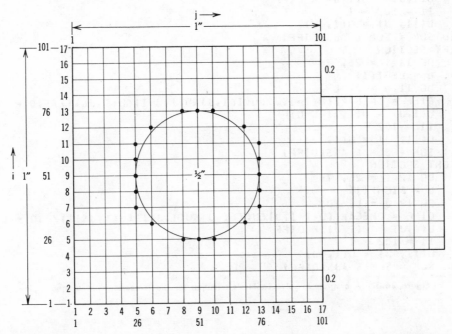

Figure 26.2. Sketch of grid of solution points laid over a cross section of Figure 26.1. The grid drawn has a 1/16 inch spacing; that used in the solution has a 1/100 inch spacing. The critical points in the numbering system of the grid used for solution are shown.

```
C CASE STUDY 26
C SOLUTION OF LAPLACE'S EQUATION TO FIND THE TEMPERATURE IN A
C PIPE WITH A HOLE AND A FIN
C
      REAL U(101, 201), ARRAY(2, 200), SYMBOL(20)
      REAL LINE(101), H(3)
      INTEGER LEFT(76), RIGHT(76)
      DATA (SYMBOL(I), I = 1, 16)/1HA, 1H , 1HB, 1H , 1HC, 1H , 1HD,
     1    1H , 1HE, 1H , 1HF, 1H , 1HG, 1H , 1HH, 1H /
      DATA H(1), H(2), H(3)/0.0, 0.666666667, 0.333333333/
      NAMELIST/INPUT/TIN, TOUT, LENGTH, MAXIT, EPS, OMEGA
    3 READ (5, INPUT)
C CLEAR ARRAY TO OUTSIDE TEMPERATURE
      DO 1 I = 1, 101
      DO 1 J = 1, 201
    1 U(I, J) = TOUT
C SET AN INSIDE SQUARE TO TEMPERATURE OF FLUID.  THIS WILL INCLUDE
C   SOME OFF-BOUNDARY POINTS, BUT IT WILL NOT MATTER.
      DO 2 I = 15, 85
      DO 2 J = 15, 85
    2 U(I, J) = TIN
C LOCATE BOUNDARY OF CIRCLE.  THE SUBSCRIPTS STORED IDENTIFY 'INTERIOR'
C POINTS, NOT 'BOUNDARY' POINTS.
      DO 4 I = 26, 76
      RIGHT(I) = 52.0 & SQRT(625.1 - FLOAT(I-51)**2)
    4 LEFT(I) = 102 - RIGHT(I)
C SET ITERATION COUNTER TO 1
      ITN = 1
C START AN ITERATION BY SETTING SUM OF RESIDUES TO ZERO
    9 SUM = 0.0
C TOP SLAB
      DO 10 I = 77, 100
      DO 10 J = 2, 100
      DIFF = OMEGA*((U(I-1,J)&U(I&1,J)&U(I,J-1)&U(I,J&1))/4.0 - U(I,J))
      U(I,J) = U(I,J) & DIFF
      I1 = 102 - I
      U(I1, J) = U(I, J)
   10 SUM = SUM & ABS(DIFF)
C LEFT SECTION
      DO 11 I = 26, 76
      L = LEFT(I)
      DO 11 J = 2, L
      DIFF = OMEGA*((U(I-1,J)&U(I&1,J)&U(I,J-1)&U(I,J&1))/4.0 - U(I,J))
      U(I,J) = U(I,J) & DIFF
      I1 = 102 - I
      U(I1, J) = U(I, J)
   11 SUM = SUM & ABS(DIFF)
C RIGHT SECTION
      DO 12 I = 26, 76
      L = RIGHT(I)
      DO 12 J = L, 100
      DIFF = OMEGA*((U(I-1,J)&U(I&1,J)&U(I,J-1)&U(I,J&1))/4.0 - U(I,J))
      U(I,J) = U(I,J) & DIFF
      I1 = 102 - I
      U(I1, J) = U(I, J)
   12 SUM = SUM & ABS(DIFF)
```

Figure 26.3. A program to solve Laplace's equation for the temperature in a square pipe.

```
C FIN
      DO 13 I = 22, 80
      L = LENGTH - 1
      DO 13 J = 101, L
      DIFF = OMEGA*((U(I-1,J)&U(I&1,J)&U(I,J-1)&U(I,J&1))/4.0 - U(I,J))
      U(I,J) = U(I,J) & DIFF
      I1 = 102 - I
      U(I1, J) = U(I, J)
   13 SUM = SUM & ABS(DIFF)
C BOTTOM SLAB
      DO 14 I = 2, 25
      DO 14 J = 2, 100
      DIFF = OMEGA*((U(I-1,J)&U(I&1,J)&U(I,J-1)&U(I,J&1))/4.0 - U(I,J))
      U(I,J) = U(I,J) & DIFF
      I1 = 102 - I
      U(I1, J) = U(I, J)
   14 SUM = SUM & ABS(DIFF)
C STORE ONE TEMPERATURE
      ARRAY(1, ITN) = U(70, 110)
C STORE SUM OF RESIDUES
      ARRAY(2, ITN) = SUM
      ITN = ITN & 1
C CHECK FOR FAILURE TO CONVERGE IN MAXIMUM ALLOWABLE ITERATIONS
      IF (ITN .GT. MAXIT) GO TO 20
C CHECK FOR CONVERGENCE
      IF (SUM .LE. EPS) GO TO 20
C NOT CONVERGED--GO BACK FOR ANOTHER ITERATION
      GO TO 9
C EITHER PROCESS HAS CONVERGED OR MAXIT HAS BEEN EXCEEDED.  WISH TO
C PLOT RESULTS EITHER WAY. IF THERE IS ANY DOUBT ABOUT WHICH IS
C WHICH, A COUNT OF THE NUMBER OF SUCCESSIVE RESIDUES WILL GIVE
C THE ANSWER
C WRITE ARRAY OF TEMPERATURES IN FIN
   20 WRITE (6, 30) (ARRAY(1, I), I = 1, ITN)
   30 FORMAT (1H1/(1H , 1P10E12.4))
C WRITE ARRAY OF SUCCESSIVE RESIDUALS
      WRITE (6, 30) (ARRAY(2, I), I = 1, ITN)
      WRITE (6, 901) TIN, TOUT, MAXIT, EPS, OMEGA
  901 FORMAT (1H1, 2F10.1, I5, 2F10.2///)
C CONVERT TEMPERATURES TO SYMBOLS
      L = 0.6 * FLOAT(LENGTH)
      DO 9011 K = 1, L
      J1 = 1 & (5 * (K-1)) / 3
      J2 = J1 & 1
      M = 1 & MOD(K-1, 3)
      DO 9012 I = 1, 101
      LINE(I) = U(I, J1) & H(M)*(U(I,J2) - U(I,J1))
 9012 IF (((I .GT. 26 .AND. I .LT. 76 .AND. J1 .GT. LEFT(I)&1 .AND.
     1    J2 .LT. RIGHT(I)-1) .OR. (J2 .GT. 101 .AND. (I .LE. 20 .OR.
     2    I .GE. 82))) LINE(I) = 401.
      DO 9013 I = 1, 101
      N = LINE(I) / 26.667 & 1.0
 9013 LINE(I) = SYMBOL(N)
 9011 WRITE (6, 9014) LINE
 9014 FORMAT (1H , 101A1)
      GO TO 3
      END
```

Figure 26.3.   (Continued)

solving the problem for a circular hole at all but for a hole made up of a set of straight lines. If the mesh spacing is fine enough, and as long as we do not need to know the temperatures near the hole accurately, this will be adequate. If more accuracy were demanded, one relatively simple attack would be to interpolate from these interior mesh points to the real boundary, which would add considerable accuracy near the boundary.*

Knowing the location of the hole, we can start iterating. This requires that at every interior point in the pipe we apply the difference equation stated earlier. Now, of course, at first we shall be using values on the right that are only approximations, but eventually, after many complete iterations, we shall find that the successive approximations differ less and less. When the changes from one iteration to the next are sufficiently small, we shall say that we have found a solution. A review of the method used in solving simultaneous equations by the Gauss-Seidel iteration method will show how close the two procedures are. The coefficients of the unknowns here are simply the numbers 1, ¼, ¼, ¼, and ¼, which we do not need to store in an array but can merely write into an assignment statement as constants.

A major difference is that we shall take measures here to speed up the rate of convergence. We compute a new approximation to the temperature at a point by using the difference equation written before. But then we reason that this new approximation is surely not the final value; why not "accelerate" the process by leaping ahead a bit? If the difference between the previous approximation and the new one is, say, X, why not multiply X by some factor greater than 1.0 to try to anticipate what succeeding iterations are going to have to do? The acceleration constant is called OMEGA in the program; the theoretical limits on its value are that it must lie between 1 and 2 or the process will diverge. Experiment showed that 1.84 was a reasonably good choice for this problem.

Now we should consider what can be done to take advantage of the symmetry of the pipe. We may, of course, ignore it, but time is going to be an issue here, and it seems wasteful not to take advantage of the rather obvious fact that the eventual solution will be completely symmetrical. We cannot just solve half the problem, which might first occur to us, because we do not know the values of the temperature needed on the new boundary that would be created by cutting the pipe in half, so to speak, along the horizontal axis of symmetry. One method sometimes used replaces the boundary condition we have stated with a condition that reflects the fact that we know the derivative perpendicular to the new "fictitious" boundary is zero. Elaborate methods of approximating this derivative with differences have been used in some studies. Here we shall take an easier way which, although quite satisfactory in terms of time and accuracy, does require that the full mesh be stored. We are thus trading some extra space for a gain in simplicity, which as we have seen is a familiar tradeoff.

The method is this: as soon as each new approximation is computed, we store it not only at the mesh point for which it was computed but at the "corresponding" point, considering symmetry. This value is not the exact solution at the symmetrical point, but it is a better guess than the old value that was there. This simple device was found, in this program, to cut the number of iterations just about in half, with much less than a doubling of the time per iteration.

A total of something like 50 to 200 iterations will be required to get convergence within reasonable variation between iterations. After convergence has been achieved we convert the temperatures to a symbolic form for ease of understanding by a process that we shall investigate in detail when we reach that point in discussing the program.

Figure 26.3 is the program developed for this task. At the beginning we have several type statements and two DATA statements for purposes that we shall explain as we proceed. Then there is a NAMELIST statement, which we have not encountered before. NAMELIST is a highly useful feature of many FORTRAN's, but it is not yet part of ASA FORTRAN and we have not emphasized it as much as might otherwise have been desirable. The idea is that we attach a name to a list of variables; here we have said that the list consisting of TIN (temperature input), TOUT, etc., is to have the name INPUT. Then in a READ statement, which just happens to be next but need not have been, we put that name where the FORMAT statement number would normally go. When this happens, FORTRAN expects to find a data card of the following type.

Column 1 is ignored and column 2 must contain a dollar sign. The name of the list must begin in

---

* For a treatment of this technique see George E. Forsythe and Wolfgang R. Wasow, *Finite Difference Methods for Partial Differential Equations*, Wiley, New York, 1960, pp. 198–202.

column 3 and be followed by one or more blanks. The data items then appear in the form of a variable name, an equal sign, the data value, and a comma. A dollar sign must follow the last data value. Any name mentioned in the NAMELIST may appear on the data card, *but not every variable so named need appear on the data card*. This is the whole point of the NAMELIST feature: we may, for instance, start with one data card on which all variables are named, then follow with others on which only one or two appear. Any variables that are not named on a NAMELIST data card simply retain their current values. This is most helpful in cases, as here, in which at the time of writing the program we do not know the variables we may wish to change.

Figure 26.4 shows the data cards that were used to produce the five plots shown later. On the first one we give values to all of the variables; then, on the others, we change the values only of LENGTH. The unnamed variables simply retain the same values.

It is also sometimes handy to name in a NAMELIST things that *might* sometimes need to be changed but to give values to them with assignment statements in the program. If we never name them on a data card, they are never changed from the values given in the program.

It is permissible to have several NAMELISTs.

The action then is that the object program will search the data deck looking for a card with the stated list on it.

The NAMELIST feature may also be used on output, in which all the variables named are written in a standard form. However, the NAMELIST feature is much more commonly used with input than with output.

Now we need to establish the boundary values: TIN for the boundary on the inside and TOUT for the boundary on the outside. While we are at it we may as well give these same values to interior points. The accuracy of the initial guesses is of little influence on the number of iterations required, but the values chosen here are probably a little better than, say, all zeros for the interior points, which would be another possibility.

Next we need to set up some way of knowing where the boundary of the inside circle is. This is a somewhat messy operation that surely should not be done on every iteration if we can establish it once and for all in a preliminary operation. The basic idea is that we know that the equation of the inside circle is

$$x^2 + y^2 = 0.0625$$

in terms of the dimensions of the actual pipe. Considering that the "origin" in this equation is at the mesh point I = J = 51, the appropriate equation is

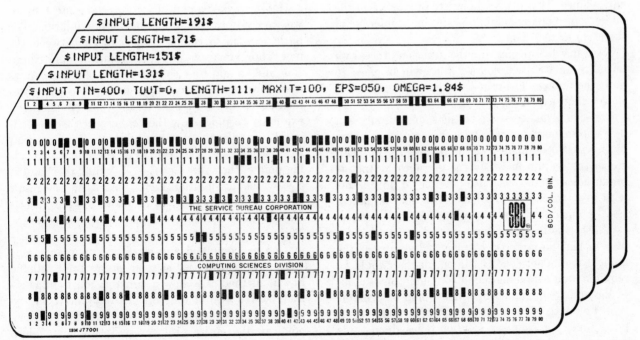

Figure 26.4.   The data deck used with the program in Figure 26.3, employing the NAMELIST feature.

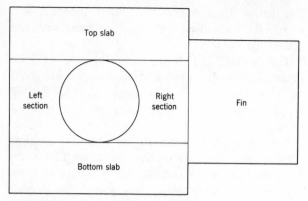

Figure 26.5. The five sections of the pipe corresponding to five parts of the program in Figure 26.3.

as shown. The extra 0.1 in the constant 625.1 is to ensure that the argument of the square root function is strictly positive; this nicety is probably not essential. The constant 52 is chosen to provide the proper starting point for a DO loop that will appear later; for that purpose we need an interior point, not a boundary point. The difference is that between 52 and 51 for this constant. Note that the right side of this assignment statement is in terms of real variables, whereas RIGHT was named in an INTEGER statement. The conversion from real to integer form will be done as a part of the action of the assignment statement. Note, too, that RIGHT and LEFT were stated as having 76 elements each, even though only elements 26 through 76 are filled with values. This is a slight waste of storage in the interest of understandability and a simpler (and therefore faster) loop later.

Now, just before entering the iteration process we initialize an iteration counter that will be used to limit the number of iterations when convergence is slower than expected.

One complete iteration consists of computing a new approximation to the temperature at every one of the 10,000 or more interior points. As we shall see in the section in which the graphical output is produced, it would be quite possible to do all this with just two nested DO loops. It would be necessary to include a quite complex IF statement to establish whether the point selected by the indices was within the material of pipe and skip over the point if it was not. The IF statement that would be required would take as much computer time as all of the actual computation, which would be prohibitive. For once we are quite concerned about computer time: one complete solution takes about three minutes on a machine that can cost more than $10

per minute, and in real life we would ordinarily want a large number of solutions for different conditions.

The answer to this little problem is to write several DO loops, each covering a section of the grid that consists entirely of interior points. Thus the logic of deciding whether a given point is interior is taken up by the program structure rather than being tested for in the object program. We shall work with five different regions, as sketched in Figure 26.5. The "top slab," "bottom slab," and the fin are all rectangular and therefore fast to scan with two nested DO loops in each case. The right and left sections both have circular arcs for part of their boundaries, but with the RIGHT and LEFT arrays specifying these arcs only one additional statement is involved.

Let us look at the program segment for the top slab to see how the computation is handled.

The I and J indexes, in every case, name interior points, and they thus name a mesh point at which a new approximation is to be computed. The four surrounding points are identified with reference to the point in the middle. With this in mind, we see that I, for instance, must range from 77 to 100, not 76 to 101. Temperatures on lines for I-76 and I-101 will enter the calculations as data—but these are boundary points, at which new approximations are never computed, and the DO parameters must be as shown.

Having picked a point by giving values to I and J, we are ready to compute a new approximation. The basic idea is to compute the change between the new approximation and the previous approximation and to add that change to the old approximation. This could, of course, be done all in one statement, except that we need to know the change itself explicitly to judge how the convergence is proceeding. Looking at the statement for DIFF, we see that we first compute the average of the four surrounding points, then subtract the old value, then multiply by the acceleration factor OMEGA. The result is then added to the old value to get the (accelerated) new value, which is stored immediately. This new value will be used when a new approximation is computed for an adjacent point; new values are always used as soon as they are available in this method. The new approximation is immediately stored also at the point corresponding to this one taking symmetry into account. Thus every point in the entire grid receives two new values during each iteration, effectively doubling the convergence rate at the slight time expense of the two added statements.

The convergence test used in this program adds up the absolute values of all the differences and stops iterating when this sum is less than a number EPS read as input. This summing operation accordingly must be done for every point. There are other ways, of course, to check for convergence; another good way is to look for the largest difference anywhere in the grid and stop when it is less than a tolerance.

After computing a new approximation for every point in the top slab (and storing these same values in the bottom slab), we move to the left section. The loop is slightly more complex because we have to get one of the DO parameters from the array LEFT. A new value is computed for each point in this section. Symmetry here means that values at the top of this section are stored at the bottom of this section and vice versa.

The right section extends from the hole to the right edge of the part that is not in the fin; this separation is required by the fact that the fin is not the same size vertically as the right section. The fin is rectangular and simple, as is the bottom slab.

No experiments were made to determine whether taking these sections in a different order would affect the number of iterations required to achieve a given degree of convergence. The details of the path to convergence would, of course, be altered, but it is not obvious that the total number of complete iterations would be changed significantly.

When all points have been dealt with, we carry out some operations that are needed at the end of each sweep. The primary output will be a picture, but we would also like to have some idea of the progress of an actual number. One point out in the fin was chosen more or less at random and the temperature at this point on every sweep entered into an array for printing later. Similarly, the sum of the differences is stored for inspection. We then increment the iteration counter and inspect the iteration counter and the sum of the differences in turn. If the iteration counter has not exceeded maximum and convergence has not been achieved, we go back for another iteration. Otherwise we break out of the iteration loop and proceed to the output. If it matters which test broke the loop, we can get the answer by looking at the number of values of temperature and the sum of differences printed.

After the last-named have been printed along with the input parameters we are ready to proceed with the graphical output. This is a relatively short but rather meaty program segment that we may profitably study quite carefully.

The primary task is to convert a temperature into a symbol, but first we must compute the temperature. The problem is this. The printer to be used has 10 printing positions per inch horizontally but only six lines per inch vertically. If we convert every temperature into a letter and print the whole array, which is readily enough done, the square pipe will not be square on the printout. The solution we shall adopt is to print only six-tenths as many lines as there are grid lines and interpolate to find the temperature at the points we need for this purpose.

It turns out that because of the size limitations of paper we prefer to print the picture turned on the axis through 90°, as may be seen in Figure 26.6. Therefore the top line on the output is the leftmost vertical line in the orientation of Figure 26.2. The next line printed will have to represent the temperatures at 101 points that are not grid points; we must interpolate between two grid points to get each of them. This process will be carried out across the grid for as many lines as there are to print, and that determination is the first action in this section of the program: there will be L lines, where L is 0.6 times LENGTH, the number of grid lines in the horizontal direction. We establish a DO loop that runs from 1 to L.

The next action necessary is to find the two J-values that will bracket a given L-value. The situation is possibly made clearer by reference to Figure 26.7, where we see how L-values line up against J-values. L = 1 coincides with J = 1. There are to be six L-values for every 10 J-values; we see that L = 4 and J = 6 coincide, as do L = 7 and J = 11. What we need, for purposes of interpolation, is to know which two J-values, J1 on the left and J2 on the right, bracket a given L-value. The formula for J1 does the job. This was chosen by judicious experimentation which we shall not attempt to reconstruct, but it is easy to demonstrate that it does what is needed.

When K = 1 it is fairly clear that J1 = 1. When K = 2, we have K − 1 is 1, which multiplied by 5 gives 5; when this is divided by 3 in integer arithmetic the quotient is 1; adding 1 gives 2, which we see in the sketch is indeed the J-value that lies just to the left of K = 1. When K = 3 we have K − 1 is 2; times 5 is 10; divided by 3 is 3; plus 1 is 4. When

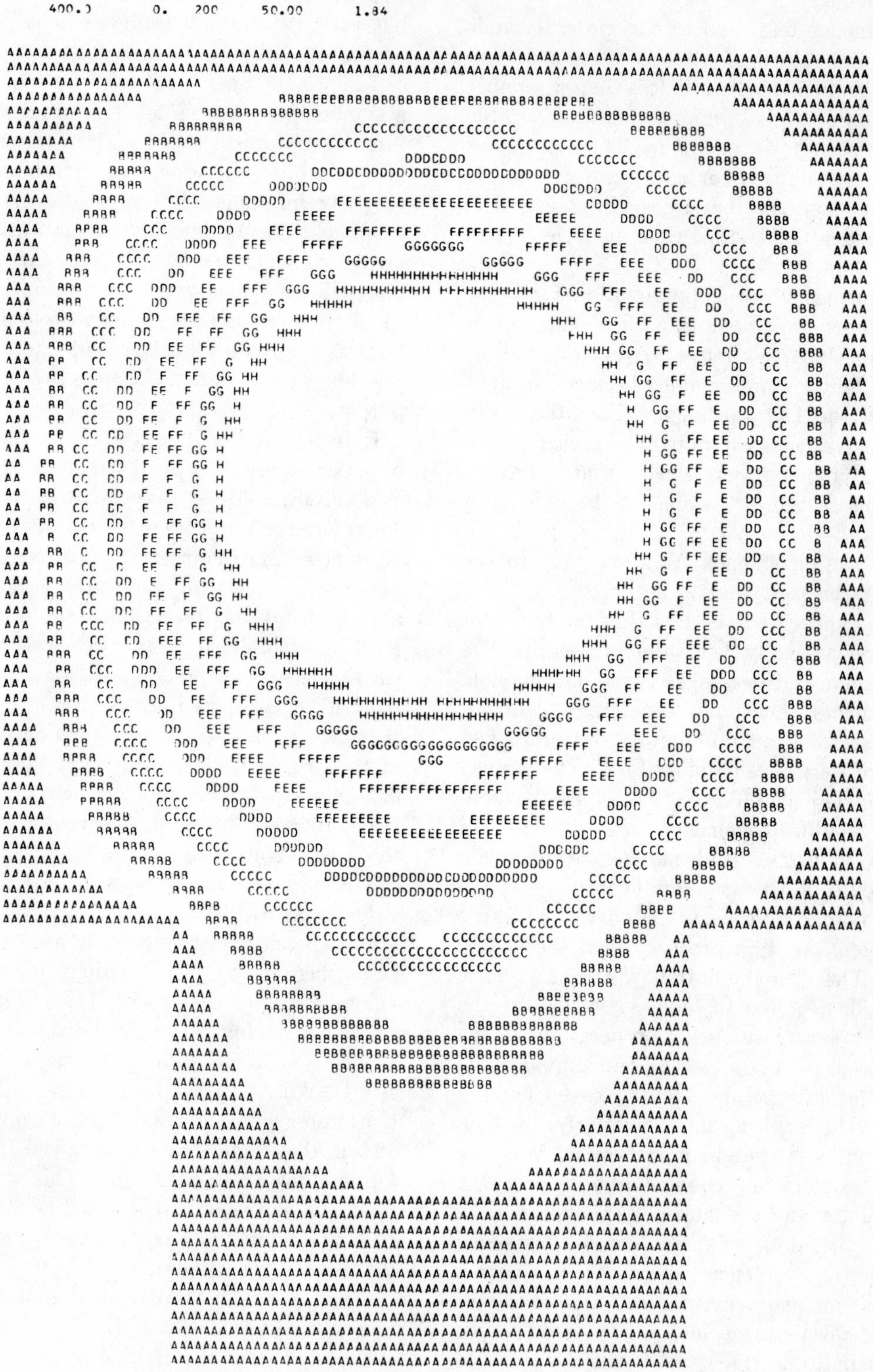

Figure 26.6. The graphical output of the program in Figure 26.3, for LENGTH = 151, corresponding to a length of ½ inch.

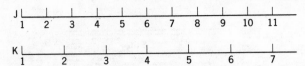

Figure 26.7. The scheme of correspondence between solution points (J, top line) and output points (K, bottom line) used in producing the graphical output in Figure 26.6 with the program in Figure 26.3.

K = 4; K — 1 is 3; times 5 is 15; divided by 3 is 5; plus 1 is 6. This time the integer division gave a result that was exact, which corresponds to the fact that for every third K-value there is an exact match with a J-value.

Now we know the two J-values that bracket K, and we are ready to interpolate. We are dealing with some fixed I-value: we wish to interpolate between two temperatures, both lying on this I-line. To interpolate in this case means:

to the value at the left point, add the product of the difference between the two values, times the fraction of the distance of the desired point from the left value.

For instance, if the given K specifies a point that is one third of the way from J1 to J2, we need to take the temperature of J1 as the base point; subtract the temperature at J1 from the temperature at J2; multiply this difference by ⅓; add the product to the temperature at J1. The reader wishing a more precise statement may review the formula presented in Case Study 14 or he may inspect the statement just before statement 9012.

The thing is, this factor by which we multiply the temperature difference will always be either zero, ⅔, or ⅓. When K = 1, 4, 7, etc., it is zero; for K = 2, 5, 8, etc., it is ⅔; for K = 3, 6, 9, etc., it is ⅓. This may be established by close study of Figure 26.6. If we call these cases 1, 2, and 3, respectively, we can simply set up an array containing the appropriate multipliers and use a simple formula to pick the right one. The formula uses the remainder function MOD. This delivers the remainder on division of the first argument by the second. The remainder on dividing any integer by 3 is zero, 1, or 2; we add 1 to get a legal FORTRAN subscript.

The actual interpolation is now a matter of one statement.

Having picked a value of K, we do the interpolation for the entire line of 101 I-values, storing the interpolated temperatures in an array named LINE. As a part of this same loop we test whether the point that we have chosen is actually in the pipe or whether it is in the hole or outside the fin. This check is made here with an elaborate IF statement, one that has parentheses in a logical expression—something we have not seen often. The parentheses around the four conditions "and-ed" together to check for a point in the hole are not strictly necessary: "and" is a stronger operator than "or" so that the action would be the same without these parentheses. In the test for a point outside the fin it is different. We wish to ask: is the point to the right of J2 = 101 *and either* below I = 20 or above

```
6.5665E-01  1.4504E 00  2.6926E 00  4.1259E 00  5.5294E 00  6.9696E 00  8.5192E 00  1.0170E 01  1.1860E 01  1.3584E 01
1.5357E 01  1.7143E 01  1.8871E 01  2.0523E 01  2.2088E 01  2.3584E 01  2.5026E 01  2.6408E 01  2.7723E 01  2.8973E 01
3.0162E 01  3.1294E 01  3.2370E 01  3.3392E 01  3.4361E 01  3.5279E 01  3.6148E 01  3.6973E 01  3.7753E 01  3.8492E 01
3.9191E 01  3.9852E 01  4.0478E 01  4.1070E 01  4.1630E 01  4.2159E 01  4.2660E 01  4.3134E 01  4.3582E 01  4.4006E 01
4.4406E 01  4.4785E 01  4.5144E 01  4.5483E 01  4.5803E 01  4.6107E 01  4.6394E 01  4.6665E 01  4.6922E 01  4.7165E 01
4.7395E 01  4.7613E 01  4.7819E 01  4.8014E 01  4.8199E 01  4.8374E 01  4.8539E 01  4.8696E 01  4.8844E 01  4.8985E 01
4.9118E 01  4.9244E 01  4.9363E 01  4.9476E 01  4.9583E 01  4.9685E 01  4.9781E 01  4.9872E 01  4.9958E 01  5.0040E 01
5.0118E 01  5.0191E 01  5.0261E 01  5.0327E 01  5.0389E 01  5.0448E 01  5.0504E 01  5.0558E 01  5.0608E 01  5.0656E 01
5.0701E 01  5.0744E 01  5.0785E 01  5.0823E 01  5.0860E 01  5.0894E 01  5.0927E 01  5.0958E 01  5.0988E 01  5.1016E 01
5.1042E 01  5.1067E 01  5.1091E 01  5.1114E 01  5.1135E 01  5.1155E 01  5.1174E 01  5.1193E 01  5.1210E 01  5.1226E 01
5.1242E 01  5.1256E 01  5.1270E 01  5.1283E 01  5.1296E 01  5.1308E 01  5.1319E 01  5.1330E 01  5.1340E 01  5.1349E 01
5.1358E 01  5.1367E 01  5.1375E 01  5.1383E 01  5.1390E 01  5.1397E 01 -0.0000E-20
```

```
1.6976E 05  1.3250E 05  9.9675E 04  7.5830E 04  5.8927E 04  4.7621E 04  4.0571E 04  3.0481E 04  2.3972E 04  1.9792E 04
1.6402E 04  1.4425E 04  1.2222E 04  1.0972E 04  9.5759E 03  8.3523E 03  7.5607E 03  7.2135E 03  6.3946E 03  5.8887E 03
5.3591E 03  4.9822E 03  4.6795E 03  4.3950E 03  4.1483E 03  3.9320E 03  3.7376E 03  3.5571E 03  3.3860E 03  3.2242E 03
3.0715E 03  2.9260E 03  2.7863E 03  2.6554E 03  2.5280E 03  2.4079E 03  2.2927E 03  2.1830E 03  2.0785E 03  1.9788E 03
1.8836E 03  1.7931E 03  1.7067E 03  1.6244E 03  1.5459E 03  1.4711E 03  1.3999E 03  1.3321E 03  1.2676E 03  1.2062E 03
1.1477E 03  1.0921E 03  1.0392E 03  9.8880E 02  9.4090E 02  8.9533E 02  8.5198E 02  8.1075E 02  7.7154E 02  7.3423E 02
6.9875E 02  6.6500E 02  6.3290E 02  6.0238E 02  5.7334E 02  5.4572E 02  5.1945E 02  4.9446E 02  4.7069E 02  4.4808E 02
4.2657E 02  4.0611E 02  3.8664E 02  3.6811E 02  3.5049E 02  3.3372E 02  3.1776E 02  3.0258E 02  2.8813E 02  2.7438E 02
2.6129E 02  2.4883E 02  2.3698E 02  2.2570E 02  2.1495E 02  2.0473E 02  1.9499E 02  1.8573E 02  1.7691E 02  1.6851E 02
1.6052E 02  1.5291E 02  1.4566E 02  1.3876E 02  1.3219E 02  1.2593E 02  1.1998E 02  1.1430E 02  1.0890E 02  1.0376E 02
9.8853E 01  9.4189E 01  8.9742E 01  8.5508E 01  8.1476E 01  7.7638E 01  7.3978E 01  7.0494E 01  6.7178E 01  6.4015E 01
6.1002E 01  5.8132E 01  5.5400E 01  5.2799E 01  5.0316E 01  4.7952E 01 -0.0000E-20
```

Figure 26.8. The numerical output of the program of Figure 26.3, with LENGTH = 151. The top half gives the successive approximations to the temperature at one point in the fin; the bottom half shows the successive values of the sum of the residues for an entire sweep.

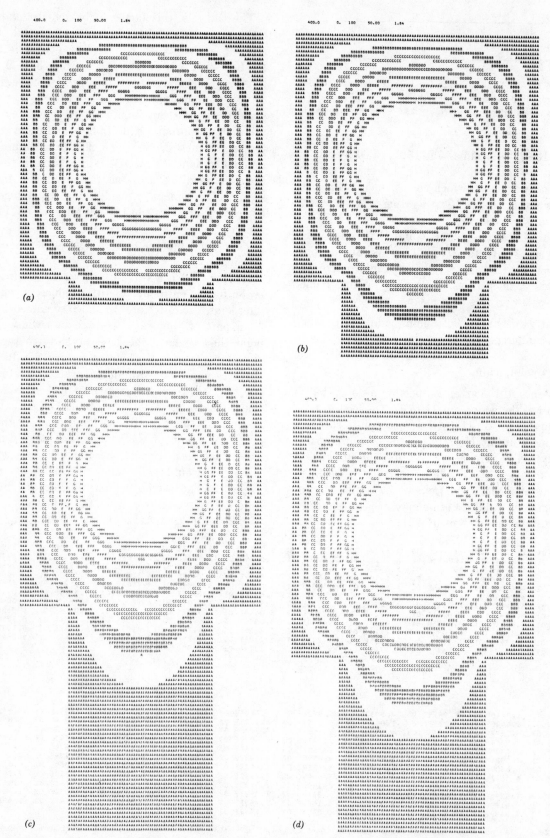

Figure 26.9. Four other solutions of the pipe temperature problem for LENGTH = 111, 131, 171, and 191.

I = 82? In other words, the question "is J2 greater than 101?" must be distributed over both of the other questions, which is the effect of the parentheses. When it is found that a point is not in the material, we set that temperature equal to 401.0, which turns out to produce a blank, as we now proceed to investigate.

The basic scheme for plotting the results is that temperatures will be translated either into a letter or a blank, according to the pattern of Table 26.1.

TABLE 26.1

Temperature	Symbol
0–26.6	A
53.3–80.0	B
106.7–133.3	C
160.0–186.7	D
213.3–240.0	E
266.7–293.3	F
320.0–346.7	G
373.3–400.0	H

The idea is that each letter will represent a band of temperatures about 26.7° wide, and between each of these bands will be a band of the same width for which a blank will be substituted. Areas of constant temperature, or "constant" to the extent of a band of temperatures, will be printed by a letter or a blank.

This is accomplished by dividing each temperature by 26.667 and adding 1, after which the result is truncated to an integer. The number will be in the range of 1 to 16 for temperatures from zero to 401.0 and will then be used as a subscript to pick out

an entry from a DATA statement that contains the eight letters we want, separated by blanks. The chosen character is replaced in the location of the temperature. When the entire array LINE has been converted from temperatures to printing characters, the line is printed and we go back to get another line. When the entire graph has been printed, we go back to read another data card and to do the whole job over.

The program was run first for LENGTH = 151, MAXIT = 200, and EPS = 50. Figure 26.8 shows the values printed for the temperature in the fin and the sum of the residues. We see that 116 interations were required to achieve convergence. The graphical representation of the solution is shown in Figure 26.6.

In Figure 26.9 we have the pictorial solution for four other pipes, with different fin lengths. We see what any reader familiar with Laplace's equation probably already has guessed: a change in the boundary of a region has little effect at points far removed from the change. *Any* change anywhere affects the solution *everywhere* — but not by much. Stated in an intuitive sort of way, the heat from the hot fluid in the hole flows out the shortest way, if the entire outside really is at the same temperature; the length of the fin just doesn't matter much. The reader who expected that lengthening the fin would provide more surface area for heat transfer should realize that in many practical situations the fluid on the outside, air for instance, would not really be held at a constant temperature right at the surface. In this more realistic case, which would be considerably harder to deal with, the length of the fin would have a greater effect.

CASE STUDY 27 FOURIER ANALYSIS OF OSCILLOSCOPE TRACE PHOTOGRAPHS

A recurring task in applied mathematics is to find an approximate representation of data. A polynomial representation is in some ways the simplest form, and we have extensive literature on ways to do the curve fitting required to find the coefficients of some "best" fit to the data. If the data refers to some physical phenomenon that follows, say, an exponential curve, then we would naturally turn to methods of finding the coefficients of a good fit of an exponential; and if the data represents a periodic function we turn immediately to a representation in terms of sines and cosines—that is, the Fourier series.

We shall restrict ourselves to the case of discrete data points, in which summations replace the integrations of the usual Fourier series presentations. In this case we have the values of some experimental function at points on the horizontal axis that have an equal spacing of unity. We then wish to represent the function by some linear combination of the functions:

$$1, \quad \cos \frac{\pi}{N} x, \quad \cos \frac{2\pi}{N} x, \quad \ldots,$$
$$\cos \frac{(N-1)\pi}{N} x, \quad \cos \frac{\pi}{N} Nx$$
$$\sin \frac{\pi}{N} x, \quad \sin \frac{2\pi}{N} x, \quad \ldots, \quad \sin \frac{(N-1)\pi}{N} x$$

that is, we assume that our arbitrary function $f(x)$ can be represented by a series of the form

$$f(x) = \frac{a_0}{2} + \sum_{k=1}^{N} \left(a_k \cos \frac{\pi}{N} kx + b_k \sin \frac{\pi}{N} kx \right) + \frac{a_n}{2} \cos \pi x$$

Our task is to determine the a_k's and b_k's to make the sum of the series exactly equal to the function values at the $2N$ sample points. (We shall work only with an even number of sample points.) As a tool in this process, we need the following *orthogonality* relations, which we state without proof:

$$\sum_{x=0}^{2N-1} \sin \frac{\pi}{N} kx \sin \frac{\pi}{N} mx = \begin{cases} 0 & \text{if } k \neq m \\ N & \text{if } k = m \neq 0 \end{cases}$$

$$\sum_{x=0}^{2N-1} \sin \frac{\pi}{N} kx \cos \frac{\pi}{N} mx = 0$$

$$\sum_{x=0}^{2N-1} \cos \frac{\pi}{N} kx \cos \frac{\pi}{N} mx = \begin{cases} 0 & \text{if } k \neq m \\ N & \text{if } k = m \neq 0, N \\ 2N & \text{if } k = m = 0, N \end{cases}$$

Now we can determine the coefficients of the expansion. To obtain a_k ($k = 1, 2, \ldots, N-1$) we multiply both sides of the assumed series by $\cos (\pi/N)mx$ and sum over all x. The result is

$$\sum_{x=0}^{2N-1} f(x) \cos \frac{\pi}{N} mx = N a_m \qquad 1 \leq m \leq N-1$$

The formula is simple precisely because of the orthogonality relations: most of the terms in the expansion of the product drop out. Using $\sin (\pi/N)mx$ in the same way, we get

$$\sum_{x=0}^{2N-1} f(x) \sin \frac{\pi}{N} mx = N b_m \qquad 1 \leq m \leq N-1$$

Finally, we have

$$\sum_{x=0}^{2N-1} f(x) = N a_0$$

$$\sum_{x=0}^{2N-1} f(x) \cos \pi x = N a_N$$

```
C CASE STUDY 27
C FOURIER ANALYSIS OF OSCILLOSCOPE TRACE PHOTOGRAPHS
C
      INTEGER X, XEND, N, NEND, KEND, N2, POINTS
      REAL NREAL, KREAL, FORM(12), F(50), A(26), B(24), IDENT(12)
      REAL PERCNT(24)
C READ AND WRITE HEADING LINE
   60 READ (5, 58) IDENT
      WRITE (6, 40) IDENT
      READ (5, 62) N
C READ OBJECT-TIME FORMAT GIVING DATA ARRANGEMENT
      READ (5, 58) FORM
      POINTS = 2 * N - 1
      READ (5, FORM) FO, (F(L), L = 1, POINTS)
C GET CONSTANT TERM
      AO = FO
      DO 63 I = 1, POINTS
   63 AO = AO + F(I)
      NREAL = N
      AO = AO / NREAL
C GET OTHER TERMS
      KEND = N - 1
      DO 64 K = 1, KEND
      A(K) = FO
      B(K) = 0.0
      KREAL = K
      DO 65 X = 1, POINTS
      XREAL = X
      ARG = 3.14159265 * KREAL * XREAL / NREAL
      A(K) = A(K) + F(X) * COS(ARG)
   65 B(K) = B(K) + F(X) * SIN(ARG)
      A(K) = A(K) / NREAL
      B(K) = B(K) / NREAL
C GET HARMONIC PERCENTAGES
   64 PERCNT(K) = SQRT((A(K)**2 + B(K)**2) / (A(1)**2 + B(1)**2))*100.0
C GET LAST TERM
      A(N) = FO
      DO 66 X = 1, POINTS
      XREAL = X
      ARG = 3.14159265 * XREAL
   66 A(N) = A(N) + F(X) * COS(ARG)
      A(N) = A(N) / NREAL
      WRITE (6, 41) N, POINTS
      WRITE (6, 42) AO
      WRITE (6, 43) A(1), B(1)
      NEND = N - 1
      DO 74 I = 2, NEND
   74 WRITE (6, 44) I, A(I), I, B(I), I, PERCNT(I)
      WRITE (6, 45) N, A(N)
      GO TO 60
   40 FORMAT (1H1, 12A6//)
   41 FORMAT (1H0, 3HN =, I4, 5X, 8HPOINTS =, I4)
   42 FORMAT (1H0, 8HA( 0) = , F10.5)
   43 FORMAT (1H0, 8HA( 1) = , F10.5, 5X, 8HB( 1) = , F10.5, 5X,
     1   14HFUNDAMENTAL = , 15H100.00 PER CENT)
   44 FORMAT (1H0, 2HA(, I2, 4H) = , F10.5, 5X, 2HB(, I2, 4H) = , F10.5,
     1 5X, 9HHARMONIC , I2, 3H = , F6.2, 9H PER CENT)
   45 FORMAT (1H0, 2HA(, I2, 4H) = , F10.5)
   58 FORMAT (12A6)
   62 FORMAT (I2)
      END
```

Figure 27.1. A program to produce a Fourier analysis of experimental data.

The last are in the same form as the others, which is the advantage of writing the assumed series form with the a_0 and a_N coefficients divided by 2.

Let us combine these results and rewrite them slightly in preparation for getting to work on the program.

$$a_k = \frac{1}{N} \sum_{x=0}^{2N-1} f(x) \cos \frac{\pi}{N} kx \qquad k = 0, 1, 2, \ldots, N$$

$$b_k = \frac{1}{N} \sum_{x=0}^{2N-1} f(x) \sin \frac{\pi}{N} kx \qquad k = 1, 2, \ldots, N-1$$

The notation and development is that of Hamming's excellent treatment.*

Turning to the programming problem, we have a not unfamiliar choice: a moderately simple program can be written by attacking these formulas directly or we can expend some additional effort and develop a program that is longer to write but much faster to execute. The latter approach would hinge on use of recursion relationships to avoid computation of all but a few of the sines and cosines

* Richard W. Hamming, *Numerical Methods for Scientists and Engineers*, McGraw-Hill, New York, 1962, pp. 67–71.

by references to the FORTRAN functions. If we were working with a slow computer or had a great many data sets to process, it would definitely be worthwhile to explore ways to buy speed with programming, but here we shall take the simple way out and program the summations just about as they are written above.*

The program in Figure 27.1 approaches the computation in a straightforward manner. We read and write an identification, read N, read an object-time format, and read the data points. For notational consistency we call the first data point FO; an alternative approach, of course, would be to put all of the data points in the array F, but then the subscripts would not be consistent with the notation of the formulas. For the sake of understandability it seemed reasonable to take this approach, which turns out not to cost anything in time.

With this assumption, then, we always have a first data point FO, followed by 2N − 1 additional points. This number 2N − 1 turns up enough times

* Readers interested in rapid calculation would do well to refer to the paper "An Algorithm for the Machine Calculation of Complex Fourier Series," James W. Cooley and John W. Tukey, *Mathematics of Computation*, **19**, No. 90 (April 1965) pp. 297–301.

```
        SIN(X) + SIN(2X+30)


    N =   12      POINTS =  23

    A( 0) =    -0.00000

    A( 1) =     0.00000    B( 1) =     1.00000    FUNDAMENTAL = 100.00 PER CENT

    A( 2) =     0.86603    B( 2) =     0.50000    HARMONIC   2 = 100.00 PER CENT

    A( 3) =    -0.00000    B( 3) =    -0.00000    HARMONIC   3 =   0.00 PER CENT

    A( 4) =    -0.00000    B( 4) =     0.00000    HARMONIC   4 =   0.00 PER CENT

    A( 5) =    -0.00000    B( 5) =    -0.00000    HARMONIC   5 =   0.00 PER CENT

    A( 6) =    -0.00000    B( 6) =    -0.00000    HARMONIC   6 =   0.00 PER CENT

    A( 7) =    -0.00000    B( 7) =     0.00000    HARMONIC   7 =   0.00 PER CENT

    A( 8) =     0.00000    B( 8) =     0.00000    HARMONIC   8 =   0.00 PER CENT

    A( 9) =    -0.00000    B( 9) =    -0.00000    HARMONIC   9 =   0.00 PER CENT

    A(10) =     0.00000    B(10) =    -0.00000    HARMONIC  10 =   0.00 PER CENT

    A(11) =     0.00000    B(11) =     0.00000    HARMONIC  11 =   0.00 PER CENT

    A(12) =    -0.00000
```

Figure 27.2.   The output of the program in Figure 27.1 for a test case based on table data.

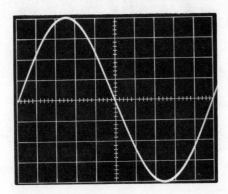

Figure 27.3. Oscilloscope trace photograph of the sine wave produced by a signal generator.

SINE WAVE FROM SIGNAL GENERATOR

N = 10   POINTS = 19

A(0) =   -0.06000

A(1) =   0.01176   B(1) =   4.02544   FUNDAMENTAL = 100.00 PER CENT

A(2) =   -0.04236   B(2) =   -0.04253   HARMONIC 2 =   1.49 PER CENT

A(3) =   -0.01902   B(3) =   -0.06017   HARMONIC 3 =   1.57 PER CENT

A(4) =   -0.01000   B(4) =   -0.01176   HARMONIC 4 =   0.38 PER CENT

A(5) =   0.00000   B(5) =   -0.00000   HARMONIC 5 =   0.00 PER CENT

A(6) =   0.00236   B(6) =   -0.02629   HARMONIC 6 =   0.66 PER CENT

A(7) =   0.01902   B(7) =   0.00027   HARMONIC 7 =   0.47 PER CENT

A(8) =   -0.01000   B(8) =   0.01902   HARMONIC 8 =   0.53 PER CENT

A(9) =   -0.01176   B(9) =   0.01466   HARMONIC 9 =   0.47 PER CENT

A(10) =   -0.02000

Figure 27.4. Fourier analysis by the program in Figure 27.1 of the sine wave in Figure 27.3.

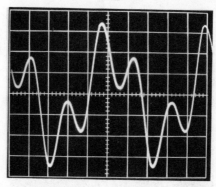

Figure 27.5. Oscilloscope trace photograph of the sum of the a-c line current with a signal generator sine wave of three times line frequency.

```
60 CPS LINE & 180 CPS FROM SIGNAL GENERATOR

N =  13      POINTS =  25

A( 0) =   -0.07692

A( 1) =   -0.96168      B( 1) =   -1.90742      FUNDAMENTAL = 100.00 PER CENT

A( 2) =    0.04065      B( 2) =    0.00104      HARMONIC  2 =    1.90 PER CENT

A( 3) =    0.03497      B( 3) =   -1.57153      HARMONIC  3 =   73.59 PER CENT

A( 4) =    0.03504      B( 4) =   -0.00315      HARMONIC  4 =    1.65 PER CENT

A( 5) =    0.02846      B( 5) =   -0.01624      HARMONIC  5 =    1.53 PER CENT

A( 6) =    0.01652      B( 6) =    0.02116      HARMONIC  6 =    1.26 PER CENT

A( 7) =   -0.02795      B( 7) =    0.01341      HARMONIC  7 =    1.45 PER CENT

A( 8) =    0.00786      B( 8) =    0.00893      HARMONIC  8 =    0.56 PER CENT

A( 9) =   -0.00901      B( 9) =    0.01067      HARMONIC  9 =    0.65 PER CENT

A(10) =   -0.01307      B(10) =   -0.00480      HARMONIC 10 =    0.65 PER CENT

A(11) =   -0.00710      B(11) =    0.03590      HARMONIC 11 =    1.71 PER CENT

A(12) =    0.00146      B(12) =    0.00393      HARMONIC 12 =    0.20 PER CENT

A(13) =   -0.01538
```

Figure 27.6. Fourier analysis of the waveform in Figure 27.5.

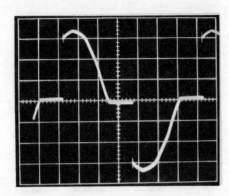

Figure 27.7. The waveform of a silicon-controlled rectifier used for dimming lights.

DIMMER SWITCH WAVEFORM

N = 12   POINTS =   23

A(0) =   -0.05000

A(1) = -0.89464	B(1) = 2.54959	FUNDAMENTAL = 100.00 PER CENT
A(2) = -0.02388	B(2) = 0.03527	HARMONIC 2 = 1.58 PER CENT
A(3) = 0.65767	B(3) = -0.58099	HARMONIC 3 = 32.48 PER CENT
A(4) = 0.01250	B(4) = 0.02165	HARMONIC 4 = 0.93 PER CENT
A(5) = 0.26196	B(5) = 0.38780	HARMONIC 5 = 17.32 PER CENT
A(6) = 0.00833	B(6) = -0.00833	HARMONIC 6 = 0.44 PER CENT
A(7) = -0.27122	B(7) = -0.13397	HARMONIC 7 = 11.20 PER CENT
A(8) = -0.01250	B(8) = -0.00722	HARMONIC 8 = 0.53 PER CENT
A(9) = 0.29233	B(9) = -0.11433	HARMONIC 9 = 11.62 PER CENT
A(10) = -0.00945	B(10) = 0.00640	HARMONIC 10 = 0.42 PER CENT
A(11) = -0.09610	B(11) = 0.23803	HARMONIC 11 = 9.50 PER CENT

A(12) =   0.00000

Figure 27.8. Fourier analysis of the waveform in Figure 27.7.

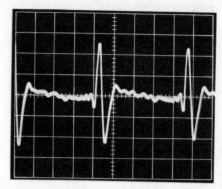

Figure 27.9. The waveform produced by the open A-string of an electric guitar.

GUITAR, OPEN A-STRING

N = 11   POINTS = 21

A(0) =   -0.10909

A(1) =   -0.14438   B(1) =   0.04342   FUNDAMENTAL = 100.00 PER CENT

A(2) =   0.14870   B(2) =   -0.18454   HARMONIC   2 = 157.19 PER CENT

A(3) =   0.33636   B(3) =   0.05035   HARMONIC   3 = 225.58 PER CENT

A(4) =   0.20437   B(4) =   0.33304   HARMONIC   4 = 259.17 PER CENT

A(5) =   -0.04055   B(5) =   0.39618   HARMONIC   5 = 264.14 PER CENT

A(6) =   -0.29110   B(6) =   0.20261   HARMONIC   6 = 235.24 PER CENT

A(7) =   -0.29404   B(7) =   0.02413   HARMONIC   7 = 195.68 PER CENT

A(8) =   -0.15732   B(8) =   -0.11034   HARMONIC   8 = 127.45 PER CENT

A(9) =   -0.03466   B(9) =   -0.11907   HARMONIC   9 = 82.26 PER CENT

A(10) =   -0.00011   B(10) =   -0.07219   HARMONIC 10 = 47.88 PER CENT

A(11) =   0.05455

Figure 27.10. Fourier analysis of the waveform in Figure 27.9.

to make it convenient to precompute it and give it a name: POINTS.

One concession to speed has been made: advantage is taken of the simplifications that occur in the formulas for a_0 and a_N. Thus a_0 requires only a simple summation, with no sines or cosines, followed by a division by N, taken as a real number rather than the integer form in which it was read. Not counting a_N, there are now $N - 1$ sine coefficients and $N - 1$ cosine coefficients to compute. We have a DO loop that runs from 1 to $N - 1$. When $x = 0$ the cosine is 1.0 and the sine is zero, which lead to the starting values for the summations of A(K) and B(K) shown. The rest of the summation process is a literal transcription of the formula.

After computing each a_k, b_k pair, we compute the percentage of the fundamental represented by the harmonic. (The lowest frequency is called the fundamental and is also called the first harmonic; the other frequencies are called the second and third and so on.) The fundamental frequency, represented by a_1, b_1, is taken to be 100%. The use of the square root of the sum of the squares is justified by the fact that

$$a \sin x + b \cos x = \sqrt{a^2 + b^2} \sin (x + \varphi)$$

where $\varphi = \tan^{-1} b/a$. This means, incidentally, that we can view the Fourier series representation as a sum of sine terms, in which each sine has a frequency that is an integral multiple of the fundamental and each has a phase offset; that is, the phase offset can be taken into account either by forming the sum of a sine and a cosine or by exhibiting the offset directly.

Getting the last term is a simple matter, after which we print the results, using slightly elaborate FORMAT statements to get a printed report that is well identified.

Figure 27.2 shows the output from a test case in which the data points were generated by using a table of sines and a desk calculator with the function

$$\sin x + \sin (2x + 30)$$

We see that the constant term is zero, which means that the average value of the function is zero. The first and second harmonics have the same magnitude, as they should, and all others are zero.

Figure 27.3 is an oscilloscope trace photograph of a waveform produced by the author's Heathkit signal generator, and Figure 27.4 is the output of the Fourier analysis program when run with the data values taken from measurements of the pattern. Considering the inherent inaccuracy of the method of measurement, the harmonics of less than 2% are quite satisfactory.

With this program in hand, shutting off the examples is about like eating only two peanuts at a cocktail party. Figures 27.5 and 27.6 tell the story of two waves added together: the a-c line waveform and a sine wave of three times the line frequency. The line waveform is rather far from sinusoidal, as we see by inspection of the size of the harmonics other than the third. Figures 27.7 and 27.8 show the waveform and its analysis for the output of a silicon-controlled rectifier used for dimming lights. Finally, Figures 27.9 and 27.10 analyze the remarkably complex waveform produced by the open A-string of guitarist Charles McCracken.

CASE STUDY 28 ION TRAJECTORIES IN A MONOPOLE MASS SPECTROMETER; OFF-LINE PLOTTING OF THE SOLUTION OF A DIFFERENTIAL EQUATION

A mass spectrometer is a device that separates ions on the basis of differences in charge/mass ratio, either to establish the presence of a particular ion or to analyze a mixture of ions. In the familiar magnetic instrument a beam of ions of the material to be analyzed is injected at right angles into a uniform magnetic field, where they follow circular arcs of curvatures that are proportional to the square roots of the charge/mass ratios. By moving a collector or varying the strength of the magnetic field, the ion currents of the various charge/mass ratios can be determined.

In this case study we shall carry out some of the calculations needed in design studies of a mass spectrometer of a rather different type, borrowing heavily from a paper by Lever.* We shall consider three aspects of the study: the integration of a second-order differential equation, plotting the results of a computation on a special plotter not attached to the computer, and, briefly, the use of computations in design studies. Let us begin with a sketch of the physical and mathematical situation.

Figure 28.1 is a sketch of the essentials of the monopole mass spectrometer. A radio-

* R. F. Lever, "Computation of Ion Trajectories in the Monopole Mass Spectrometer by Numerical Integration of Mathieu's Equation," *IBM J. Res. Develop.*, **10**, No. 1, January 1966.

frequency electric field is maintained in the space between the cylindrical and the Vee electrodes. A beam of monoenergetic ions is injected into this field at an angle in the design proposed by Lever; such an instrument may also be designed to inject the ions parallel to the Z-axis. The electric field is described by

$$\phi = \frac{x^2 - y^2}{r_0{}^2} \quad (U + V \cos \omega t)$$

This field has the desirable property, from the computational standpoint, that its y-component is independent of x and vice versa, so that we may study the trajectories in the x- and y-directions separately. These trajectories are determined from Mathieu's equation:

$$\frac{d^2V}{dZ^2} + (a + 2q \cos 2Z)\, V = 0$$

in which V is either x or y, and a and q are constants that depend on the physical dimensions of the device, the field strength, the frequency, and the charge and mass of the ion. For $V = x$, a and q are positive, and for $V = y$ they are negative.

The path followed by an ion traversing such a field is a complex one, as we shall see. The purpose of the initial phase of the study is to determine whether a and q can be chosen as to guarantee that an ion injected into

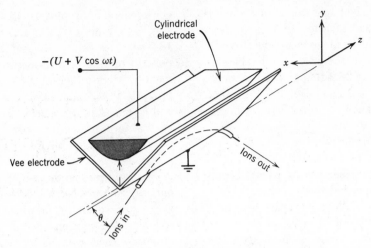

Figure 28.1. Schematic representation of the design of a proposed monopole mass spectrometer.

this field will "focus" at the collector, which would be placed at the point marked "ions out" in Figure 28.1. If we cannot operate the instrument so that ions of a specified charge/mass ratio converge to a fairly sharp point, the device is of little value.

This issue is further complicated by the fact that the ions are being injected in a continuous stream, so that they enter in all phases of the electric field. One ion may enter just as the field goes through zero, another at its maximum, and so on. The major question we shall consider is, do ions of all phases, under suitable operating conditions, arrive in focus at the collector?

Fortunately, blind groping is not required. Analysis that we shall not reproduce shows that only certain combinations of values of a and q can possibly produce a focus; the derivation of these combinations is well presented in Lever's paper. We shall simply accept these results and choose a pair of values to use in illustrating the program.

We are now ready to attack the matter of integrating the differential equation. Let us review, for the sake of a clear understanding of what we are doing, the meaning of the differential equation at hand.

The first derivative of a function gives its slope and the second its curvature. In general, a second-order differential equation will have both derivatives, but ours has not: the first derivative is missing. Our situation therefore is this: at a given starting point, Z_0, we know the value of the function and its first derivative; knowing these values and the equation for the second derivative, what is the value of the function and of the first derivative a short distance from the starting point? Having found

the new values, what are the values a further short distance away? We proceed to ask and answer these questions as far down the Z-axis as interests us.

The heart of the program is the procedure for stepping along the Z-values and finding the function values at each new point. There are many numerical techniques for doing this; the one we shall use is the Runge-Kutta method, modified for use with a second-order equation having no first derivative. The procedure is this. Rewrite the equation as

$$\frac{d^2V}{dZ^2} = -(a + 2q \cos 2Z)\,V = DERIV(Z, V)$$

where we have written on the right the name of the arithmetic statement function that will be used in the program to represent the equation. Now suppose that we have found the solution at a point Z_i, where the value of the function is V_i and of the first derivative is DV_i. Then to find the values of the function and derivative at a point $Z_{i+1} = Z_i + h$, apply the following formulas in the order written:

$a_1 = h * DERIV(Z_i, V_i)$

$a_2 = h * DERIV(Z_i + h/2, V_i + h/2 * DV_i + h/8 * a_1)$

$a_3 = h * DERIV(Z_i + h, V_i + h * DV_i + h/2 * a_2)$

Then the desired new values of V and DV at $Z_{i+1} = Z + h$ are

$$V_{i+1} = V_i + h * (DV_i + (a_1 + 2a_2)/6)$$

$$DV_{i+1} = DV_i + (a_1 + 4a_2 + a_3)/6$$

Often these values would simply be printed as they were computed, requiring very little storage. For our purposes, however, we need to work with the complete set of solution values after they have been found, so we place all the values in arrays.

We can see this much of the procedure in the part of the program of Figure 28.2 that is headed "solve differential equation." The equation will, in fact, be solved 16 times for each combination of a and q, varying the starting values of Z over a range of zero to $15\pi/16$, in order to study the effect of variations in phase of the electric field on the ion trajectories. The solution section therefore begins with a DO loop that generates these starting values. Then the initial values of V and DV are assigned; frequently these values would be read from a data card because in a complete design study they would need to be varied. Similarly, we would normally provide for varying h, but here we wish to concentrate on other things and simply assign a suitable value with an assignment statement.

Now we have the actual equation solution, set up to be evaluated at 600 points. With the value of h chosen, this will carry the solution out to 12π; from theoretical analyses presented in Lever's paper, we know that a focus should occur at 10π. The computation of V and DV is a literal transcription of the statements given above. Note, however, the techniques for incrementing Z: we convert the index to floating point and multiply by h, rather than simply adding h in a statement that might read

$$Z(I+1) = Z(I) + H$$

The purpose is to minimize the roundoff error that would accumulate in the latter method—a problem to which we trust the reader is properly sensitive by now.

When all 600 values have been found, we are almost ready to plot them, but not quite. Recall that we introduced the factor of phase variations by starting the solutions at values between zero and $15\pi/16$, so that the value in $Z(1)$ is not zero except for the first curve; but we wish to plot the vibrations in the x- and y-directions as a function of *position*, not as a function of their relation to the field; assigning $Z(1)$ a nonzero value was a convenient way to "trick" the equation into accounting for the variations of phase on ions entering at varying times; there was no suggestion that the point of injection moves physically. In short, to get the picture of the focusing properties that is our goal, we must now subtract the starting value from all entries in the Z-array.

Now we must back away from the solution and consider how to plot it. The reader has already seen one kind of plot of the solution in Figure 18.5 of Case Study 18; but that was only a temporary, preliminary measure taken in the development of this program to see if it was behaving properly in a gross sense. Now we would like to see a plot that is quite accurate, and we would like all 16 curves for one set of operating conditions to be plotted on one graph in order to see the focussing properties.

The solution is to use an off-line plotter. The one used in our case was a Gerber Series 600, which accepts a specially prepared magnetic tape that specifies the motion of the plotting arm in two dimensions, as well as the control of the plotting pen. The format of the tape that controls the plotter must conform precisely to the requirements laid down in the appropriate manual; the conversion from the form in which our results appear in the computer storage to the form the plotter needs is a rather complex task and one that the individual programmer should not have to do himself. The answer, of course, is a set of subroutines, in this case prepared by systems programmers at the Service Bureau Corporation, where the programs in this book were tested.

Not every reader concerned with plotting will have the same plotter and subroutine package, so we shall be content with sketching the general features of the subroutines.

The first call, of PLOTS, mostly establishes BUFFER as the name of an array of 500 locations in which plotter commands can be assembled until they are written on a tape. PLOT establishes the reference point for all subsequent operations that involve a table position; we have said that it is one inch above and one inch to the right of the position at which the plotter operator initially locates the pen, and the −2 says that the move to that location should be made with the pen up.

Now we are ready to draw axes. The AXIS call says, reading across the parameters: this axis begins at the origin established in the PLOT call; it is 4 inches long; it is at an angle of 90° to the horizontal; the label to be written is the letter X; there is one character in this label, and the minus sign says that the label and scale should be on the negative side of the axis; the starting point on this axis is associated with a value of −20; one inch on this axis corresponds to 10 units in the variable that is measured along this axis; the −1, finally, means that the scale is to be printed as integers without a decimal point.

```
C CASE STUDY 28
C ION TRAJECTORIES IN A MONOPOLE MASS SPECTROMETER
C INTEGRATION OF A SECOND ORDER DIFFERENTIAL EQUATION
C OUTPUT IS WRITTEN ON TAPE AND PLOTTED WITH AN OFF-LINE PLOTTER
C
      DIMENSION Z(1200), V(1200), DV(1200), BUFFER(500), XX(2), YY(2)
      INTEGER SWITCH
      DATA XX(1), XX(2), YY(1), YY(2)/0.0, 12.0, 2.0, 2.0/
      DERIV(Z,V) = - (A + 2.0*Q*COS(2.0*Z))*V
C SET PARAMETERS FOR X PLOT
      A = 0.225041
      Q = 0.699745
C INITIALIZE PLOTTING ROUTINES
      CALL PLOTS (BUFFER, 500, 1)
C SET PLOTTING TABLE ORIGIN
      CALL PLOT (1.0, 1.0, -2)
C DRAW VERTICAL AXIS FOR X PLOT
      CALL AXIS (0.0, 0.0, 4.0, 90.0, 1HX, -1, -20.0, 10.0, -1)
C SET SWITCH
      SWITCH = 1
C DRAW HORIZONTAL AXIS
    2 CALL AXIS (0.0, 0.0, 12.0, 0.0, 20HZ - Z0,  UNITS OF PI, -20, 0.0,
     1    1.0, -1)
C DRAW 'REAL' AXIS--SEE TEXT
      CALL LINE (XX, YY, 2, 1, 1)
C SOLVE DIFFERENTIAL EQUATION
      DO 70 K = 1, 16
      D = K - 1
      Z(1) = 3.14159265 * D / 16.0
      V(1) = 0.0
      DV(1) = 1.0
      H = 3.14159265 / 50.0
      H2 = H / 2.0
      DO 20 I = 1, 600
      A1 = H * DERIV(Z(I), V(I))
      A2 = H * DERIV(Z(I) + H2, V(I) + H2*DV(I) + H/8.0*A1)
      A3 = H * DERIV(Z(I) + H, V(I) + H*DV(I) + H2*A2)
      V(I+1) = V(I) + H*(DV(I) + (A1 + 2.0*A2)/6.0)
      DV(I+1) = DV(I) +(A1 + 4.0*A2 + A3)/6.0
   20 Z(I+1) = Z(1) + H*FLOAT(I)
C TRANSLATE FOR PLOTTING
      DO 69 I = 2, 600
   69 Z(I) = Z(I) - Z(1)
      Z(1) = 0.0
C CONVERT SOLUTION FROM POSITION COORDINATES TO INCREMENTS
      CALL SCALE (Z, -600, 1, 12.0, 0.0, 3.14159265)
      CALL SCALE (V, -600, 1, 4.0, -20.0, 10.0)
C PLOT THE CURVE CONNECTING THESE POINTS
   70 CALL LINE (Z, V, 600, 1, 1)
C USE SWITCH TO DETERMINE WHETHER THIS COMPLETES FIRST OR SECOND PLOT
      GO TO (100, 200), SWITCH
C RESET THE PLOTTING TABLE ORIGIN
  100 CALL PLOT (20.0, 1.0, -2)
      SWITCH = 2
C SET PARAMETERS FOR Y PLOT
      A = -0.225041
      Q = -0.699745
C DRAW VERTICAL AXIS FOR Y
      CALL AXIS (0.0, 0.0, 4.0, 90.0, 1HY, -1, -20.0, 10.0, -1)
      GO TO 2
C WRAPUP OUTPUT ACTIONS FOR PLOTTER TAPE
  200 CALL PLOTE
      STOP
      END
```

Figure 28.2. A program to produce a magnetic tape for plotting the trajectories of ions in the mass spectrometer in Figure 28.1, using an off-line plotter.

Now we set a switch. This program will be used to find and plot the solutions for both the x- and the y-vibrations; we wish to go through the program twice and need to be able to tell at the end of the equation solution whether we are in the first or the second pass through.

Drawing the horizontal axis is much the same, but with certain more or less obvious changes. The axis is longer, the label is different, and the scale is different. We have said that one inch along this axis corresponds to a value of π and have written an appropriate label. The call of LINE draws a line across the graph, parallel to what has been called the Z-axis. In fact, in terms of usual terminology, the real axis is the line we now draw, whereas the thing earlier called the axis merely exhibits the scale. This was necessary because otherwise the graphs would obliterate the scale indications.

Now we solve the differential equation, as already discussed, ending with two arrays, Z and V, that contain 600 position coordinates for points through which we wish to plot a curve. The plotter requires that all position information be expressed not in terms of position relative to the origin but of incremental position from the present location. This conversion is the main function of the SCALE routine. Reading across the parameters, we have said, in the case of the first call, the array is named Z; we have 600 values to convert, and the scaling information is supplied by this call; the increment in storage location between Z-values is 1; the axis with reference to which these values will be plotted is 12 inches long; the starting value is zero; and one inch along this axis is worth π units in the computed values. If the 600 had been written with a plus sign, the routine would have searched the entire array for the smallest value and the largest

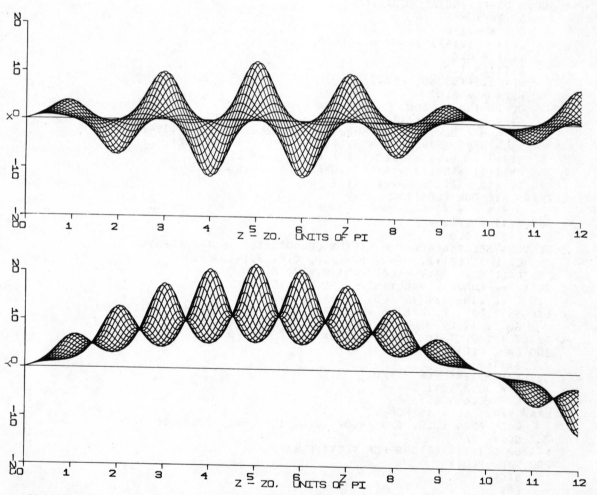

Figure 28.3.   The plots produced by a Gerber Series 600 plotter under control of the magnetic tape produced by the program in Figure 28.2. The plotter produced the plots in their entirety, including axis identification.

value and used what it found for scaling rather than the values we supplied. The point of the specification about the increment between storage locations is that we might wish to store several kinds of numbers in one array and then dictate that this routine is to pick up every second or third entry. The second call of SCALE is not greatly different.

Now that the values have been converted to the form the plotter requires, we tell the LINE routine to write a tape with plotter commands to draw a continuous curve through these points. We name the two arrays for the two dimensions and state that there are 600 points and the storage increment for each is 1. The tape is written as a result of this call. The tape that is written, and this applies to all routines that write plotter commands, is tape 6 on channel A. We have no choice in the matter and accordingly do not have to name it. In the instructions to the operator of the computer we simply specify that a scratch tape (one that is either blank or contains nothing of further value) should be mounted on A6 and saved for us.

This tape is then taken to the plotter at our leisure, mounted on the plotter's tape reader, a suitable piece of paper is taped to the table, and the button is pushed. Some length of time later the graphs of Figure 28.3 are produced. The precision and quality of the results are remarkable.

Returning to the purpose of the whole application, we see that at 10π the graphs of all 16 curves do indeed cross the Z-axis together in the cases of both the x- and y-vibrations.

So much for the question of focussing. The researcher studying the possible design of such an instrument would now be able to use the program for a number of additional studies of the operation of the mass spectrometer. How sensitive would it be to variations in the operating conditions, namely frequency and voltage of the electric field? In some cases a study of the equations provides the answer, but if not it is a simple matter to try it: vary a and q from the values that were expected from theory to provide a focus and see if there is a great change in the focus or if perhaps there is a good focus somewhere else. How sensitive would the operation be to fluctuations in the field along the Z-direction, caused by unavoidable irregularities in construction? This can be studied by rewriting the program to make a and q functions of Z and resolving the equations for typical operating points. Which is better, to inject the ions parallel to the Z-axis or at an angle? Lever's conclusion, based in large part on the results provided by his program, was that better focus and less dependence on accuracy of operating conditions is provided by injection at an angle.

These and other questions are readily attacked by the techniques we have seen. Thus a quite complete analysis of how the proposed instrument would work can be made, without the time and cost of physical experimentation. Naturally, computer-aided analysis cannot answer all the questions, but if after thorough studies physical experimentation seems justified a great deal of guidance is available from the computer studies.

This kind of attack on problems in research and development is highly typical of the application of computers.

CASE STUDY 29 FINDING THE INVERSE LAPLACE TRANSFORM OF A QUOTIENT OF TWO POLYNOMIALS

The method of the Laplace transform is heavily used in several branches of engineering, especially electrical. The technique is most readily described in terms of transformations between the "time domain" and the "frequency domain," although the technique is by no means limited to such applications.

We begin with a problem formulated in terms of some function of time, $f(t)$, which is then transformed into the frequency domain by the Laplace transformation

$$F(s) = \mathfrak{L}[f(t)] = \int_0^\infty f(t)e^{-st}\,dt$$

where $s = \alpha + \beta i$ is the complex Laplace operator. Tables of Laplace transform pairs give the transforms of a great many common functions, and actual integration is seldom required in practice. For instance, Table 29.1 lists a few pairs that we shall need subsequently.

One major reason for using the Laplace transform is that many common mathematical operations on the original function $f(t)$ have parallel but much simpler operations on the result function $F(s)$. For instance, differentiation of the original function $f(t)$, corresponds to multiplication of the result function, $F(s)$, by s and the addition of a constant [i.e., $\mathfrak{L}(df/dt) = s\,F(s) + \text{constant}$]; a translation of the original function by a distance b corresponds to multiplying the result function e^{-bs}.

Many problems are commonly formulated

TABLE 29.1

$f(t)$	$F(s)$
1	$\dfrac{1}{s}$
t	$\dfrac{1}{s^2}$
$\dfrac{t^2}{2}$	$\dfrac{1}{s^3}$
e^{at}	$\dfrac{1}{s-a}$
te^{at}	$\dfrac{1}{(s-a)^2}$
$Ae^{at}\sin(\omega t + \alpha)$	$\dfrac{s+d}{(s-a)^2+\omega^2}$

$$A = \frac{1}{\omega}[(a+d)^2 + \omega^2]^{1/2}$$

$$\alpha = \arctan\frac{\omega}{a+d}$$

initially in terms of the transform, so powerful are the methods that have been developed for dealing with the transformed equations. In some applications it is never necessary to "get back" to the original time function, that is, to make the inverse Laplace transformation. This will be true, for example, if what is needed is simply the response of a system to signals of varying frequencies.

In other cases, however, what is needed is the time response of the system. If the transform corresponds to some simple original function that we can find in a handbook of Laplace transform pairs, well and good. Frequently, however, the transform is too complex for this approach and we must find another. One common case is that the transform is given as the quotient of two polynomials of moderate degree, say 5 to 15. Our attack in that case will be to find the partial fraction representation of the quotient. We have a theorem that the inverse transform of a sum is the sum of the inverse transforms, and the individual partial fractions can then be dealt with separately. Barring denominator roots of high multiplicity, which would be uncommon in ordinary design work, the partial fractions can take only a limited number of forms, and the problem is reduced to manageable dimensions.

In this case study we shall begin with a quotient of polynomials given by a prior design analysis and knowing very little else. We do not know the roots of the denominator, for instance, which are necessary for several reasons, and we shall have to compute them. Having the roots, we compute the partial fraction parameters and from them the inverse Laplace transform parameters. Finally, we produce a plot of the time function, using the inverse transform and employing a technique for picking the time increments for plotting. The straightforward approaches to these requirements will turn out to have some highly frustrating but very educational pitfalls, which we shall have to work around.

This case study leans heavily on papers by Titus* and Bashkow,† and on personal discussions with Professor Bashkow.

The given quotient is, in fact, stated as the product of two fractions:

$$g_1(s) = \frac{472.5}{s^5 + 15s^4 + 105s^3 + 420s^2 + 945s + 945}$$

$$g_2(s) = \frac{-s^7 + 28s^6 - 378s^5 + 3150s^4 - 17325s^3 + 62370s^2 - 135135s + 135135}{s^7 + 28s^6 + 378s^5 + 3150s^4 + 17325s^3 + 62370s^2 + 135135s + 135135}$$

The biggest part of the job is to find the partial fraction representation of the product of these two polynomial quotients by a process that we shall consider carefully. To do this, however, it is essen-

---

*C. K. Titus, A General Card-Program for the Evaluation of the Inverse Laplace Transform, *J. Assoc. Computing Machinery*, **2**, No. 1, January 1958.

† T. R. Bashkow, A "Curve Plotting" Routine for the Inverse Laplace Transform of Rational Functions," *J. Assoc. Computing Machinery*, **5**, No. 1, January 1958.

tial to know the roots of the denominator of the product, and we turn first to that problem.

Now the product of two polynomials has the same roots as the individual expressions. We can therefore apply a root-finding procedure to the two denominators separately and thus avoid *part* of the roundoff error difficulty that is almost sure to arise with polynomials of high degree with large coefficients.

Writing a complete program for root-finding is no small chore, and we turn without a second thought to the library of preprogrammed routines. In the selection available to the present author the best choice seemed to be a SHARE program library routine written by H. S. Schecter of the Republic Aviation Corporation. This subroutine obtains the roots of an arbitrary polynomial with *complex* coefficients

$$f(x) = c_1 x^n + c_2 x^{n-1} + \cdots + c_n x + c_{n+1}$$

Note the assumptions on the order in which the coefficients are stored; this assumption must be known, of course, for all we shall enter is the coefficients without any identification other than the order in which they appear in storage. The sequence used here is a natural one for the purpose.

Note, too, that this subroutine expects complex coefficients, whereas ours are real. We shall simply store zeros for the imaginary parts of the coefficients. There is, no doubt, a program somewhere to be found that finds the complex roots of a polynomial with real coefficients. Such a subroutine did not present itself when needed, however, and one can spend only so long looking for the ideal when something merely good enough is at hand. The reader will do well to be prepared for many such compromises.

The subroutine is named NEWRA and is called with the statement

CALL NEWRA (N, C, R, MAX, EPS, ZERO, G, KE)

where

N   = degree of polynomial

C   = complex array of coefficients $c_1, \ldots, c_{n+1}$

R   = complex array of roots $R_1, \ldots, R_n$

MAX = maximum number of iterations allowed

EPS = relative convergence criterion

ZERO = test constant used to see if roots satisfy the equation $f(R)/(c_{n+1}/c_1) \leq$ ZERO

G   = complex guess for a root

KE = -0, -1, or m, according as all, no, or m roots respectively, were found.

The brief writeup did not state how the program searches for starting values for the Newton-Raphson approximation process, which can be a problem with complex roots. The suggestion is that we have to supply guesses with the variable G. However, experimentation showed that the program could find the roots of our polynomials, regardless of what initial guesses were used, and in the program of Figure 29.1 we simply enter zero as an initial guess. The choice of the convergence criterion constants is something of a problem too. We are going to need all the accuracy we can get, yet it is quite possible to ask for accuracy that is impossible and simply not get any roots back. The program is set up, accordingly, to try a series of values for EPS. ZERO, on the other hand, we have rather little interest in; this is presumably intended to provide a warning of a root that meets the convergence criterion but does not satisfy the equation. Such can happen in certain circumstances. We are grate-

ful for the warning but not greatly concerned about it.

The program in Figure 29.1 enters the coefficients in a long DATA statement, which is appropriate in a problem that is going to be done exactly once after it has been checked out. Note the zeros for all the imaginary parts. The calls of NEWRA for each of the polynomials are set up in a DO loop to permit several values of EPS, which are also entered via the DATA statement, to be tried.

Figure 29.2 shows the subroutine itself for any reader who may wish to study it. SHARE standards for submission of programs do not require the use of explanatory comments.

The exact form of the output is of no particular interest and is not reproduced. For the two smallest values of EPS the program was able to find only two roots of the seventh-degree polynomial. Using EPS = 0.0000005, we found all roots of both. The best values for the roots of the denominator of $g_1(x)$ were

$$-3.6467384$$
$$-3.3519569 \pm 1.7426618$$
$$-2.3246745 \pm 3.5710230$$

For $g_2(x)$ the best values were

$$-2.6856769 \pm 5.4206941$$
$$-4.9717872$$
$$-4.0701382 \pm 3.5171752$$
$$-4.7582876 \pm 1.7392822$$

```
C CASE STUDY 29
C FINDING THE INVERSE LAPLACE TRANSFORM OF A QUOTIENT OF TWO POLYNOMIALS
C THIS PROGRAM USES A SHARE SUBROUTINE TO FIND THE ROOTS OF TWO POLYNOMIALS
C THE PROGRAM TRIES A SERIES OF VALUES OF THE CONVERGENCE CRITERION,
C SEEKING THE CLOSEST POSSIBLE APPROXIMATION TO THE ROOT
C
      COMPLEX P1(8), P2(6), X(7)
      REAL EPS(5)
      DATA (P1(I), I = 1, 8)/(1.0, 0.0), (28.0, 0.0), (378.0, 0.0),
     1  (3150.0, 0.0), (17325.0, 0.0), (62370.0, 0.0), (135135.0, 0.0),
     2  (135135.0, 0.0)/, (P2(I), I = 1, 6)/(1.0, 0.0), (15.0, 0.0),
     3  (105.0, 0.0), (420.0, 0.0), (945.0, 0.0), (945.0, 0.0)/,
     4  (EPS(I), I = 1, 5)/0.000001, 0.0000008, 0.0000005, 0.0000002,
     5  0.0000001/
      DO 26 I = 1, 5
      CALL NEWRA (7, P1, X, 40, EPS(I), 0.00001, (0.0, 0.0), KE)
      WRITE (6, 27) KE, EPS(I), (X(J), J = 1, 7)
   27 FORMAT (1H0, I10, F10.7/(1H , 1P8E15.7))
      CALL NEWRA (5, P2, X, 40, EPS(I), 0.00001, (0.0, 0.0), KE)
   26 WRITE (6, 27) KE, EPS(I), (X(J), J = 1, 5)
      STOP
      END
```

Figure 29.1. A short main program to call the root-finding subroutine in Figure 29.2. The data is included in the program, in the DATA statement, rather than being read.

```
      SUBROUTINE NEWRA(M,C,R,MAX,EPS,ZERO,GG,KE)
      DIMENSION C(1),R(1),GG(2),G(4),S(2),T(2),X(2)
      DATA G(3),G(4)/.1,.1/
      KE=0
      N=M
      S(1)=C(1)
      S(2)=C(2)
      Q=S(1)**2+S(2)**2
      IF(N.NE.1) GO TO 5
      R(1)=-(S(1)*C(3)+S(2)*C(4))/Q
      R(2)=(S(2)*C(3)-S(1)*C(4))/Q
      GO TO 200
    5 L=0
      KR=0
      KT=0
      EN=N
      G(1)=GG(1)
      G(2)=GG(2)
      CM=(C(2*N+1)**2+C(2*N+2)**2)/Q
      IF(CM.NE.0.) GO TO 10
      CM=1.
      R(1)=0.
      R(2)=0.
      G(1)=.1
      L=1
   10 DO 15 I=1,N
      S(1)=S(1)+C(2*I+1)
   15 S(2)=S(2)+C(2*I+2)
      IF(SQRT((S(1)**2+S(2)**2)/CM)-ZERO)20,20,25
   20 R(2*L+1)=1.
      R(2*L+2)=0.
      G(1)=1.1
      G(2)=0.
      L=L+1
      IF(L-N)25,200,200
   25 IF(MOD(N,2))30,35,30
   30 S(1)=C(3)-C(1)
      S(2)=C(4)-C(2)
      K=3
      GO TO 40
   35 S(1)=C(1)
      S(2)=C(2)
      K=2
   40 DO 45 I=K,N,2
      S(1)=S(1)-C(2*I-1)+C(2*I+1)
   45 S(2)=S(2)-C(2*I)+C(2*I+2)
      IF(SQRT((S(1)**2+S(2)**2)/CM)-ZERO)50,50,55
   50 R(2*L+1)=-1.
      R(2*L+2)=0.
      G(1)=-1.1
      G(2)=0.
      L=L+1
      IF(L-N)55,200,200
   55 Q=G(1)**2+G(2)**2
      IF(Q-1.)75,75,60
   60 G(1)=G(1)/Q
      G(2)=(-G(2))/Q
      KR=-1
   65 CM=1./CM
      K=(N+1)/2
      DO 70 I=1,K
      J=N-I
      T(1)=C(2*I-1)
      T(2)=C(2*I)
      C(2*I-1)=C(2*J+3)
      C(2*I)=C(2*J+4)
      C(2*J+3)=T(1)
   70 C(2*J+4)=T(2)
   75 IG=1

   80 X(1)=G(2*IG-1)
      X(2)=G(2*IG)
      DO 110 J=1,MAX
      P=EN
      S(1)=C(1)*X(1)-C(2)*X(2)+C(3)
      S(2)=C(2)*X(1)+C(1)*X(2)+C(4)
      T(1)=P*C(1)
      T(2)=P*C(2)
      DO 85 I=2,N
      Q=S(1)*X(1)-S(2)*X(2)+C(2*I+1)
      S(2)=S(2)*X(1)+S(1)*X(2)+C(2*I+2)
      S(1)=Q
      P=P-1.
      Q=T(1)*X(1)-T(2)*X(2)+P*C(2*I-1)
      T(2)=T(2)*X(1)+T(1)*X(2)+P*C(2*I)
   85 T(1)=Q
      P=S(1)**2+S(2)**2
      Q=(T(1)*S(1)+T(2)*S(2))/P
      T(2)=(T(2)*S(1)-T(1)*S(2))/P
      T(1)=Q
      IF(P.EQ.0.) T(1)=1.E16
      IF(L)90,100,90
   90 DO 95 I=1,L
      S(1)=X(1)-R(2*I-1)
      S(2)=X(2)-R(2*I)
      Q=S(1)**2+S(2)**2
      T(1)=T(1)-S(1)/Q
   95 T(2)=T(2)+S(2)/Q
  100 Q=T(1)**2+T(2)**2
      X(1)=X(1)-T(1)/Q
      X(2)=X(2)+T(2)/Q
      IF(1./SQRT(Q*(X(1)**2+X(2)**2))-EPS)105,105,110
  105 IF(SQRT(P/CM)-ZERO)160,160,110
  110 CONTINUE
      IG=IG+1
      IF(IG-2)115,80,115
  115 IF(KT)120,155,120
  120 IF(L)125,175,125
  125 KE=L
  130 IF(KR)135,200,135
  135 KT=0
  140 DO 145 I=1,L
      Q=R(2*I-1)**2+R(2*I)**2
      R(2*I-1)=R(2*I-1)/Q
  145 R(2*I)=(-R(2*I))/Q
  150 IF(KT)65,200,65
  155 KT=1
      KR=KR+1
      Q=X(1)**2+X(2)**2
      G(1)=X(1)/Q
      G(2)=(-X(2))/Q
      IF(L)140,65,140
  160 IF(ABS(X(2)/X(1))-ZERO)165,165,170
  165 X(2)=0.
  170 R(2*L+1)=X(1)
      R(2*L+2)=X(2)
      G(1)=1.1*X(1)
      G(2)=1.1*X(2)
      L=L+1
      IF(L-N)75,130,130
  175 KE=-1
  200 RETURN
      END
```

Figure 29.2. A subroutine to find the roots of an arbitrary polynomial with complex coefficients.

With the roots known we can now proceed with the partial fraction expansion. Recall that the quotient of two polynomials can always be expressed as a sum of terms of the forms

$$\frac{k}{s} \quad \frac{k}{s^2} \quad \frac{k}{s^3} \qquad \text{roots at the origin}$$

$$\frac{k}{s-a} \quad \frac{k}{(s-a)^2} \qquad \text{real roots, not at the origin}$$

$$\frac{k(s+d)}{(s-a)^2+\omega^2} \qquad \text{complex root}$$

The k's here represent any real constant multipliers. Another theorem about the Laplace transform tells us that multiplying the original function by a constant corresponds to the same action on the result function. In principle, there is, of course, no limit to the multiplicity of the roots. We have indicated that a root at the origin may not be of higher order than 3, a real root may not be of higher order than 2, and a complex root must be first order. These are the assumptions that are built into the program and represent design practice that covers a wide range of applications in electrical engineering. For physical systems of interest the real parts of these roots are *always* negative, and in some of what follows we shall assume that they are so and write them without a sign.

As a simple example of the reduction of a quotient to partial fractions, observe that

$$\frac{3s^4 + 9s^3 + 10s^2 + 4s + 2}{s^5 + 3s^4 + 4s^3 + 2s^2}$$

$$= \frac{1}{s^2} + \frac{2}{s+1} + \frac{s+3}{s^2+2s+2}$$

The roots here are a second-order root at the origin, a first-order real root at -1, and a first-order complex root at $-1 \pm i$.

A note about terminology. Consider the *equation*

$$(x-3)\ (x-5) = 0$$

This has *roots* $x = 3$ and $x = 5$; for these values the equation is satisfied. Now consider the *function* $f(x) = (x-3)\ (x-5)$. This has *zeros* at 3 and 5, the values of x for which the function $f(x)$ is zero. Finally, the function

$$\frac{1}{f(x)} = \frac{1}{(x-3)(x-5)}$$

has *poles* at 3 and 5, the values for which the reciprocal of the function $f(x)$ is infinite.

The roots of an equation, the zeros of a function, and the poles of a function are, of course, three different things. We have followed the practice, no doubt indefensible, of using the terms root, zero, and pole virtually interchangeably where the meaning is clear from the context.

The reduction to partial fractions is developed in most texts on advanced calculus. We shall present as much as we need for the cases to be found here, following the presentation in Titus' paper.

First-Order (Nonrepeated) Pole At Origin

For notational convenience we assume that the polynomial quotient has been written in the form

$$\frac{N(s)}{D(s)} = \frac{\cdots + N_4 s^4 + N_3 s^3 + N_2 s^2 + N_1 s + N_0}{\cdots + D_4 s^4 + D_3 s^3 + D_2 s^2 z + D_1 s + D_0}$$

The partial fraction is of the form

$$\frac{1}{s}$$

The inverse transform is

$$b$$

where

$$b = \frac{N_0}{D_1}$$

Second-Order Pole At Origin

Partial fraction form:

$$\frac{1}{s^2}$$

Inverse transform:

$$b + ct$$

where

$$b = \frac{N_0}{D_2}$$

$$c = \frac{D_2 N_1 - D_3 N_0}{D_2{}^2}$$

First-Order Real Pole Not At Origin

Partial fraction form:

$$\frac{1}{s-a}$$

Inverse transform:

$$be^{-\alpha t}$$

where

$$b = \frac{N(-\alpha)}{D'(-\alpha)}$$

This means to evaluate the numerator at $-\alpha$ and divide by the value of the derivative of the denominator evaluated at $-\alpha$. For physical systems of practical interest the real parts of the roots are always negative, and we may as well let the computer attach the required sign; we shall enter the roots written with the real parts plus.

Second-Order Real Pole Not at Origin

Partial fraction form:

$$\frac{1}{(s - a)^2}$$

Inverse transform:

$$be^{-\alpha t} + cte^{-\alpha t}$$

where

$$b = \frac{2N(-\alpha)}{D''(-\alpha)}$$

$$c = \frac{\dfrac{D''(-\alpha)N'(-\alpha)}{2} - \dfrac{N(-\alpha)D'''(-\alpha)}{6}}{\dfrac{D''(-\alpha)^2}{4}}$$

First-Order Complex Pole

Partial fraction form:

$$\frac{s + d}{(s - a)^2 + \omega^2}$$

Inverse transform:

$$ae^{-\alpha t} \sin(\rho t + \theta)$$

where, if we write

$$\frac{N(-\alpha + \beta i)}{D'(-\alpha + \beta i)} = \text{Re} + \text{Im } i$$

$$a = 2\sqrt{\text{Re}^2 + \text{Im}^2}$$

$$\theta = \arctan \frac{\text{Re}}{-\text{Im}}$$

We note that each of these cases can be characterized by, at most, five numbers: the real part of the root, the imaginary part of the root, the multiplicity of the root, and either one or two constants for the inverse transform. Bearing in mind that at the origin the real part of the root is zero, so that e^{-t} is one, we can describe the inverse transform constants as follows. For the real poles there is a constant b which is multiplied by an exponential, plus a constant c which is multiplied by t and by an exponential. For a complex pole there is a constant a which is multiplied by an exponential and a sine, plus a constant, which is an offset in the sine. (The notation here is that of Bashkow's paper.) Any one pole is, of course, either real or complex, and it appears that we could call the inverse form parameters AORB (A or B) and THORC (theta or C) to suggest the meanings. These same five parameters characterize the task of the plotting routine that we shall discuss later.

Now we may turn to the program of Figure 29.3 to see how all this is accomplished. The program begins with a set of definitions of most of the program variables, which may be some help in study.

One of the main tasks of the program will be the evaluation of various polynomials. There are lots of ways to set up polynomial evaluation for a computer, and a way chosen here would not always be the best choice if we were faced with some different mixture of the objectives of programming simplicity, ease of understanding, accuracy, speed, and storage space. The method used here is quite adequate for our purposes and has no really important drawbacks.

The idea is to begin by deciding what is the maximum size polynomial that will be accepted by the program; this has to be done with any method. The decision here was that the numerator would be limited to twentieth degree, for which there are 21 coefficients; the denominator would be limited to twenty-first degree. (For systems of interest the denominator is always of higher degree.) Next we decide that if there are fewer than the maximum the coefficients will be stored at the "high" end of the array and in decreasing order of the power of x to which they correspond. Turning quickly to an example of this none-too-lucid statement, take our numerator: it is of the seventh degree, with eight coefficients. If we write the polynomial as

$$c_7 x^7 + c_6 x^6 + \cdots + c_1 x + c_0$$

```
C CASE STUDY 29
C FINDING THE INVERSE LAPLACE TRANSFORM OF A QUOTIENT OF TWO POLYNOMIALS
C THIS PROGRAM FINDS THE INVERSE TRANSFORM PARAMETERS
C
C    N = COEFFICIENTS OF NUMERATOR
C    NP = COEFFICIENTS OF DERIVATIVE OF NUMERATOR
C    D = COEFFICIENTS OF DENOMINATOR
C    DP = COEFFICIENTS OF FIRST DERIVATIVE OF DENOMINATOR
C    DPP = COEFFICIENTS OF SECOND DERIVATIVE OF DENOMINATOR
C    DPPP = COEFFICIENTS OF THIRD DERIVATIVE OF DENOMINATOR
C    DN = DEGREE OF NUMERATOR, AS ELEMENT NUMBER OF COEFFICIENT OF HIGHEST POWER
C    DD = DEGREE OF DENOMINATOR, DITTO
C    DNP1 = DEGREE OF NUMERATOR PLUS 1
C    DDP1 = DEGREE OF DENOMINATOR PLUS 1
C    NO = VALUE OF NUMERATOR, EVALUATED AT A REAL ROOT
C    N1 = VALUE OF DERIVATIVE OF NUMERATOR, EVALUATED AT A REAL ROOT
C    LIKEWISE FOR D1, D2, D3
C    G = FORTRAN COMPLEX VALUE OF A COMPLEX ROOT
C    NOC = VALUE OF NUMERATOR, EVALUATED AT A COMPLEX ROOT G
C    SIMILARLY FOR D1C
C    KA = COMPLEX QUOTIENT OF NOC AND D1C
C    RE = REAL PART OF KA
C    IM = IMAGINARY PART OF KA
C
      REAL N(21), NP(20), D(22), DP(21), DPP(20), DPPP(19)
      REAL COEF, NO, N1, D1, D2, D3, RE, IM
      COMPLEX G, NOC, D1C, KA
      INTEGER DN, DNP1, DD, DDP1, I, K
C READ THE COEFFICIENT OF THE HIGHEST POWER OF S IN NUMERATOR
   99 READ (5, 1) K, COEF
    1 FORMAT (I2, 8X, F10.0)
      I = 21 - K
C WHAT IS CALLED THE DEGREE OF THE NUMERATOR (DN) IS IN FACT THE
C ELEMENT NUMBER OF THE COEFFICIENT OF THE HIGHEST POWER OF S IN
C THE NUMERATOR
      DN = I
      DNP1 = DN + 1
      N(I) = COEF
C READ THE REST OF THE COEFFICIENTS OF THE NUMERATOR
    2 READ (5, 1) K, COEF
      I = 21 - K
      N(I) = COEF
C CHECK FOR COEFFICIENT NUMBER OF ZERO, INDICATING LAST COEFFICIENT
      IF (K .NE. 0) GO TO 2
C READ COEFFICIENT OF HIGHEST POWER OF S IN DENOMINATOR
      READ (5, 1) K, COEF
      I = 22 - K
C GET DEGREE OF DENOMINATOR
      DD = I
      DDP1 = DD + 1
      D(I) = COEF
C READ REST OF COEFFICIENTS OF DENOMINATOR
    3 READ (5, 1) K, COEF
      I = 22 - K
      D(I) = COEF
```

Figure 29.3. A program to find the inverse Laplace transform parameters of a quotient of two polynomials.

```
      IF (K .NE. 0) GO TO 3
C GET DERIVATIVE OF DENOMINATOR
      DO 4 I = DD, 21
      AI = 22 - I
    4 DP(I) = AI * D(I)
C GET 2ND DERIVATIVE OF DENOMINATOR
      DO 5 I = DD, 20
      AI = 21 - I
    5 DPP(I) = AI * DP(I)
C GET 3RD DERIVATIVE OF DENOMINATOR
      DO 6 I = DD, 19
      AI = 20 - I
    6 DPPP(I) = AI * DPP(I)
C GET DERIVATIVE OF NUMERATOR
      DO 7 I = DN, 20
      AI = 21 - I
    7 NP(I) = AI * N(I)
C TEST FOR POLES AT ORIGIN
      IF (D(22) .EQ. 0.0 .AND. D(21) .NE. 0.0) GO TO 11
      GO TO 10
C FIRST ORDER POLE AT ORIGIN
   11 ALPHA = 0.0
      BETA = 0.0
      K = 1
      AORB = N(21) / D(21)
      THORC = 0.0
      WRITE (6, 9) ALPHA, BETA, K, AORB, THORC
    9 FORMAT (1H0, 1P2E18.7, I4,2E18.7)
      GO TO 40
   10 IF (D(22) .EQ. 0.0 .AND. D(21) .EQ. 0.0) GO TO 13
C NO POLES AT ORIGIN - GO TO READ A POLE CARD
      GO TO 40
C SECOND ORDER POLE AT ORIGIN
   13 ALPHA = 0.0
      BETA = 0.0
      K = 2
      AORB = (D(20) * N(20) - D(19) * N(21)) / D(20)**2
      THORC = N(21) / D(20)
      WRITE (6, 9) ALPHA, BETA, K, AORB, THORC
C READ A POLE CARD
   40 READ (5, 20) ALPHA, BETA, K
   20 FORMAT (2F12.0, I1)
C CHECK FOR BLANK CARD SIGNALLING THE LAST POLE CARD
C GO TO READ NEW POLYNOMIAL IF SO
      IF (ALPHA .EQ. 0.0) GO TO 99
      IF (K .NE. 2) GO TO 30
C SECOND ORDER REAL, NOT AT ORIGIN
C EVALUATE NUMERATOR
      NO = N(DN)
      DO 21 I = DNP1, 21
   21 NO = - ALPHA * NO + N(I)
C EVALUATE DERIVATIVE OF NUMERATOR
      N1 = NP(DN)
      DO 22 I = DNP1, 20
   22 N1 = - ALPHA * N1 + NP(I)
C EVALUATE 2ND DERIVATIVE OF DENOMINATOR
```

Figure 29.3.   (Continued)

```
          D2 = DPP(DD)
          DO 23 I = DDP1, 20
       23 D2 = - ALPHA * D2 + DPP(I)
C EVALUATE 3RD DERIVATIVE OF DENOMINATOR
          D3 = DPPP(DD)
          DO 24 I = DDP1, 19
       24 D3 = - ALPHA * D3 + DPPP(I)
          AORB = (D2 / 2.0 * N1 - NO * D3 / 6.0) / (D2**2 / 4.0)
          THORC = 2.0 * NO / D2
          WRITE (6, 9) ALPHA, BETA, K, AORB, THORC
          GO TO 40
       30 IF (BETA .NE. 0.0) GO TO 50
C FIRST ORDER REAL, NOT AT ORIGIN
          NO = N(DN)
          DO 31 I = DNP1, 21
       31 NO = - ALPHA * NO + N(I)
          D1 = DP(DD)
          DO 32 I = DDP1, 21
       32 D1 = - ALPHA * D1 + DP(I)
          AORB = NO / D1
          THORC = 0.0
          WRITE (6, 9) ALPHA, BETA, K, AORB, THORC
          GO TO 40
C COMPLEX (ALL COMPLEX POLES ARE FIRST ORDER, BY ASSUMPTION)
       50 G = CMPLX(-ALPHA, BETA)
C EVALUATE NUMERATOR
          NOC = CMPLX(N(DN), 0.0)
          DO 51 I = DNP1, 21
       51 NOC = G * NOC + N(I)
C EVALUATE DERIVATIVE OF DENOMINATOR
          D1C = CMPLX(DP(DD), 0.0)
          DO 52 I = DDP1, 21
       52 D1C = G * D1C + DP(I)
C EVALUATE INVERSE LAPLACE PARAMETERS
          KA = NOC / D1C
          RE = REAL(KA)
          IM = AIMAG(KA)
          AORB = 2.0 * SQRT(RE**2 + IM**2)
          THORC = ATAN2(RE, -IM)
C IF THE ANGLE IS IN THE FIRST OR FOURTH QUADRANT, LEAVE IT AS IS
C OTHERWISE ADD OR SUBTRACT PI TO GET SUCH AN ANGLE, AND REVERSE SIGN OF
C AORB TO COMPENSATE
          IF (ABS(THORC) .LT. 1.57079632) GO TO 453
          IF (THORC) 450, 451, 452
      450 THORC = THORC + 3.14159265
          GO TO 451
      452 THORC = THORC - 3.1415926
      451 AORB = - AORB
      453 WRITE (6, 9) ALPHA, BETA, K, AORB, THORC
          GO TO 40
          END
```

Figure 29.3. (Continued)

the coefficients would be stored in the array N in this way:

$N(1)$ through $N(13)$: immaterial

$N(14) = c_7$

$N(15) = c_6$

$N(16) = c_5$

$N(17) = c_4$

$N(18) = c_3$

$N(19) = c_2$

$N(20) = c_1$

$N(21) = c_0$

To simplify later programming operations the way we characterize the degree of the polynomial is not to store the degree itself but the element number of the coefficient of the highest power.

In finding the roots, we worked with the two denominator polynomials separately. This is quite impossible here; it is (sadly) not true that the partial fraction reduction of a product is the product of their reductions. It was necessary to multiply out the two polynomials "by hand" to enter the product into this program.*

The coefficient cards each contain the power with which they are associated. The cards must be in order, from highest power to lowest. The degree of the polynomial is deduced from the first coefficient card, and the corresponding element number is stored. The numerator is read first, followed by the denominator.

The next problem is to prepare for evaluating the various derivatives that are required. There is also a variety of ways to do this, of which we have chosen one that is adequate, although it would not always be the first choice: we simply compute the coefficients of the derivative polynomial and store them in a separate array. This can be done without regard to the degree of the polynomial; we *assume* that the N and D arrays are full and rely on the

* A process that led to a series of wasted computer runs because the multiplication was done incorrectly. Recourse was finally made to FORMAC, one of several systems that permits literal multiplication of polynomials and the analytical differentiation of fairly complex expressions, among other fascinating operations. For an introduction to the capabilities of programs of this kind see Jean E. Sammet, "Survey of Formula Manipulation," *Communications Assoc. Computing Machinery*, **9**, No. 8, 555, August 1966.

fact that any missing high-order coefficients will be correctly handled as zeros by the differentiation process. All that is required is to multiply the coefficient of x^n by n and store the product in the appropriate location in the derivative array. This must be done with due regard for the way the coefficients were stored: the element in location I in the N array, for instance, is the coefficient of x to the 21-I power. The various other derivatives are handled similarly.

Now we can get to work on the poles. All of the nonzero roots will, of course, be entered on cards, but there is no real need to enter the roots at the origin, for their existence can be readily determined by inspecting the denominator: if D_0 and D_1 are both zero, we have a second-order pole at the origin; if D_0 is zero but D_1 is not, we have a first-order pole at the origin. The appropriate tests are made, the real and imaginary parts of the roots are set equal to zero, the multiplicity of the pole is written as the value of K, and the parameters AORB and THORC are computed according to the formulas given earlier.

Now we are ready to read the cards that give the locations of the roots (poles). These cards can be dealt with one at a time; the partial fraction reduction never requires us to have all the roots in storage at the same time. The program is set up to expect a blank card at the end of the pole card deck; this enables us to use the program to work with several polynomial quotients in one computer run. This would not be possible, of course, if we used the end of the deck to signify the completion of the set of pole cards.

The first action on reading a pole card therefore is to check whether the real part of the root is zero; this should never happen with actual data and thus signals a blank card when it does appear. Next we check the multiplicity of the root; if it is 2, we have a real pole, for all complex poles are assumed to be simple (first-order). If the multiplicity is 1, we distinguish between real and complex by looking at the imaginary part of the root.

For a second-order real pole, not at the origin, we have to evaluate the numerator, the derivative of the numerator, and the second and third derivatives of the denominator. All of the evaluations are done by the procedure known as *Horner's rule*, which we may explain by noting, for instance, that

$$c_5 x^5 + c_4 x^5 + c_3 x^3 + c_2 x^2 + c_1 x + c_0$$
$$= (((((c_5)x + c_4)x + c_3)x + c_2)x + c_1)x + c_0$$

From the appearance of the latter expression, the procedure is also often called *nesting*. It is quite readily programmed: we set a variable equal to the coefficient of the highest power, then repeatedly multiply that variable by x and add another coefficient.

This process is carried out for all the polynomial evaluations in the program. It is not only simple to program and rapid to execute but also has significant advantages in accuracy in many cases of practical interest (including this case study).

We read the real part of the root with the understanding that it will have been punched as positive; hence the minus signs that appear throughout the program.

The complex poles present only slight additional problems. We have to set up the computation to use complex arithmetic, but that is a rather small project in FORTRAN. After dividing the complex value of the numerator by the complex value of the denominator we separate the quotient into real and imaginary parts and go into the evaluation of the inverse transform coefficients. This is more or less straightforward, with the exception of the arctangent function. In order to get the phase angle exactly right it is necessary to use a plus sign on the imaginary part of the root, a minus sign on the imaginary part of the quotient KA, and the function ATAN2 that keeps track of the quadrant information of the two coordinate values. (Getting this right took some days of off-and-on fighting with the program.) Then, to make the results match up with the results given in Bashkow's paper, we make a simple transformation based on the identity

$$\sin(\alpha + \pi) = -\sin \alpha$$

to reduce the phase to an angle less than π. If this reduction is carried out, that is, if the angle as first computed is in the second or third quadrant, we have to reverse the sign of AORB to compensate for the change.

The appropriate actions, depending on real or complex and first- or second-order real, are carried out for all pole cards. For our example all poles are first-order and there are no poles at the origin. The output of the program is shown in Figure 29.4. Everything seems reasonable enough and in *fairly* close agreement with the results in Bashkow's paper, in which the roots were given as being just slightly different from those found by our SHARE subroutine.

So we turn to the routine that computes the actual inverse transform, that is, produces the time response of the system described in the frequency domain by the quotient of polynomials that we started with. As we have noted, the inverse transform of the quotient is just the sum of the transforms of the separate partial fractions to which the quotient is equivalent. In Bashkow's notation it is just

$$f(t) = \sum_{i=1}^{5} a_i e^{-\alpha_i t} \sin(\beta_i t + \theta_i) + \sum_{i=6}^{7} b_i e^{-\alpha_i t}$$

where the subscripts denote the various complex and real roots. All we really have to do is pick a set of values of t and evaluate these expressions for them.

To make the problem more interesting, however, we shall introduce another element into the job: we shall ask the program to determine the initial and final values of t that should be used and the spacing between values to produce the data for a smooth plot of the time response. We know from theoretical considerations that when $t = 0$ the time response must be zero, even though it is not obvious from anything available here, and it is clear that in our case the response drops to zero for large time.

We initially evaluate $f(t)$ for $t = 0$, then ask what value of t should be used next. This is the

2.6856769E 00	5.4206941E 00	1	3.6202672E 01	1.1820550E 00
3.6467384E 00	0.	1	1.5417162E 04	0.
3.3519569E 00	1.7426618E 00	1	-1.1017863E 04	-1.4312222E 00
2.3246745E 00	3.5710230E 00	1	-3.1614408E 02	-1.3677133E 00
4.9717872E 00	0.	1	-1.8331984E 04	0.
4.0701382E 00	3.5171752E 00	1	-2.3169281E 03	-5.8406407E-01
4.7582876E 00	1.7392822E 00	1	-1.8995255E 04	5.3140447E-01

Figure 29.4.   The output of the program in Figure 29.3 operating on the quotient of polynomials given in the text.

same as asking what the initial value should be for Δt, the increment in t between evaluation points. We should like to choose Δt_1 so that $f(\Delta t_1)$ will not have changed appreciably from its value at zero, which requires that the fastest exponential (the one with the largest α), and the highest frequency sine wave (the one with the largest β) will not have changed much at Δt_1.

Let us denote by α_l and β_l the largest values of α and β. Bashkow then suggests choosing $\Delta t_1 = 0.1/\alpha_l$ or $0.1/\beta_l$, whichever is larger. This will give

$$e^{-\alpha_l \Delta t_1} \simeq 1.0$$

and

$$\sin \beta_l \Delta t_1 \simeq 0.0$$

We would like to choose the final value of t, t_{end}, so that $f(t_{\text{end}})$ is essentially steady-state. We need for the slowest exponential to damp out and for the lowest frequency sine wave to complete a cycle. Bashkow's analysis shows that this goal can be reached by letting

$$t_{\text{end}} = \frac{\log\left(|B_s| + |C_s|\right) + 10}{\alpha_s} \quad \text{or} \quad \frac{10}{\beta_s}$$

whichever is larger. B_s and C_s mean the parameters associated with α_s, and α_s and β_s are the smallest value of α and β, not including the β values of zero for real roots.

With starting and ending values in hand, we can investigate the method of correcting Δt to keep it within the bounds we want. The primary consideration is that we want the time values to be closely spaced when the function is changing rapidly but spaced out when it is changing slowly. The precise procedure is as follows:

We evaluate the derivative of $f(t)$, $f'(t)$, at the "current" value of t and at $t + \Delta t$. These are then the tangents of the slopes at the points; we convert to angles by taking the arctangents. Denote by ϕ the absolute value of the difference between the slopes: this is a measure of the rate of change of the slope of the function in the vicinity of time t.* We then apply the following rules:

---

* It is also, of course, an estimate of the absolute value of the second derivative of the function—which is the curvature. This process can thus be described alternatively in terms of bounds on the curvature, and the program could likewise have been written to evaluate the second derivative. But the latter is a slightly messy expression; this way is probably as good as any.

1. If ϕ is less than 0.05 radian, approximately 3°, then the value of $f(t + \Delta t)$ is printed, $t + \Delta t$ is used as the new value of t, and Δt is doubled. The function is changing too slowly; we can safely increase the distance between time values.

2. If ϕ is between 0.05 and 0.15, the value of $f(t + \Delta t)$ is printed, $t + \Delta t$ becomes the new value of t and is left unchanged. This is about how we wanted it.

3. If ϕ is greater than 0.15, the function is changing too rapidly; we may be missing points that we need in order to draw a good graph. In fact, $t + \Delta t$ might be too far away from t, so we do *not* print the value of $f(t + \Delta t)$ but rather cut Δt in half and go back to try again, unless, that is, Δt is less than $t_{\text{end}}/50$, in which case Δt is left alone. The point of this is to prevent the interval-choosing routine from bogging down indefinitely at a point at which the function is changing rapidly.

The values are printed, but while we are at it we may as well produce a plot of the graph ourselves. This is a relatively simple matter of using the time and function values to pick a point in an array, as we did in Case Study 18, except that here we shall produce two lines of dots for the axes.

The program in Figure 29.5 has a few interesting points that we may note.

The selection of α_s, α_e, β_s, and β_e makes use of the FORTRAN functions for finding maxima and minima, functions we have not used before this. In the case of α_s we not only need the real part of the root but also the values of AORB and THORC associated with it; thus we need to know the subscript of the smallest root as well. We set IS to 1 and ALPHAS to the absolute value of ALPHA(1). Then we inspect all the other values of looking for a smaller one. Each time we find one that is smaller, we set ALPHAS equal to it and IS equal to its element number. The basic logic is much the same for the other values, with the exception that in the case of β_s we must not accept the zero values of the real roots as being the smallest.

The statements for $\Delta t_1 = $ DELTAT and $t_{\text{end}} = $ TEND also use the maximum and minimum functions to good advantage.

The evaluation of the function presents no special problems. We simply evaluate the inverse transform of each partial fraction and add all such values.

It was discovered by experimentation that for this function the algorithm for choosing Δt did not

```
C CASE STUDY 29
C FINDING THE INVERSE LAPLACE TRANSFORM OF A QUOTIENT OF TWO POLYNOMIALS
C THIS PROGRAM CHOOSES THE TIME INTERVAL AND GRAPHS THE RESULTS
C
      REAL ALPHA(20), BETA(20), AORB(20), THORC(20), E, F, T, TF, FP
      INTEGER K(20), I, L
      REAL ALPHAS, ALPHAL, BETAS, BETAL, DELTAT, TEND, DELTA
      INTEGER IS
      REAL GRAPH(60, 120)
      DATA BLANK/1H /, DOT/1H./, X/1HX/
C BLANK OUT ARRAY FOR GRAPH
 9999 DO 612 I = 1, 60
      DO 612 J = 1, 120
  612 GRAPH(I, J) = BLANK
C PUT IN DOTS FOR AXES
      DO 667 I = 1, 60
  667 GRAPH(I, 1) = DOT
      DO 668 J = 1, 120
  668 GRAPH(6, J) = DOT
C WRITE HEADING LINE
      WRITE (6, 9998)
 9998 FORMAT (1H1, 6X, 1HT, 6X,3H2*F, 5X, 5HDELTA, 3X, 6HDELTAT////)
      DO 63 L = 1, 21
      READ (5, 62) ALPHA(L), BETA(L), K(L), AORB(L), THORC(L)
   62 FORMAT (2F12.0, I1, 2F12.0)
      IF (K(L) .EQ. 0) GO TO 64
   63 CONTINUE
      STOP
   64 L = L - 1
C GET ALPHAS, ALPHAL, IS, BETAS, BETAL
      IS = 1
      ALPHAS = ABS(ALPHA(1))
      ALPHAL = ABS(ALPHA(1))
      BETAS = ABS(BETA(1))
      BETAL = ABS(BETA(1))
      DO 23 I = 2, L
      IF (ALPHAS .LT. ABS(ALPHA(I))) GO TO 24
      ALPHAS = ABS(ALPHA(I))
      IS = I
   24 ALPHAL = AMAX1(ALPHAL, ABS(ALPHA(I)))
      IF (BETA(I) .EQ. 0.0) GO TO 23
      BETAS = AMIN1(BETAS, ABS(BETA(I)))
      BETAL = AMAX1(BETAL, ABS(BETA(I)))
   23 CONTINUE
C GET  DELTAT, TEND
      DELTAT = AMIN1(0.1/ALPHAL, 0.1/BETAL)
      TEND = AMAX1((ALOG(ABS(AORB(IS)) + ABS(THORC(IS))) + 10.0)/ALPHAS,
     1      10.0/BETAS)
C START EVALUATION AT T = 0.0
      T = 0.0
      N = 1
C EVALUATE FUNCTION
   67 F = 0.0
      DO 78 I = 1, L
      IF (BETA(I)) 80, 81, 80
C COMPLEX - BETA NOT ZERO
   80 F = F + AORB(I) * EXP(-ALPHA(I) * T) * SIN(BETA(I) * T + THORC(I))
      GO TO 78
```

Figure 29.5. A program to compute and plot the time response of the system represented by the inverse Laplace transform given in the parameters in Figure 29.4 and represented in the frequency domain by the original polynomial quotient.

```
C REAL - BETA ZERO
   81 E = EXP(-ALPHA(I) * T)
      F = F + AORB(I) * E
      IF (K(I) .EQ. 2) F = F + THORC(I) * T * E
   78 CONTINUE
      TF = 2.0 * F
C END  OF FUNCTION EVALUATION
C
      WRITE (6, 72) T, TF, DELTA, DELTAT
   72 FORMAT (1H , 4F9.4)
      I = 50.0*TF + 6.0
      J = 12.0*T + 1.0
      IF (I.GE.1.AND.I.LE.60.AND.J.GE.1.AND.J.LE.120) GRAPH(I,J)=X
      IF (T .GT. TEND) GO TO 666
      N = N + 1
      IF (N .GT. 100) GO TO 666
C EVALUATE DERIVATIVE AT T = T AND T = T + DELTA T
   49 FP = 0.0
      DO 178 I = 1, L
      E = EXP(-ALPHA(I) * T)
      IF (BETA(I)) 180, 181, 180
C COMPLEX - BETA NOT ZERO
  180 FP = FP + AORB(I) * E * COS(BETA(I) * T + THORC(I)) * BETA(I)
     1    - AORB(I) * ALPHA(I) * E * SIN(BETA(I)*T + THORC(I))
      GO TO 178
C REAL - BETA ZERO
  181 FP = FP - ALPHA(I) * AORB(I) * E
      IF (K(I) .EQ. 2) FP = FP + THORC(I) * E * (1.0 - ALPHA(I) * T)
  178 CONTINUE
      AX = 2.0 * FP
      Y = T + DELTAT
      FP = 0.0
      DO 278 I = 1, L
      E = EXP(-ALPHA(I) * Y)
      IF (BETA(I)) 280, 281, 280
  280 FP = FP + AORB(I) * E * COS(BETA(I) * Y + THORC(I)) * BETA(I)
     1    - AORB(I) * ALPHA(I) * E * SIN(BETA(I)*Y +THORC(I))
      GO TO 278
  281 FP = FP - ALPHA(I) * AORB(I) * E
      IF (K(I) .EQ. 2) FP = FP + THORC(I) * E * (1.0 - ALPHA(I) * Y)
  278 CONTINUE
      FP = 2.0 * FP
      DELTA = ABS(ATAN(AX) - ATAN(FP))
C END OF DERIVATIVE EVALUATION
C
      IF (DELTA .GE. 0.05) GO TO 50
      T = T + DELTAT
      DELTAT = 2.0 * DELTAT
      GO TO 67
   50 IF (DELTA .GE. 0.15 .AND. DELTAT     .GE. TEND/50.0) GO TO 60
      T = T + DELTAT
      GO TO 67
   60 DELTAT = DELTAT / 2.0
      GO TO 49
  666 WRITE (6, 28) (GRAPH(60, J), J = 1, 120)
   28 FORMAT (1H1, 120A1)
      DO 29 I1 = 2, 60
      I = 61 - I1
   29 WRITE (6, 30) (GRAPH(I, J), J = 1, 120)
   30 FORMAT (1H , 120A1)
C GO BACK FOR ANOTHER SET OF DATA
      GO TO 9999
      END
```

Figure 29.5. (Continued)

```
C CASE STUDY 29
C FINDING THE INVERSE LAPLACE TRANSFORM OF A QUOTIENT OF TWO POLYNOMIALS
C THIS PROGRAM FINDS THE PARTIAL FRACTION COEFFICIENTS
C MODIFIED TO USE DOUBLE PRECISION
C
C    N = COEFFICIENTS OF NUMERATOR
C    NP = COEFFICIENTS OF DERIVATIVE OF NUMERATOR
C    D = COEFFICIENTS OF DENOMINATOR
C    DP = COEFFICIENTS OF FIRST DERIVATIVE OF DENOMINATOR
C    DPP = COEFFICIENTS OF SECOND DERIVATIVE OF DENOMINATOR
C    DPPP = COEFFICIENTS OF THIRD DERIVATIVE OF DENOMINATOR
C    DN = DEGREE OF NUMERATOR, AS ELEMENT NUMBER OF COEFFICIENT OF HIGHEST POWER
C    DD = DEGREE OF DENOMINATOR, DITTO
C    DNP1 = DEGREE OF NUMERATOR PLUS 1
C    DDP1 = DEGREE OF DENOMINATOR PLUS 1
C    N0 = VALUE OF NUMERATOR, EVALUATED AT A REAL ROOT
C    N1 = VALUE OF DERIVATIVE OF NUMERATOR, EVALUATED AT A REAL ROOT
C    LIKEWISE FOR D1, D2, D3
C    G = FORTRAN COMPLEX VALUE OF A COMPLEX ROOT
C    NOC = VALUE OF NUMERATOR, EVALUATED AT A COMPLEX ROOT G
C    SIMILARLY FOR D1C
C    KA = COMPLEX QUOTIENT OF NOC AND D1C
C    RE = REAL PART OF KA
C    IM = IMAGINARY PART OF KA
C
      INTEGER DN, DNP1, DD, DDP1, I, K
      DOUBLE PRECISION N(21), NP(20), D(22), DP(21), DPP(20), DPPP(19)
      DOUBLE PRECISION COEF, N0, N1, D1, D2, D3, RE, IM
      DOUBLE PRECISION AR, AI, BR, BI, NOCR, NOCI, D1CR, D1CI, TEMP
      DOUBLE PRECISION CMPYR, CMPYI, CDIVR, CDIVI
      DOUBLE PRECISION RRP, RIP, RR, RI, DOCR, DOCI
      DOUBLE PRECISION DOD, D1D, DOUBLE
      CMPYR(AR, AI, BR, BI) = AR*BR - AI*BI
      CMPYI(AR, AI, BR, BI) = AR*BI + AI*BR
      CDIVR(AR, AI, BR, BI) = (AR*BR + AI*BI)/(BR**2 + BI**2)
      CDIVI(AR, AI, BR, BI) = (AI*BR - AR*BI)/(BR**2 + BI**2)
```

~~~

```
C FIRST ORDER REAL, NOT AT ORIGIN
C GO THROUGH ONE ITERATION OF THE NEWTON-RAPHSON METHOD TO IMPROVE ROOT
C EVALUATE DENOMINATOR AT GIVEN ROOT
      DOD = D(DD)
      DO 753 I = DDP1, 22
  753 DOD = -ALPHA * DOD + D(I)
C EVALUATE DERIVATIVE OF DENOMINATOR AT GIVEN ROOT
      D1D = DP(DD)
      DO 754 I = DDP1, 21
  754 D1D = -ALPHA * D1D + DP(I)
C GET IMPROVED VALUE
      DOUBLE = -ALPHA - DOD / D1D
      WRITE (6, 755) DOUBLE
  755 FORMAT (1H0, 1PD25.15)
C USE IMPROVED ROOT IN GETTING PARTIAL FRACTION COEFFICIENTS FOR INVERSE
      N0 = N(DN)
      DO 31 I = DNP1, 21
   31 N0 = DOUBLE * N0 + N(I)
      D1 = DP(DD)
      DO 32 I = DDP1, 21
   32 D1 = DOUBLE * D1 + DP(I)
      AORB = N0 / D1
      THORC = 0.0
C CONVERT TO SINGLE PRECISION FOR PRINTING AND USE IN PLOTTING ROUTINE
      ALPHA = -DOUBLE
      WRITE (6, 9) ALPHA, BETA, K, AORB, THORC
      GO TO 40
C COMPLEX (ALL COMPLEX POLES ARE FIRST ORDER, BY ASSUMPTION)
C REFINE THE APPROXIMATIONS TO THE ROOTS BY USING THE NEWTON-RAPHSON METHOD
C START WITH THE VALUES SUPPLIED BY THE ROOT-FINDER SUBROUTINE
   50 RRP = -ALPHA
      RIP = BETA
```

Figure 29.6. Two sections of a modified version of the program in Figure 29.3, using double precision methods.

210

```
C INITIALIZE THE ITERATION COUNTER
      ICTR = 1
C EVALUATE THE DENOMINATOR POLYNOMIAL
  152 DOCR = D(DD)
      DOCI = 0.0
      DO 153 I = DDP1, 22
      TEMP = CMPYR(RRP, RIP, DOCR, DOCI) + D(I)
      DOCI = CMPYI(RRP, RIP, DOCR, DOCI)
  153 DOCR = TEMP
C EVALUATE THE DERIVATIVE OF THE DENOMINATOR
      D1CR = DP(DD)
      D1CI = 0.0
      DO 52 I = DDP1, 21
      TEMP = CMPYR(RRP, RIP, D1CR, D1CI) + DP(I)
      D1CI = CMPYI(RRP, RIP, D1CR, D1CI)
   52 D1CR = TEMP
C GET NEW APPROXIMATION
      RR = RRP - CDIVR(DOCR, DOCI, D1CR, D1CI)
      RI = RIP - CDIVI(DOCR, DOCI, D1CR, D1CI)
C CHECK FOR CONVERGENCE
      IF (DABS(RR - RRP) + DABS(RI - RIP) .LT. 1.D-11) GO TO 154
C CHECK ITERATION COUNTER
      IF (ICTR .GT. 40) GO TO 155
C GO BACK FOR ANOTHER ITERATION
      ICTR = ICTR + 1
      RRP = RR
      RIP = RI
      GO TO 152
  155 WRITE (6, 156)
  156 FORMAT (1H0, 5HNOCON)
  154 WRITE (6, 157) RR, RI, ICTR
  157 FORMAT (1H0, 1P2D25.15, I10)
C NOW USE THESE NEW VALUES FOR THE ROOT TO FIND THE INVERSE LAPLACE PARAMETERS
C EVALUATE NUMERATOR
  250 NOCR = N(DN)
      NOCI = 0.0
      DO 251 I = DNP1, 21
      TEMP = CMPYR(RR, RI, NOCR, NOCI) + N(I)
      NOCI = CMPYI(RR, RI, NOCR, NOCI)
  251 NOCR = TEMP
C EVALUATE DERIVATIVE OF DENOMINATOR
      D1CR = DP(DD)
      D1CI = 0.0
      DO 252 I = DDP1, 21
      TEMP = CMPYR(RR, RI, D1CR, D1CI) + DP(I)
      D1CI = CMPYI(RR, RI, D1CR, D1CI)
  252 D1CR = TEMP
C COMPUTE PARAMETERS
      RE = CDIVR(NOCR, NOCI, D1CR, D1CI)
      IM = CDIVI(NOCR, NOCI, D1CR, D1CI)
      AORB = 2.0 * DSQRT(RE**2 + IM**2)
      THORC = DATAN2(RE, -IM)
C CONVERT ROOT TO SINGLE PRECISION FOR PRINTING AND USE IN PLOTTING ROUTINE
      ALPHA = - RR
      BETA = RI
C IF ANGLE IS IN FIRST OR FOURTH QUADRANTS, LEAVE AS IS
C OTHERWISE ADD OR SUBTRACT PI TO GET SUCH AN ANGLE, AND REVERSE SIGN
C OF AORB TO COMPENSATE
      IF (ABS(THORC) .LT. 1.57079632) GO TO 453
      IF (THORC) 450, 451, 452
  450 THORC = THORC + 3.14159265
      GO TO 451
  452 THORC = THORC - 3.14159265
  451 AORB = - AORB
  453 WRITE (6, 9) ALPHA, BETA, K, AORB, THORC
      GO TO 40
      END
```

Figure 29.6. (Continued)

space the values quite closely enough; therefore the program is written to multiply the function (and the derivative) by 2.

The value of the function at $t = 0$ is printed and plotted; we then go immediately into the routine that corrects Δt; for the initial value of Δt might conceivably have been too large. The derivatives are not too complicated, and the segment for modifying Δt is a direct translation of the procedure described.

We shall not reproduce the output of this program because one glance showed it to be nonsense. The maximum value of the function (as multiplied by 2) turns out to be just a little greater than 1, and we noted earlier that the value at $t = 0$ should be zero. Instead the computed value was -18!

A careful check establishes that there are no outright blunders in the programs. Two further possibilities come to mind. The first is that the roots supplied by the root-finder are not sufficiently accurate and the second is our old friend roundoff. The coefficients of the denominator polynomial are quite large, and a twelfth degree polynomial causes the experienced numerical analyst twinges of apprehension before the computer ever enters the scene.

The road back to the drawing board seems to point toward refining the roots and double precision for the polynomial evaluations.

Refining the roots is a relatively simple process. We have approximations that are surely quite close, if not exact. An extra iteration or two of Newton-Raphson, carried out in double precision, surely ought to remove that source of error from culpability.

Figure 29.6 shows two modified sections of the program of Figure 29.3, in which we have converted most of the type statements to double precision and have rewritten the evaluation of the inverse transform parameters for the nonzero cases.

The first modification includes, as well as the new DOUBLE PRECISION statements, several arithmetic statement functions. The problem is that we now want to do double precision complex arithmetic. There is no such thing by automatic FORTRAN actions, and we program the complex operations in terms of their real and imaginary parts, all performed in double precision.

Looking at some of these operations in the second part of the modified program of Figure 29.6 we see that the Newton-Raphson method has been used to find a better approximation to the real root not at the origin. Very little added effort is required, actually. We evaluate the denominator and the derivative of the denominator at the approximate root supplied by the root-finder. The variable DOUBLE applies the Newton-Raphson correction

```
    -2.6856768789430770 00      5.4206941307167380 00           2

     2.6856769E 00      5.4206941E 00    1    3.6202390E 01      1.1820568E 00

    -3.6467385953268280 00

     3.6467386E 00     0.             1    1.5413955E 04      0.

    -3.3519563991542950 00      1.7426614161839800 00           2

     3.3519564E 00      1.7426614E 00    1    -1.1017029E 04    -1.4318495E 00

    -2.3246743031814490 00      3.5710229203374350 00           2

     2.3246743E 00      3.5710229E 00    1    -3.1615409E 02    -1.3676901E 00

    -4.9717868585113520 00

     4.9717869E 00     0.             1    -1.8314144E 04      0.

    -4.0701391636358660 00      3.5171740477104330 00           3

     4.0701391E 00      3.5171740E 00    1    -2.3166758E 03    -5.8427823E-01

    -4.7582905281539940 00      1.7392860611453590 00           6

     4.7582905E 00      1.7392861E 00    1    -1.8997845E 04      5.3168750E-01
```

Figure 29.7. The double-precision output of the program in Figure 29.6, corresponding to the single-precision results in Figure 29.4.

T	2*F	DELTA	DELTAT
0.	0.0007	-0.0000	0.0184
0.0184	0.0010	0.0010	0.0369
0.0553	-0.	0.0088	0.0738
0.1291	-0.0015	0.0029	0.1476
0.2767	0.0034	0.0512	0.1476
0.4243	0.0009	0.1289	0.1476
0.5719	-0.0110	0.0593	0.1476
0.6457	-0.0089	0.1027	0.0738
0.7195	-0.0001	0.0694	0.0738
0.7933	0.0111	0.0074	0.1476
0.8670	0.0187	0.0856	0.0738
0.9408	0.0182	0.1251	0.0738
1.0146	0.0088	0.1039	0.0738
1.0884	-0.0060	0.0360	0.1476
1.1622	-0.0205	0.0477	0.1476
1.2360	-0.0285	0.1177	0.0738
1.3098	-0.0264	0.1451	0.0738
1.3836	-0.0143	0.1160	0.0738
1.4574	0.0042	0.0462	0.1476
1.5312	0.0231	0.0389	0.1476
1.6050	0.0358	0.1190	0.0738
1.6788	0.0374	0.1677	0.0738
1.7525	0.0266	0.1585	0.0738
1.8263	0.0057	0.0942	0.0738
1.9001	-0.0190	0.0032	0.1476
1.9739	-0.0388	0.1242	0.0738
2.0477	-0.0440	0.2579	0.0738
2.1215	-0.0260	0.3381	0.0738
2.1953	0.0212	0.2974	0.0738
2.2691	0.0996	0.1992	0.0738
2.3429	0.2074	0.1178	0.0738
2.4167	0.3384	0.0632	0.0738
2.4905	0.4836	0.0250	0.1476
2.6380	0.7721	0.0480	0.2952
2.7118	0.8934	0.0871	0.0738
2.7856	0.9873	0.1614	0.0738
2.8594	1.0481	0.2857	0.0738
2.9332	1.0730	0.4355	0.0738
3.0070	1.0624	0.4499	0.0738
3.0808	1.0193	0.3067	0.0738
3.1546	0.9488	0.1741	0.0738
3.2284	0.8575	0.0939	0.0738
3.3022	0.7525	0.0459	0.1476
3.4497	0.5285	0.0035	0.2952
3.5973	0.3235	0.0850	0.1476
3.6711	0.2375	0.0741	0.0738
3.7449	0.1648	0.0949	0.0738
3.8187	0.1060	0.1130	0.0738
3.8925	0.0604	0.1247	0.0738
3.9663	0.0268	0.1264	0.0738
4.0401	0.0037	0.1172	0.0738
4.1139	-0.0109	0.0996	0.0738
4.1877	-0.0187	0.0780	0.0738
4.2614	-0.0216	0.0563	0.0738
4.3352	-0.0211	0.0370	0.1476
4.4828	-0.0146	0.0302	0.2952
4.7780	-0.0005	0.0233	0.5903
5.3683	0.0018	0.0396	1.1807
6.5490	0.0000	0.0087	2.3613
8.9103	0.0000	0.0005	4.7226

Figure 29.8. The numerical output of the program in Figure 29.5, using the results shown in Figure 29.7.

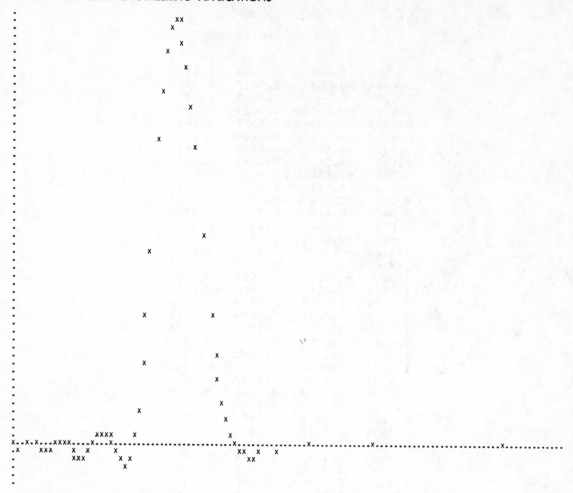

Figure 29.9. The graphical output of the program in Figure 29.5, showing the time response of the system under study.

to this approximate root; DOUBLE is used in the subsequent evaluation of the inverse transform parameters.

The improvement of the complex roots is another matter. First of all, there is reason to suspect that they may be less accurate to begin with and harder to improve. Then there is the matter of programming the complex double-precision operations. This is done by setting up separate double-precision variables for the real and imaginary parts of everything that interests us. RRP is meant to stand for "root, real, previous," and RIP for "root, imaginary, previous," the "previous" referring to the Newton-Raphson approximation process. All of the complex arithmetic operations are handled with the four arithmetic statement functions, with the exception of addition, which is taken into account automatically by the compiler. The mechanics of the iteration process are much as we have seen them before in other case studies.

The output this time is shown in Figure 29.7, where we see that the values of AORB differ from the single precision versions by enough to explain the problems in the evaluation of the time response. Whether the discrepancy was caused by the inaccuracy in the roots or by the roundoff troubles unfortunately is not made clear. The indications gathered during the development of the programs suggest that the roots supplied by the root-finder were close enough and that the real trouble lay mostly in the inability of the single-precision version to retain all the digits of the coefficients of the twelfth-degree polynomial.

The values of the roots and the inverse transform parameters produced by this program were fed into the plotting routine (unmodified), which produced the values of Figure 29.8 and the plot of Figure 29.9. We see that the value of the function at $t = 0$ is 0.0007, which, according to Bashkow, is reasonably close to the theoretically expected value of zero.

APPENDIX 1 STATEMENT PUNCTUATION; OPERATOR FORMATION RULES

The page number given for each statement refers to the location in which a discussion of the statement may be found.

Statement	Samples	Page
a = b (arithmetic or logical assignment statement)	A = B + C*D	4
	L = X .GT. 12.0 .OR. M .AND. N	73
GO TO n (Unconditional GO TO)	GO TO 12	19
GO TO n, $(n_1, n_2, \ldots, n_m)$ (Assigned GO TO)	GO TO N, (1, 123, 12, 4321)	
ASSIGN i TO n	ASSIGN 4321 TO N	
GO TO $(n_1, n_2, \ldots, n_m)$, i (Computed GO TO)	GO TO (1, 123, 12, 4321), N	33
IF (e) n_1, n_2, n_3 (Arithmetic IF)	IF (X — 12.0) 34, 54, 21	30
IF (e) s (Logical IF)	IF (X .GE. 12.0) Y = 23.6 + R	24
STOP	STOP	5
PAUSE	PAUSE	
END	END	5
DO n i = m_1, m_2	DO 69 I = 1, LAST	86
or		
DO n i = m_1, m_2, m_3	DO 54321 INDEX = INIT, LAST, 5	89
CONTINUE	CONTINUE	91
CALL Name $(a_1, a_2, \ldots, a_n)$	CALL MATMPY (A, B, C, 4, 6, 10)	129
SUBROUTINE Name $(a_1, a_2, \ldots, a_n)$	SUBROUTINE MATMPY (A, B, C, I, J, K)	129
FUNCTION Name $(a_1, a_2, \ldots, a_n)$	FUNCTION BESSEL (X, N)	127
RETURN	RETURN	127
READ f, list	READ 819, X, Y, Z	
READ (u) list	READ (3) A, B, N	117
READ (u, f) list	READ (5, 69) FIRST, LAST	4
WRITE (u) list	WRITE (4) (A(I), I = 1, N)	117
WRITE (u, f) list	WRITE (6, FMT) A(1), A(2), R	5
PRINT f, list	PRINT 26, (A(I), I = 1, 5)	
PUNCH f, list	PUNCH 900, DELT, EPS	
FORMAT (specifications)	12 FORMAT (1H0, F10.0/1PE16.7, I6)	4
REWIND u	REWIND 8	117
BACKSPACE u	BACKSPACE 3	117
ENDFILE u	ENDFILE 7	117
DIMENSION v, v, v, . . .	DIMENSION A(10), B(2, 5), C(3, 3, 4)	80

Statement	Samples	Page
COMMON /x₁/a₁/ . . . /xₙ/aₙ	COMMON /BLOCK1/A, I/BLOCK2/B, C, K	134
	COMMON A, B, C, M, N	133
EQUIVALENCE (k₁), (k₂), . . . , (kₙ)	EQUIVALENCE (A, B, C), (I, J)	131
EXTERNAL v₁, v₂, . . . , vₙ	EXTERNAL SIN, SQRT, OTHER	144
INTEGER v₁, v₂, . . . , vₙ	INTEGER X	36
REAL v₁, v₂, . . . , vₙ	REAL A, B, M, LAST, ARRAY(23)	3
DOUBLE PRECISION v₁, v₂, . . . , vₙ	DOUBLE PRECISION A, B(2), CUBERT	64
COMPLEX v₁, v₂, . . . , vₙ	COMPLEX OMEGA, TRANS	53
LOGICAL v₁, v₂, . . . , vₙ	LOGICAL A1, B1, RESULT(100)	72
DATAk₁/d₁/, k₂/d₂/, . . . , kₙ/dₙ/	DATA A/12.0/, I, J, K, L/2, 3*19/	110
NAMELIST/A/B/C, . . . , D/E/F, G, . . . , H	NAMELIST/INPUT/TIN, LENGTH, EPS	174
ENTRY NAME a₁, a₂, . . . , aₙ	ENTRY YBES (1, XM, FXM, $500)	147

Table A1.1 indicates the types of operand that may be combined by the four arithmetic operators other than exponentiation to form valid arithmetic expressions. Table A1.2 gives the same information for exponentiation. Table A1.3 indicates how logical operands may be combined by the six relational operands to form valid logical expressions.

In all three tables the entries show the type of the resulting expression if valid; an X indicates an invalid combination.

TABLE A1.1
Type of Right Operand

+ − * /	Integer	Real	Double	Complex	Logical
Integer	Integer	X	X	X	X
Real	X	Real	Double	Complex	X
Double	X	Double	Double	X	X
Complex	X	Complex	X	Complex	X
Logical	X	X	X	X	X

(Type of Left Operand)

TABLE A1.2
Type of Exponent

**	Integer	Real	Double	Complex	Logical
Integer	Integer	X	X	X	X
Real	Real	Real	Double	X	X
Double	Double	Double	Double	X	X
Complex	Complex	X	X	X	X
Logical	Integer	X	X	X	X

(Type of Left Operand)

TABLE A1.3
Type of Right Operand

.EQ. .NE. .GT. .GE. .LT. .LE.	Integer	Real	Double	Complex	Logical
Integer	Logical	X	X	X	X
Real	X	Logical	Logical	X	X
Double	X	Logical	Logical	X	X
Complex	X	X	X	X	X
Logical	X	X	X	X	X

(Type of Left Operand)

APPENDIX 2 BASIC SUPPLIED FUNCTIONS

Table A2.1 lists the characteristics of the external functions that may be expected to be supplied with every FORTRAN IV compiler; Table A2.2 does the same for intrinsic functions. The difference is that between closed and open subroutines, plus the fact that the external functions may be named in an EXTERNAL statement, whereas the intrinsic functions may not.

TABLE A2.1 EXTERNAL FUNCTIONS

External Function	Definition	Number of Arguments	Symbolic Name	Type of Argument	Type of Function
Exponential	e^a	1	EXP	Real	Real
		1	DEXP	Double	Double
		1	CEXP	Complex	Complex
Natural logarithm	$\log_e (a)$	1	ALOG	Real	Real
		1	DLOG	Double	Double
		1	CLOG	Complex	Complex
Common logarithm	$\log_{10} (a)$	1	ALOG10	Real	Real
			DLOG10	Double	Double
Trigonometric sine	$\sin (a)$	1	SIN	Real	Real
		1	DSIN	Double	Double
		1	CSIN	Complex	Complex
Trigonometric cosine	$\cos (a)$	1	COS	Real	Real
		1	DCOS	Double	Double
		1	CCOS	Complex	Complex
Hyperbolic tangent	$\tanh (a)$	1	TANH	Real	Real
Square root	$(a)^{1/2}$	1	SQRT	Real	Real
		1	DSQRT	Double	Double
		1	CSQRT	Complex	Complex
Arctangent	$\arctan (a)$	1	ATAN	Real	Real
		1	DATAN	Double	Double
	$\arctan (a_1/a_2)$	2	ATAN2	Real	Real
		2	DATAN2	Double	Double
Remaindering*	$a_1 \pmod{a_2}$	2	DMOD	Double	Double
Modulus		1	CABS	Complex	Real

* The function DMOD (a_1, a_2) is defined as $a_1 - [a_1/a_2]a_2$, where $[x]$ is the integer whose magnitude does not exceed the magnitude of x and whose sign is the same as the sign of x.

TABLE A2.2 INTRINSIC FUNCTIONS

Intrinsic Function	Definition	Number of Arguments	Symbolic Name	Type of Argument	Type of Function
Absolute value	$\lvert a \rvert$	1	ABS	Real	Real
			IABS	Integer	Integer
			DABS	Double	Double
Truncation	Sign of a times largest integer $\leq \lvert a \rvert$	1	AINT	Real	Real
			INT	Real	Integer
			IDINT	Double	Integer
Remaindering*	$a_1 \,(\mathrm{mod}\ a_2)$	2	AMOD	Real	Real
			MOD	Integer	Integer
Choosing largest value	Max $(a_1, a_2, \ldots)$	$\geqq 2$	AMAX0	Integer	Real
			AMAX1	Real	Real
			MAX0	Integer	Integer
			MAX1	Real	Integer
			DMAX1	Double	Double
Choosing smallest value	Min $(a_1, a_2, \ldots)$	$\geqq 2$	AMIN0	Integer	Real
			AMIN1	Real	Real
			MIN0	Integer	Integer
			MIN1	Real	Integer
			DMIN1	Double	Double
Float	Conversion from integer to real	1	FLOAT	Integer	Real
Fix	Conversion from real to integer	1	IFIX	Real	Integer
Transfer of sign	Sign of a_2 times $\lvert a_1 \rvert$	2	SIGN	Real	Real
			ISIGN	Integer	Integer
			DSIGN	Double	Double
Positive difference	$a_1 - \text{Min}\,(a_1, a_2)$	2	DIM	Real	Real
			IDIM	Integer	Integer
Obtain most significant part of double precision argument		1	SNGL	Double	Real
Obtain real part of complex argument		1	REAL	Complex	Real
Obtain imaginary part of complex argument		1	AIMAG	Complex	Real
Express single precision argument in double precision form		1	DBLE	Real	Double
Express two real arguments in complex form	$a_1 + a_2\sqrt{-1}$	2	CMPLX	Real	Complex
Obtain conjugate of a complex argument		1	CONJG	Complex	Complex

* The function MOD or AMOD (a_1, a_2) is defined as $a_1 - [a_1/a_2]a_2$, where $[x]$ is the integer whose magnitude does not exceed the magnitude of x and whose sign is the same as x.

ANSWERS TO SELECTED EXERCISES

There are several acceptable answers to many of the exercises. The one shown here is sometimes "better" than other possibilities, but only occasionally is the criterion of "goodness" stated. In other cases there are several equally "good" answers; for instance it makes no difference whether we write

$A = B + C$ or $A = C + B$. In short, the answers given here are correct but not ordinarily *uniquely* correct. If another answer can be shown to be equivalent, it must be accepted unless other criteria have been stated.

CASE STUDY 1 (page 6)

1.   REAL A, B, H
        READ (5, 1) B, H
      1 FORMAT (8F10.0)
        A = B * H/2.0
        WRITE (6, 2) B, H, A
      2 FORMAT (1H , 1P10E13.5)
        STOP
        END

3.   REAL I, V, R
        READ (5, 1) R, V
      1 FORMAT (8F10.0)
        I = V/R
        WRITE (6, 2) V, R, I
      2 FORMAT (1H , 1P10E13.5)
        STOP
        END

5.   REAL C, A, S
        READ (5, 1) A, S
      1 FORMAT (8F10.0)
        C = 0.000008855 * A/S
        WRITE (6, 2) A, S, C
      2 FORMAT (1H , 1P10E13.5)
        STOP
        END

CASE STUDY 2 (page 9)

1. 256.   2.56   −43000.   1.0E12   4.92E − 7   −10.0   −1.E − 16

3. 87,654.3 (comma not permitted); +987 (has no decimal point);
    9.2E + 87 (too large in most FORTRANS); 7E − 9 (has no decimal point).

5. Yes.

6. −234. (decimal point not permitted); 23,400 (comma not permitted); 1E12 (exponent form not permitted);
    +1000000000000 (too large in most Fortrans).

8. REAL F, M, T, A, X
    READ (5, 1) F, M, T
    1 FORMAT (8F10.0)
    A = F/M
    X = 0.5 * A * T**2
    WRITE (6, 2) T, F, M, A, X
    2 FORMAT (1H , 1P10E13.5)
    STOP
    END

10. REAL F, W, L, X
    READ (5, 1) F, L
    1 FORMAT (8F10.0)
    W = 6.2831853 * F
    X = 1.E — 6 * W * L
    WRITE (6, 2) F, L, W, X
    2 FORMAT (1H , 1P10E13.5)
    STOP
    END

CASE STUDY 3 (page 12)

1. Integer: I, IJK, LARGE, KAPPA
    Real: G, GAMMA, BT07TH, CDC160, DELTA, AlP4, ALGOL
    Unacceptable: GAMMA423 (too many characters); IJK* (* not permitted);
    J79—14 (— not permitted); R(2)16 (parentheses not permitted); 2N173
    (does not begin with a letter); 6CA7 (does not begin with a letter); EPSILON
    (too many characters); A1.4 (decimal point not permitted); FORTRAN
    (too many characters).

3. b. $(X + 2.0)/(Y + 4.0)$. Constants may be written in
    any other equivalent form, such as $(X + 2.)/(Y + 4.)$.
    e. $((X + A + 3.1415927)/(2.0*Z))**2$
    g. $(X/Y)**(R — 1.0)$
    j. $A + X*(B + X*(C + D*X))$

4. a. $X + Y**3$
    d. $A + B/C$
    f. $A + B/(C + D)$
    h. $((A + B)/(C + D))**2 + X**2$
    j. $1.0 + X + X**2/2.0 + X**3/6.0$
    k. $(X/Y)**(G — 1.0)$

5. REAL R1, R2, R
    READ (5, 1) R1, R2
    1 FORMAT (8F10.0)
    R = 1.0/(1.0/R1 + 1.0/R2)
    WRITE (6, 2) R1, R2, R
    2 FORMAT (1H , 1P10E13.5)
    STOP
    END
    Assignment statement could also be
    R = R1 * R2 /(R1 + R2)

8. REAL X, XBAR
    READ (5, 1) X
    1 FORMAT (8F10.0)
    XBAR = (X**2 + 3.0 * X + 5.0) / (2.0 * X + 3.0)
    WRITE (6, 2) X, XBAR
    2 FORMAT (1H , 1P10E13.5)
    STOP
    END

CASE STUDY 4 (page 16)

2.   a. DELTA = BETA + 2.0
     c. C = SQRT(A**2 + B**2) or
        C = SQRT(A*A + B*B)
     d. R = 1.41421356
     g. Y = COS(2.0*X)*SQRT(X/2.0)
     h. G = G + 2.0

3.   a. AREA = 2.0*P*R*SIN (3.14159265/P)
     c. ARC = 2.0*SQRT(Y**2 + 1.3333333*X**2)
     e. S = − COS(X)**(P + 1.0)/(P + 1.0)
     f. G = 0.5*ALOG((1.0 + SIN(X))/(1.0 − SIN(X)))
        Preferably written as two statements to avoid
        computing the sine twice:
        S = SIN (X)
        G = 0.5*ALOG((1.0 + S)/(1.0 − S))
     i. E = X*ATAN(X/A) − A/2.0*ALOG(A**2 + X**2)
     l. Q = (2./(3.1415927*X))**0.5*SIN(X)
        Since $(2/\pi)^{1/2}\ (1/X)^{1/2} = \sqrt{2/\pi}/\sqrt{X} = 0.7978846/\sqrt{X}$,
        this can be written more compactly and thus requires less
        time in the object program:
        Q = 0.7978846/SQRT(X)*SIN(X)
     n. Y = 2.5066283*X**(X + 1.0)*EXP(− X)

4.   REAL T, V
     READ (5, 1) T
   1 FORMAT (8F10.0)
     V = 50.0 − 40.0 * EXP(− 20.0*T)
     WRITE (6, 2) T, V
   2 FORMAT (1H , 1P10E13.5)
     STOP
     END

6.   REAL L, THETA, TAU
     READ (5, 1) L, THETA
   1 FORMAT (8F10.0)
     TAU = 6.2831853 * SQRT(L*COS(THETA/57.29578)/32.2)
     WRITE (6, 2) L, THETA, TAU
   2 FORMAT (1H , 1P10E13.5)
     STOP
     END

CASE STUDY 5 (page 22)

1.   a. Two extra commas in REAL, two extra commas in READ, two missing
       parentheses and a missing asterisk in assignment statement, extra comma
       in WRITE, no FORMAT for WRITE, no END.

2.   REAL X, Q
   9 READ (5, 1) X
   1 FORMAT (8F10.0)
     Q = SIN(X) / X
     WRITE (6, 2) X, Q
   2 FORMAT (1H , 1P10E13.5)
     GO TO 9
     END

6.   REAL T, V
   39 READ (5, 1) T
   1 FORMAT (8F10.0)
     V = 50.0 − 40.0 * EXP(− 20.0*T)
     WRITE (6, 2) T, V
   2 FORMAT (1H , 1P10E13.5)
     GO TO 39
     END

9.   REAL I, K1, K2, S1, S2, T, V, L, R, C, D
        READ (5, 1) V, R, L, C
      1 FORMAT (8F10.0)
        WRITE (6, 2) V, R, L, C
      2 FORMAT (1H , 1P10E13.5)
      7 READ (5, 1) T
        D = SQRT(R**2/(4.0*L**2) — 1.0/(L*C))
        K1 = —V/(2.0*L*D)
        K2 = —K1
        S1 = —R/(2.0*L) — D
        S2 = —R/(2.0*L) + D
        I = K1*EXP(SI*T) + K2*EXP(S2*T)
        WRITE (6, 2) T, I
        GO TO 7
        END

CASE STUDY 6 (page 25)

1. a.   IF (A .GT. B) X = 16.9
        IF (A .LE. B) X = 56.9
        Another way is to set X to 16.9 unconditionally, then
        change it if necessary:
        X = 16.9
        IF (A .LE. B) X = 56.9
        A third way is to use GO TO statements:
        IF (A .GT. B) GO TO 77
        X = 56.9
        GO TO 76
     77 X = 16.9
     76

   c.   IF (RHO + THETA .LT. 1.0E — 6) GO TO 156
        GO TO 762
   d.   IF (X .LT. Y) BIG = Y
        IF (X .GE. Y) BIG = X
        This, too, can be done with only one IF by first setting
        BIG equal to Y unconditionally:
        BIG = Y
        IF (X .GE. Y) BIG = X
   g. 400 IF (THETA .LT. 6.2831853) GO TO 402
        THETA = THETA — 6.2831853
        GO TO 400
     402
   h.   IF (G .LT. 0.0 .AND. H .LT. 0.0) SIGNS = —1
        IF (G .GT. 0.0 .AND. H .GT. 0.0) SIGNS = +1
        IF (G * H .LT. 0.0) SIGNS = 0
   j.   IF ((A .LT. 0.0 .AND. B .GT. 0.0) .OR. C .EQ. 0.0) OMEGA = COS(X + 1.2)
   l.   First way:
        IF (0.999 .LE. X .AND. X .LE. 1.001) STOP
        GO TO 639
        Second way:
        IF (ABS(X — 1.000) .LE. 0.001) STOP
        GO TO 639
   m.   IF (ABS(XREAL) .LT. 1.0 .AND. ABS(XIMAG) .LT. 1.0) SQUARE = 1

2. a.
```
        READ (5, 20) ANNERN
    20 FORMAT (F10.0)
        IF (ANNERN − 2000.00) 40, 40, 50
    40 TAX = 0.0
        GO TO 100
    50 IF (ANNERN − 5000.00) 60, 60, 70
    60 TAX = 0.02*(ANNERN − 2000.00)
        GO TO 100
    70 TAX = 60.00 + 0.05*(ANNERN − 5000.00)
   100 WRITE (6, 30) ANNERN, TAX
    30 FORMAT (1P2E20.7)
        STOP
        END
```
c.
```
        X = 1.0
    61 Y = 16.7*X + 9.2*X**2 − 1.02*X**3
        WRITE (6, 62) X, Y
    62 FORMAT (1P2E20.7)
        IF (X .GE. 9.9) STOP
        X = + 0.1
        GO TO 61
        END
```
d.
```
        I = 10
    61 X = I
        X = X/10.0
        Y = 16.7*X + 9.2*X**2 − 1.02*X**3
        WRITE (6, 62) X, Y
    62 FORMAT (1P2E20.7)
        IF (I .EQ. 99) STOP
        I = I + 1
        GO TO 61
        END
```

CASE STUDY 7 (page 32)

1. a.　　　IF (A − B) 69, 69, 96
　　　69 X = 56.9
　　　　　GO TO 70
　　　96 X = 16.9
　　　70

　　　Note that the statement numbers in the arithmetic
　　　IF need not be distinct.

c.　　　IF (RHO + THETA − 1.0E − 6) 156, 762, 762
d.　　　IF (X − Y) 11, 11, 12
　　　11 BIG = Y
　　　　　GO TO 13
　　　12 BIG = X
　　　13

g.　　400 IF (THETA − 6.2831853) 402, 401, 401
　　401 THETA = THETA − 6.2831853
　　　　GO TO 400
　　402

h.　　　IF (G) 100, 200, 200
　　100 IF (H) 300, 400, 400
　　200 IF (H) 400, 500, 500
　　300 SIGNS = −1
　　　　GO TO 600
　　400 SIGNS = 0
　　　　GO TO 600
　　500 SIGNS = +1
　　600

j.　　　IF (C) 98, 100, 98
　　98 IF (A) 101, 99, 99
　　101 IF (B) 99, 99, 100
　　100 OMEGA = COS(X + 1.2)
　　99

l.　　　IF (X − 0.999) 67, 63, 21
　　21 IF (X − 1.001) 63, 63, 67
　　63 STOP
　　67

　　　IF (ABS(X − 1.0) − 0.001) 63, 63, 67
　　63 STOP
　　67

m.　　　IF (ABS(XREAL) − 1.0) 39, 45, 45
　　39 IF (ABS(XIMAG) − 1.0) 40, 45, 45
　　40 SQUARE = 1.0
　　45

2.　　　REAL A, Q
　　　READ (5, 1) A
　　1 FORMAT (8F10.0)
　　　IF (A) 12, 123, 1234
　　12 Q = −1.5707963
　　　　GO TO 12345
　　123 Q = 0.0
　　　　GO TO 12345
　　1234 Q = 1.5707963
　　12345

```
4.          REAL R, S
         23 READ (5, 1) R
          1 FORMAT (8F10.0)
            IF (R − 120.0) 33, 44, 44
         33 S = 1.7E4 − 0.485*R**2
            GO TO 22
         44 S = 1.8E4/(1.0 + R*R/1.8E4)
         22 WRITE (6, 2) R, S
          2 FORMAT (1H , 1P10E13.5)
            GO TO 23
            END
```

CASE STUDY 8 (page 35)

1.   GO TO (250, 250, 251, 249, 249, 249, 251, 250), N

3.   GO TO (10, 15, 20, 25), CODE
```
         10 BONDS = 0.0
            GO TO 30
         15 BONDS = 18.75
            GO TO 30
         20 BONDS = 37.50
            GO TO 30
         25 BONDS = 0.1 * GROSS
         30 NET = GROSS − BONDS
```

CASE STUDY 9 (page 42)

2. a.   M = bb12X = bb407.8Y = bb − 32.9

3.   READ (5, 238) BOS, EWR, PHL, DCA
        238 FORMAT (4F8.0)

6.   READ (5, 233) LGA, JFK, BAL, TPA
        233 FORMAT (2I3, 2E14.7)

7. a.   FORMAT (1H , 2I7, 2F8.2)
    b.   FORMAT (1H , 2I7, 2F6.0)
    c.   FORMAT (1H , I6, I8, 2E13.5)
    d.   FORMAT (1H , I6, I8, 1P2E12.4)
    e.   FORMAT (3HbI =, I6, 4HbbJ =, I6, 4HbbR =, F6.1, 4HbbS =, F6.1)
    f.   FORMAT (3HbI =, I6/3HbJ =, I6/3HbR = , F6.1/3HbS = , F6.1)

9.   WRITE (6, 49) A, B, X, Z
         49 FORMAT (2F12.4, 2E20.8)

CASE STUDY 11 (page 57)

1. b.   (2.0, 4.0)
    d.   (5.0, 16.0)
    g.   24.0

2.
```
        COMPLEX A11, A12, A21, A22, B1, B2, X, Y
        A11 = CMPLX(2., 3.)
        A12 = CMPLX(4., −2.)
        A21 = CMPLX(4., 1.)
        A22 = CMPLX(−2., 3.)
        B1 = CMPLX(5., −3.)
        B2 = CMPLX(2., 13.)
        FACT = A21/A11
        A22 = A22 − FACT*A12
        B2 = B2 − FACT * B1
        Y = B2/A22
        X = (B1 − A12*Y)/A11
        WRITE (6, 23) X, Y
     23 FORMAT (1H , 4F12.5)
        STOP
        END
```

The method of solution is called Gauss elimination; the goal is to reduce A21 to zero, which it would be if we actually carried out the arithmetic, after which the second equation involves only one unknown.

4.
```
        COMPLEX Z, EXPZ
    239 READ (5, 240) Z
    240 FORMAT (8F10.0)
        EXPZ = CEXP(Z)
        WRITE (6, 241) Z, EXPZ
    241 FORMAT (1H , 4F10.4)
        GO TO 239
        STOP
        END
```

11.
```
        REAL K, FIRST, LAST, INC, W
        COMPLEX T, N1, N2, D1, D2, D3, D4
        READ (5, 500) K, FIRST, LAST, INC
    500 FORMAT (8F10.0)
        W = FIRST
      6 N1 = CMPLX(1.0, 0.4*W)
        N2 = CMPLX(1.0, 0.2*W)
        D1 = CMPLX(1.0, W)
        D2 = CMPLX(1.0, 2.5*W)
        D3 = CMPLX(1.0, 1.43*W)
        D4 = CMPLX(1.0, 0.02*W)
        T = K*N1*N2/(D1*D2 * D3 * D4**2)
        WRITE (6, 501) W, T
    501 FORMAT (1H , 3F12.6)
        W = W * INC
        IF (W .LE. LAST) GO TO 6
        STOP
        END
```

CASE STUDY 12 (page 70)

3. a. Yes
   d. No
   e. Yes
   h. Yes

CASE STUDY 13 (page 77)

1. a. Yes
   c. Yes
   e. Yes
   f. Yes

CASE STUDY 14 (page 84)

1. DIMENSION X(3)
    DIST = SQRT(X(1)**2 + X(2)**2 + X(3)**2)

3. DIMENSION A(2,2), C(2,2), B(2,2)
    C(1,1) = A(1,1)*B(1,1) + A(1,2)*B(2,1)
    C(1,2) = A(1,1)*B(1,2) + A(1,2)*B(2,2)
    C(2,1) = A(2,1)*B(1,1) + A(2,2)*B(2,1)
    C(2,2) = A(2,1)*B(1,2) + A(2,2)*B(2,2)

5. DIMENSION A(30), B(30)
    I = 1
    D = 0.0
    456 D = D + (A(I) − B(I))**2
    IF (I .EQ. 30) GO TO 455
    I = I + 1
    GO TO 456
    45 D = SQRT(D)

6. DIMENSION X(50), DX(49)
    I = 1
    9 DX(I) = X(I + 1) − X(I)
    IF (I .EQ. 49) GO TO 7
    I = I + 1
    GO TO 9
    7

9. DIMENSION Y(50)
    S = Y(I) + U*(Y(I + 1) − Y(I − 1))/2.0 +
        U**2/2.0*(Y(I + 1) − 2.0*Y(I) + Y(I − 1))

11. DIMENSION A(7), B(7)
    21 FORMAT (7F10.0)
    READ (5, 21) A
    READ (5, 21) B
    I = 1
    SUM = 0.0
    4 SUM = SUM + A(I)*B(I)
    I = I + 1
    IF (I .LE. 7) GO TO 4
    ANORM = SQRT(SUM)
    WRITE (6, 22) ANORM
    22 FORMAT (1PE20.7)

13. Either add the statement
    DOUBLE PRECISION A, B
    or replace the DIMENSION statement with
    DOUBLE PRECISION A(30), B(30)

15. COMPLEX COMPLX(30)
    I = 1
    SUMABS = 0.0
    81 SUMABS = CABS(COMPLX(L)) + SUMABS
    I = I + 1
    IF (I .LE. 30) GO TO 81

17. LOGICAL TRUTH (40)
    INTEGER TRUE, FALSE
    I = 1
    TRUE = 0
    FALSE = 0
    66 IF(TRUTH(I)) GO TO 29
    FALSE = FALSE + 1
    GO TO 36
    29 TRUE = TRUE + 1
    36 I = I + 1
    IF (I .NE. 41) GO TO 66

CASE STUDY 15 (page 91)

1.
```
      DO 67 I = 100, 300
      X = I
      X = X/100.0
      Y = 41.298*SQRT(1.0 + X**2) + X**0.33333333*EXP(X)
   67 WRITE (6, 68) X, Y
   68 FORMAT (1P2E20.7)
```

CASE STUDY 16 (page 101)

1.
```
      DIMENSION A(30), B(30)
      D = 0.0
      DO 23 I = 1, 30
   23 D = D + (A(I) − B(I))**2
      D = SQRT(D)
```

4.
```
      DIMENSION M(20)
      DO 92 I = 1, 20
   92 M(I) = I * M(I)
```

5.
```
      DIMENSION R(40), S(40), T(40)
      DO 3 I = 1, M
    3 T(I) = R(I) + S(I)
```

7.
```
      DIMENSION F(50)
      MM1 = M − 1
      DO 692 I = 2, MM1
  692 F(I) = (F(I − I) + F(I) + F(I + I))/3.0
```

8.
```
      DIMENSION B(50)
      BIGB = B(1)
      NBIGB = 1
      DO 42 I = 2, 50
      IF (BIGB .GE. B(I)) GO TO 42
      BIGB = B(I)
      NBIGB = I
   42 CONTINUE
```

10.
```
      DIMENSION A(15, 15), X(15), B(15)
      DO 61 I = 1, 15
      B(I) = 0.0
      DO 61 J = 1, 15
   61 B(I) = B(I) + A(I, J)*X(J)
```

12.
```
      DIMENSION RST(20, 20)
      DPROD = RST(1, 1)
      DO 1 I = 2, 20
    1 DPROD = DPROD*RST(I, I)
```

CASE STUDY 17 (page 106)

1.
```
      Somewhere at the beginning of the program add the
      statements
      DIMENSION FORM(12)
      READ (5, 9999) (FORM(I), I = 1, 12)
 9999 FORMAT (12A6)
      Make statement 42 read
   42 WRITE (6, FORM) I, X(I)
      Delete statement 200.
```

3.
```
      READ (5, 99) N, (DATA(I), I = 1, N)
   99 FORMAT (I2, 10F7.0)
```

5. a.   One data card will be read. It should have seven
numbers punched in it, with four columns for each number;
if decimal points are not punched, the numbers will be taken
to be integers and and converted to real form. The first of
the numbers should begin in column 1.

b.   One data card will be read. It should have a two-
digit integer punched in columns 1–2, and a maximum of 10
four-digit numbers punched in the columns following.
There should be as many numbers as the value of the integer in
columns 1–2.

c.   Two cards will be read. The first should
contain an integer in columns 1–2. The second should contain
as many four-digit numbers as the value of the integer in the
first card.

d. $N + 1$ cards will be read. The first should contain an integer
in columns 1–2. There should be as many cards following
as the value of this integer, each containing a four-digit number
in columns 1–4.

8.
```
    FORMAT (2I2, 1PE20.7)
    DO 14 I = 1, M
    DO 14 J = 1, N
 14 PUNCH 13, I, J, STL(I, J)
```

9.
```
    READ (5, 29) (MONTH(I), I = 1, 12)
 29 FORMAT (12A3)

    WRITE (6, 30) MONTH(I)
 30 FORMAT (1Hb, A3)
```

11. a.   (10F10.2)

b.   (1P5E20.6) or (1P5E20.6/1P5E20.6)

c.   (F10.2//(P3E20.6)

CASE STUDY 18 (page 115)

2.
```
    DIMENSION G(48, 80)
    INTEGER A
    DATA BLANK, DOT, XPRINT/1Hb. 1H., 1HX/
    DO 42 I = 1, 48
    DO 42 J = 1, 80
 42 G(I, 40) = BLANK
    DO 64 I = 1, 48
 64 G(I, 40) = DOT
    DO 65 J = 1, 80
 65 G(24, J) = DOT
    DO 69 A = 9, 360, 9
    T = A
    T = T/(180.0/3.14159265)
    X = COS(T)
    Y = SIN(T)
    I = 6.0*(Y + 4.0) + 0.5
    J = 10.0*(X + 4.0) + 0.5
 69 G(I, J) = XPRINT
    I = 48
 71 WRITE (6, 70) (G(I, J), J = 1, 80)
 70 FORMAT (80A1)
    I = I − 1
    IF (I .NE. 0) GO TO 71
    STOP
```

CASE STUDY 19 (page 119)

```
  1.      REAL ARRAY A(8)
          REWIND 3
          REWIND 8
        2 READ (8, 1) (A(I), I = 1, 8)
        1 FORMAT (8F10.0)
          IF (A(1) .EQ. 0.0) GO TO 10
          WRITE (3) (A(I), I = 1, 8)
          GO TO 2
       10 END FILE 3
          REWIND 3
          REWIND 8
          STOP
          END
```

CASE STUDY 20 (page 125)

```
  1.      DENOM(X) = X**2 + SQRT(1.0 + 2.0*X + 3.0*X**2)
          ALPHA = (6.9 + Y)/DENOM(Y)
          BETA = (2.1*Z + Z**4)/DENOM(Z)
          GAMMA = SIN(Y)/DENOM(Y**2)
          DELTA = 1.0/DENOM(SIN(Y))

  3.      LOGICAL EXOR, A, B, ANS1, ANS2
          EXOR(A, B) = A .AND. .NOT. B .OR. .NOT. A .AND. B
          ANS1 = EXOR(EXOR(A, B), C)
          ANS2 = EXOR(A, EXOR(B, C))

  5.      LOGICAL A(40), B(40), C(41), K, SUM, CARRY
          C(1) = SUM(A(1), B(1), .FALSE.)
          K = CARRY(A(1), (B1), .FALSE.)
          DO 28 I = 2, N
          C(I) = SUM(A(I), B(I), K)
       28 K = CARRY(A(I), B(I), K)
          C(N + 1) = K
```

CASE STUDY 21 (page 140)

```
  1.      FUNCTION Y(X)
          IF (X) 10, 11, 12
       10 Y = 1.0 + SQRT(1.0 + X*X)
          RETURN
       11 Y = 0.0
          RETURN
       12 Y = 1.0 - SQRT(1.0 + X*X)
          RETURN
          END

          F = 2.0 + Y(A + Z)
          G = (Y(X(K)) + Y(X(K + 1)))/2.0
          H = Y(COS(6.2831853*X)) + SQRT(1.0 + Y(6.2831853*X))
  5.      FUNCTION SUMABS(A, M, N)
          DIMENSION A(M, N)
          SUMABS = 0.0
          DO 27 I = 1, M
          DO 27 J = 1, N
       27 SUMABS = SUMABS + A(I, J)
          RETURN
          END
```

7.
```
      SUBROUTINE AVERNZ(A, N, AVER, NZ)
      DIMENSION A(50)
      AVER = 0.0
      NZ = 0
      DO 19 I = 1, N
      AVER = AVER + A(I)
      IF (A(I) .EQ. 0.0) NZ = NZ + 1
   19 CONTINUE
      AN = N
      AVER = AVER/AN
      RETURN
      END
```

```
      CALL AVERNZ(ZETA, 20, ZMEAN, NZCNT)
```

8.
```
      SUBROUTINE LARGE(A, M, N, SUM, ROWN)
      DIMENSION A(M, N)
      SUM = AVER(ROW, 1, A, M, N)
      ROWN = 1
      DO 39 I = 2, M
      BIG = AVER(ROW, I, A, M, N)
      IF (BIG .LE. SUM) GO TO 39
      SUM = BIG
      ROWN = I
   39 CONTINUE
      RETURN
      END
```

```
      CALL LARGE(OMEGA, 15, 29, OMEGAL, NROW)
```

11. a. A(1) A(2) A(3)
          B(1) B(2) B(3) B(4)

   b.   C(1,1) C(2,1) C(1,2) C(2,2) C(1,3) C(2,3)
          D(1,1) D(2,1) D(3,1) D(1,2) D(2,2) D(3,2)

12. a. C(1) C(1) C(2) C(2) C(3) C(3)
          I(1) I(2) R(1) R(2) R(3) R(4)

   b. D(1) D(1) D(2) D(2) D(3) D(3) D(4) D(4)
          R1   R2   I(1) I(2) I(3) I(4) I(5)

INDEX